Cholesterol-Lowering Therapy

Evaluation of Clinical Trial Evidence

Cholesterol-Lowering Therapy

Evaluation of Clinical Trial Evidence

edited by

Scott M. Grundy

University of Texas Southwestern Medical Center at Dallas
Dallas, Texas

CRC Press
Taylor & Francis Group
Boca Raton London New York

CRC Press is an imprint of the
Taylor & Francis Group, an **informa** business

CRC Press
Taylor & Francis Group
6000 Broken Sound Parkway NW, Suite 300
Boca Raton, FL 33487-2742

First issued in paperback 2019

© 2000 by Taylor & Francis Group, LLC
CRC Press is an imprint of Taylor & Francis Group, an Informa business

No claim to original U.S. Government works

ISBN-13: 978-0-8247-8216-0 (hbk)
ISBN-13: 978-0-367-39932-0 (pbk)

A CIP record for this book is available from the British Library.

Library of Congress Cataloging-in-Publication Data available on application

Visit the Informa Web site at
www.informa.com

and the Informa Healthcare Web site at
www.informahealthcare.com

Foreword

That arteries can "degenerate into bones" has been known since the latter part of the sixteenth century (Gabriel Fallopius, 1575). However, only since the beginning of the twentieth century has cholesterol achieved a central position in basic and clinical research to understand the "ossification process" of the arteries, which by then was known as atherosclerosis. Studies of the metabolism and role of the blood lipoproteins intensified greatly in the second half of this century, largely because the association between atherosclerosis and coronary heart disease became irrefutable.

A great deal of the work done in the 1960s and 1970s came to a head with Brown and Goldstein's fundamental discovery of the LDL receptors and of the mechanism of cellular uptake and transport of LDL in the cell. For this, they were awarded the Nobel Prize in Physiology or Medicine in 1985.

This extraordinary discovery, and others that followed, led to a surge of optimism that perhaps soon we would conquer coronary heart disease and halt its epidemic. This hope was articulated in a 1996 editorial in *Science* by Brown and Goldstein (1) titled "Heart Attacks: Gone with the Century?" In it, they wrote:

> Exploitation of recent breakthroughs on proof of the cholesterol hypothesis, discovery of effective drugs, and better definition of genetic susceptibility facctors may well end coronary disease as a major public health problem early in the next century.

As this volume appears on the eve of the next century, we must ask: "Have we fulfilled the prediction of Brown and Goldstein?" Unfortunately, for the United States the answer is no, although we have made significant progress, as attested to by a continuous decline in the coronary heart disease death rate. Elsewhere, beyond our borders, the problems has worsened. The World Health Organization predicts that in 2020 ischemic heart disease will rank first on the list of global disease burdens. How can this be the case, when discovery of effective drugs and better definition of genetic susceptibility factors have certainly occurred? Indeed, in the last few years we have witnessed the publication of superb studies demonstrating the effectiveness of the statins in primary and secondary prevention of coronary heart disease.

This is a complex question that has no single answer, but at least one aspect of this situation is well recognized: statins are not prescribed as much as they should be. This then raises another question: Why are physicians not prescribing these medications?

This volume is not an answer to that question, but it is intended to provide busy physicians with the information they need to make the best possibile decisions for their patients. For the first time, we bring together in one place the rationale and results of all the major clinical trials initiated to assess the efficacy of the newest and best cholesterol-lowering agents.

The editor, Scott Grundy, has a worldwide reputation as an expert in the cholesterol-lowering field. The chapters, each describing a different study, are written by the researchers who designed and coordinated the trials. So much information is given to practicing physicians nowadays that it often makes their job *more* difficult. This authoritative compendium can make it easier and, in so doing, will greatly benefit patients. Heart attacks may not be gone with the century, but they soon could be if they are prevented and treated appropriately. This book will help.

Claude Lenfant, M.D.
Director, National Heart, Lung and Blood Institute
National Institutes of Health

REFERENCE

1. Brown, MS, Goldstein, JL. Heart attacks: gone with the century? [editorial]. Science 1996; 272(5262):629.

Preface

Clinical trials of cholesterol-lowering therapy constitute a triumph of modern medicine. No question has been more contentious than that of the relationship between serum cholesterol and coronary heart disease (CHD). In contrast, the medical community readily accepted the importance of cigarette smoking and high blood pressure as risk factors. Part of this acceptance grew out of the multiple dangers of smoking and hypertension. Cigarette smoking causes chronic lung disease and lung cancer as well as CHD. Hypertension leads to stroke and chronic renal failure as well as to CHD. The dangers of high serum cholesterol, however, are limited to the development of atherosclerotic disease, particularly CHD. Furthermore, the database that defines the impact of high serum cholesterol on CHD risk was acquired more slowly than that for the other risk factors. Slowly but surely, nonetheless, a broad base of evidence has emerged that links an elevation of serum cholesterol levels to development CHD. The driving force for this evidence was the "cholesterol hypothesis." This is a two-part hypothesis postulating that a rising serum cholesterol level enhances the risk of CHD, whereas a falling level reduces the risk.

The first line of support for the hypothesis that an elevated serum cholesterol causes atherosclerosis came from studies in laboratory animals. For example, the feeding of excess cholesterol to rabbits produced marked hypercholesterolemia and caused deposition of cholesterol in arteries. This deposition closely resembled human atherosclerosis. Since those early studies almost 100 years ago, a great many investigations have been carried out in animals. This research extended to a wide variety of animals, recently including genetically modified animals. Taken in aggregate, this body of research provides a strong base of support for the concept that elevated serum cholesterol—particularly when the cholesterol is carried in low-density lipoproteins (LDLs) and remnant lipoproteins—is highly atherogenic. These early studies in laboratory animals were the foundation of the "cholesterol hypothesis," and a vast number of publications in animal research strongly supports this hypothesis.

The second line of support for the cholesterol hypothesis comes from epidemiological investigations. Prospective and cross-sectional studies have examined the correlation between serum cholesterol levels and incidence (or prevalence) of CHD. Notable examples include the Seven Countries Study and the Framingham Heart Study, but there are many others. On the whole these studies reveal a positive correlation between serum cholesterol levels and risk for CHD. Such investigations have the advantage of comparing the influence of high serum cholesterol to the other major risk factors for CHD.

Three important conclusions emerge from these investigations. First, in countries in which serum cholesterol levels are very low, rates of CHD are also low, even in the presence of other risk factors. This confirms the finding in animal studies that cholesterol-rich lipoproteins are the primary atherogenic factor among all the known risk factors.

The second conclusion is that multivariate analysis of epidemiological investigations indicates that elevated serum cholesterol in high-risk populations accounts for approximately one-third of all CHD. Indeed, after correcting the data for a factor called *regression-dilution bias*, prospective studies reveal that the dangers of serum cholesterol are even greater than suggested by earlier analyses.

Finally, multiple prospective studies demonstrate that serum cholesterol concentrations are positively correlated with CHD risk over a broad range of cholesterol levels, from very low to high. These prospective studies provide useful information about the optimal level of serum cholesterol. Earlier and smaller prospective studies suggested that the optimal level of serum cholesterol is a concentration anywhere below 200 mg/dL. Later investigations, however, reveal that up to one-quarter of major coronary events occur in persons whose total cholesterol concentrations are in the range of 150 to 200 mg/dL; in contrast, CHD events are less frequent when cholesterol levels are ≤150 mg/dL. These findings strongly imply that the optimal level of serum total cholesterol is ≤150 mg/dL. Indeed, populations in which mean levels of total cholesterol are below 150 mg/dL have a very low prevalence of CHD.

A third line of support for the "cholesterol hypothesis" comes from genetic disorders of lipoprotein metabolism. Foremost among these disorders are genetic forms of elevated serum cholesterol (namely, familial hypercholesterolemia) and familial defective apolipoprotein B-100. When these disorders are present, coronary atherosclerosis develops rapidly, leading to early onset of CHD. Genetic forms of hypercholesterolemia can produce premature CHD even in the absence of other risk factors. The occurrence of premature CHD in persons with genetic hypercholesterolemia provides the strongest evidence that a high level of serum cholesterol is a direct cause of atherosclerosis in humans.

Thus, the past century leaves an enormous body of evidence that higher serum cholesterol levels promote the development of CHD. This evidence derives from animal studies, epidemiological investigations, and genetic disorders of lipoprotein metabolism; all of them support the "cholesterol hypothesis." In spite of this support, a critical question remains: Does lowering serum cholesterol reduce the risk of CHD? According to most investigators, ultimate proof of the "cholesterol hypothesis" depends on the demonstration of benefit in clinical outcomes from reducing cholesterol levels. Such proof can be obtained only through controlled clinical trials. The history of controlled clinical trials of cholesterol-lowering therapy goes back almost 40 years. The first clinical trials of this type were initiated in the 1960s. These clinical trials can be divided into earlier end-point trials, angiographic trials, and recent trials with HMG-CoA reductase inhibitors (statins). To introduce this monograph on cholesterol-lowering trials, these categories are examined briefly. They are considered in more detail in the first chapter, and the key trials are described and analyzed thoroughly in subsequent chapters.

Between the late 1960s and 1990, a sizable number of smaller clinical trials were carried out to determine whether cholesterol-lowering therapy would reduce the risk of CHD. These trials employed either diet or drugs to reduce serum cholesterol levels. The studies have been of two types: primary prevention trials (prevention of new-onset CHD) and secondary prevention trials (prevention of recurrent coronary morbidity and mortality in patients with established CHD). Most of these clinical trials suggested some benefit from lowering cholesterol. In several trials, the favorable result was statistically significant; in others, strong nonsignificant trends were recorded. In spite of these results, however, the evidence from single trials was not robust enough to convince the medical community of the benefit of cholesterol-lowering therapy in high-risk patients. More recently, these multiple trials have been subjected to meta-analysis with more encouraging results. Meta-analysis strengthened the conclusion that lowering of serum cholesterol levels reduces risk for CHD. On the other hand, meta-analysis also raised a disturbing question. It failed to show that lowering cholesterol reduces total mortality. Although this failure likely reflected a lack of statistical power in the analysis, the possibility could not be excluded that reducing serum cholesterol concentrations induces side effects that offset the benefit of reducing serum cholesterol on coronary risk. This latter possibility created a widespread skepticism in the medical community as to the benefit of cholesterol-lowering therapy.

A secondary category of clinical trials included those that have addressed the question of whether cholesterol-lowering therapy will slow the progression or promote the regression of coronary plaques. This question was studied by coronary angiographic techniques. In most of these studies aggressive measures, often including drug combinations, were employed to lower cholesterol levels. The result of these trials were strikingly consistent. They showed that lowering cholesterol levels does indeed retard progression and promote regression of coronary lesions. On the other hand, the observed changes in lesion size were small and would not be expected to affect clinical outcomes. Surprisingly, however, the groups of patients receiving cholesterol-lowering therapy manifested a substantial reduction in the incidence of acute coronary syndromes (unstable angina and acute myocardial infarction). This discrepancy between small changes in coronary lesions as observed grossly and large reductions in acute coronary syndromes raises the intriguing questions of mechanism. The hypothesis currently under consideration is that reducing cholesterol levels "stabilizes" coronary lesions and prevents plaque rupture, the major cause of acute coronary syndromes. Because of their support of the plaque-stabilizing hypothesis, the angiographic trials made an important contribution to the current rationale for aggressive cholesterol-lowering therapy in high-risk patients.

The final category of clinical trials includes studies that have tested statin therapy. The statin are powerful cholesterol-lowering drugs. They reduce all atherogenic lipoproteins, both LDLs and remnant lipoproteins. Since 1993, five major clinical trials have been reported in which statin monotherapy was the critical variable. Three of these trials were secondary prevention trials, and two were primary prevention trials. The outcomes of all five were dramatic. Statin therapy consistently reduced major coronary events by about one-third, and in three of the trials, statin treatment lowered total mortality. These favorable outcomes occurred in the absence of significant side effects. No increase in noncardiovascular mortality emerged in any of the trials; thus, the possibility that lowering cholesterol levels is inherently dangerous was largely ruled out. The outcomes of these trials with statins have greatly affected the thinking of the medical community about the role of serum cholesterol in development of CHD and about the benefit of lowering cholesterol in prevention of CHD. In recent years the use of statins in clinical practice has proliferated and so has the number of different statins available to physicians.

The availability of statins and their efficacy as proven through clinical trials have been a boon to physicians and patients. Even so, the very existence of an effective cholesterol-lowering drug brought to the fore new issues for both primary and secondary prevention. For the latter, a pressing issue is how aggressive the lowering of LDL cholesterol should be in patients with established CHD. Moreover, the benefit derived from reducing LDL concentrations, as shown in the statin trials, naturally leads to the question of whether further benefit can be obtained by modifying other lipid risk factors, e.g., raised triglycerides and low levels of high-density lipoproteins (HDLs). Turning to primary prevention, the critical question at present is how to identify and select patients for aggressive intervention with cholesterol-lowering drugs. This question is complex and requires a balancing of absolute risk of patients with efficacy, safety, and costs of drug therapy. A central issue in the selection of patients for aggressive cholesterol-lowering therapy is how best to identify the high-risk patient through global risk assessment.

These questions have become key topics for future research. Indeed, several new clinical trials have been initiated to address these timely issues. New cholesterol-lowering trials are summarized briefly in the first chapter of this book. Although most current trials will not be completed for another three to five years, they promise to provide important and incremental information that will modify current practice.

In conclusion, this volume brings together up-to-date information on all the recent clinical trials of cholesterol-lowering therapy. Emphasis of this monograph rightly belongs on the large clinical trials of statin therapy. These statin trials can be viewed as the culmination of almost 40 years of effort to test the efficacy and safety of reducing cholesterol levels. Their documentation that cholesterol-lowering therapy reduces total mortality, as well as coronary morbidity, represents a great advance and gives a final stamp of approval to the benefit of reducing serum cholesterol levels in high-risk patients.

Scott M. Grundy

Contents

Contributors

Eugene Braunwald, M.D. Department of Medicine, Brigham and Women's Hospital and Harvard Medical School, Boston, Massachusetts

B. Greg Brown, M.D., Ph.D. Division of Cardiology, Department of Medicine, University of Washington School of Medicine, Seattle, Washington

Henry Buchwald, M.D., Ph.D. Department of Surgery, The University of Minnesota, and the POSCH Group, Minneapolis, Minnesota

Robert P. Byington, M.P.H., Ph.D. Department of Public Health Sciences, Wake Forest University School of Medicine, Winston-Salem, North Carolina

Michael Clearfield, D.O. Department of Internal Medicine, University of North Texas Health Science Center, Fort Worth, Texas

Thomas G. Cole, Ph.D. Department of Medicine, Washington University, St. Louis, Missouri

Barry R. Davis, M.D., Ph.D. Department of Biometry, University of Texas School of Public Health, Houston, Texas

Lt. Col. John R. Downs, M.D., U.S.A.F., M.C. Internal Residency Training Program, Wilford Hall Medical Center, Lackland Air Force Base, Lackland, Texas

Melissa Ferraro-Borgida, M.D. Department of Medicine, University of Connecticut School of Medicine, Farmington, and Hartford Hospital, Hartford, Connecticut

Allan Gaw, M.D., Ph.D.   Institute of Biochemistry, Royal Infirmary, Glasgow, Scotland

David J. Gordon, M.D., Ph.D.   Division of Heart and Vascular Diseases, National Heart, Lung, and Blood Institute, National Institutes of Health, Bethesda, Maryland

Antonio M. Gotto, Jr., M.D. D. Phil.   Department of Medicine, Weill Medical College of Cornell University, New York, New York

Scott M. Grundy, M.D., Ph.D.   Departments of Internal Medicine and Clinical Nutrition, Center for Human Nutrition, University of Texas Southwestern Medical Center at Dallas, Dallas, Texas

C. Morton Hawkins, D.Sc.   Department of Biometry, University of Texas School of Public Health, Houston, Texas

Cheryl L. Holmes, M.D.   Department of Medicine, Vancouver Hospital and Health Sciences Centre, University of British Columbia, Vancouver, British Columbia, Canada

David Hunt, M.B.B.S., M.D., F.R.A.C.P., F.A.C.C.   Department of Cardiology, Royal Melbourne Hospital, Melbourne, Victoria, and The LIPID Study Group, Australia

G. B. John Mancini, M.D., F.R.C.P.C., F.A.C.C.   Division of Cardiology, Department of Medicine, Vancouver Hospital and Health Sciences Centre, University of British Columbia, Vancouver, British Columbia, Canada

Lemuel A. Moyé, M.D., Ph.D.   Department of Biometry, University of Texas School of Public Health, Houston, Texas

Terje R. Pedersen, M.D.   Department of Cardiology, Aker University Hospital, Oslo, Norway

Marc A. Pfeffer, M.D., Ph.D.   Department of Medicine, Cardiovascular Division, Brigham and Women's Hospital and Harvard Medical School, Boston, Massachusetts

Jean L. Rouleau, M.D., F.R.C.P.   Department of Medicine, Division of Cardiology, University Health Network and Mount Sinai Hospital, Toronto, Ontario, Canada

Frank M. Sacks, M.D.   Department of Nutrition, Harvard School of Public Health, and Department of Medicine, Brigham and Women's Hospital and Harvard Medical School, Boston, Massachusetts

Michael Schulzer, M.D., Ph.D.   Departments of Medicine and Statistics, Vancouver Hospital and Health Sciences Centre, University of British Columbia, Vancouver, British Columbia, Canada

James Shepherd, M.D., Ph.D.   Institute of Biochemistry, Royal Infirmary, Glasgow, Scotland

Andrew M. Tonkin, M.D.   National Heart Foundation of Australia, Melbourne, Victoria, and The LIPID Study Group, Australia

David Waters, M.D.   Division of Cardiology, Department of Medicine, San Francisco General Hospital, San Francisco, California

Stephen Weis, D.O.   Department of Internal Medicine, The University of North Texas Health Science Center, Fort Worth, Texas

Xue-Qiao Zhao, M.D.   Division of Cardiology, Department of Medicine, University of Washington School of Medicine, Seattle, Washington

1

Cholesterol-Lowering Clinical Trials

A Historical Perspective

Scott M. Grundy
University of Texas Southwestern Medical Center at Dallas, Dallas, Texas

The "cholesterol hypothesis" holds that increasing concentrations of serum cholesterol will raise the risk for coronary heart disease (CHD), whereas decreasing of serum cholesterol levels will reduce the risk (Fig. 1). This hypothesis is rooted in studies in experimental animals that were carried out early in the twentieth century. In these studies, feeding large amounts of cholesterol to animals led to marked hypercholesterolemia and to cholesterol accumulation within the arterial wall. This accumulation resembled the first stages of human atherosclerosis. Later, premature atherosclerotic disease was observed in people who had very high concentrations of serum concentrations. These congruent findings created increasing interest in the cholesterol hypothesis and laid the foundation for a great expansion of research in the second half of the 20th century. Between 1950 and 1975, newly acquired data pertained mainly to the ascending aim of the cholesterol hypothesis, i.e., increasing cholesterol levels impose greater risk (Fig. 1A); the last quarter of a century has witnessed a growing body of data supporting a reversal of risk by the lowering of serum cholesterol (Fig. 1B).

Experimental evidence undergirding the concept that a rising serum cholesterol increases CHD risk is of several types. Abundant data (1,2) in laboratory animals reveal that elevated serum cholesterol, induced by various means, promotes atherosclerosis, the major cause of CHD. Epidemiological surveys in humans (3,4) further document a positive relationship between serum cholesterol levels and incidence of CHD. More support comes from the high incidence of premature CHD in patients with genetic forms of hypercholesterolemia (5,6). Finally, a growing body of literature from in vitro studies provides insight as to the mechanisms whereby a high serum cholesterol level promotes atherogenesis

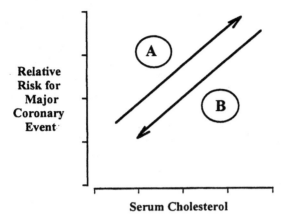

Figure 1   The cholesterol hypothesis. (A) Increasing serum cholesterol levels raises the relative risk for major coronary events. (B) Reducing serum cholesterol lowers risk for major coronary events.

(7). These multiple lines of evidence leave little doubt that increasing serum cholesterol levels will raise the risk for CHD (Fig. 1A).

Although the finding that elevated concentrations of serum cholesterol predispose to CHD makes it likely that reducing cholesterol levels will decrease the risk for CHD, this theoretical probability alone did not justify a major national commitment to lower serum cholesterol concentrations in the general public. Several important questions had to be answered before a massive public health intervention to reduce serum cholesterol could be advocated. Foremost, was the question of whether a lowering of serum cholesterol *actually* will reduce risk for CHD. Early investigators (8) speculated that the association between high cholesterol levels and CHD may not be causal. The abundance of different lines of evidence linking serum cholesterol to atherosclerosis strongly supported a causal connection, but for a long time this link remained uncertain in the minds of skeptical investigators (9,10). Equally germane were questions pertaining to the magnitude of risk reduction that can be realized from cholesterol-lowering efforts and how this benefit compares to that derived from management of other coronary risk factors. Another critical issue is whether cholesterol-lowering interventions are safe, and particularly, whether benefits in CHD risk reduction outweigh the untoward effects of the intervention (11,12). This issue was closely tied to that of all-cause mortality, i.e., does cholesterol lowering reduce total mortality? Finally, it has been asked whether widespread intervention on elevated serum cholesterol is cost-effective (13,14). In a society suffering from escalating costs of medical care, any major program of intervention must be justified on financial grounds as well as on efficacy of intervention. Many investigators have held that the only way to resolve these issues is through controlled clinical trials. First efforts to carry out cholesterol-lowering trials started in the 1950s and 1960s. These trials failed to answer the key questions, but they laid the foundation for future trials. Moreover, they were important in a negative sense; i.e., they uncovered the necessity for more effective cholesterol-lowering therapies, for larger study groups, and for more prolonged periods of intervention.

The purpose of this chapter is to review briefly the history of clinical trials of cholesterol-lowering therapies. The available trials can be categorized in several ways—e.g., diet versus drug, primary versus secondary prevention, or angiographic versus clinical endpoints. In this chapter, the major division will be made between trials of primary prevention and secondary prevention; other categories will be subsumed under these major headings. First, earlier primary and secondary prevention trials will be reviewed. Included for review will be those studies leading up to trials using HMG CoA reductase inhibitors (statins). Earlier studies will be examined to illustrate how issues of trial design have evolved in the cholesterol field. Second, cholesterol-lowering trials in which the major endpoints were changes in coronary plaque size, as determined by coronary angiography, will be reviewed. Third, the recent trials of statin therapy will be summarized. Finally, new trials that are currently under way will be previewed. At the end of each section, the unresolved issues and unanswered questions will be considered. In general, inadequacies of earlier studies set the stage for more ambitious trials to follow.

I. EARLIER PRIMARY PREVENTION TRIALS

A. Diet-Heart Theory

Few issues have been more controversial in the cholesterol field than that of the diet–heart question. This question typically has been posed in a deceptively simple form, namely, whether modifying the composition of the diet of the general public will produce a significant reduction in risk for CHD. The emphasis on dietary prevention had its origins in animal studies in which high intakes of dietary cholesterol led to hypercholesterolemia and atherosclerosis. Dietary cholesterol itself became the first potential villain. In humans, however, in contrast to some animals, high intakes of cholesterol were found not to cause a marked rise in serum cholesterol levels (15,16); this negative result shifted the focus to other dietary factors. Both clinical investigation and epidemiological studies pointed the finger at "saturated fats" as being a major dietary cause of elevated serum cholesterol concentrations. Feeding studies (17–19) in humans revealed that diets rich in saturated fatty acids raise serum cholesterol levels when compared to diets high in unsaturated fatty acids. The earliest feeding studies (17–19) further suggested that the greatest differential between serum total cholesterol levels occurs when polyunsaturated fatty acids are substituted for saturated fatty acids. Prospective epidemiological studies (3) likewise implicated saturated fatty acids as the major-culprit dietary component for raising serum cholesterol levels.

The latter studies (3) further linked high intakes of saturated fatty acids to an increased incidence of CHD. Thus, on the basis human feeding studies (17–19) and epidemiological data (3), most earlier researchers agreed that the control arm of any diet-heart trial should consist of a diet relatively high in saturated fatty acids. Human feeding studies (17–19) further suggested that linoleic acid, the predominant polyunsaturated fatty acid, is a *cholesterol-lowering* fatty acid; hence the preferred "treatment" arm for cholesterol-lowering trials was a diet rich in linoleic acid. In retrospect, the choice of linoleic acid is interesting. The decision was based largely on human feeding studies, not on epidemiological surveys. Among populations of the world that have low rates of CHD, none consume a diet high in polyunsaturated fatty acids. One population having a low prevalence of CHD is that living in the Mediterranean region, notably Greece, Crete, and southern Italy (3); the diet in this region is rich in olive oil, and hence high in monounsaturated fatty acids (mainly oleic acid). Other low-risk populations are those of the Far East (e.g., Japan and China); in this region the diet traditionally has been low in total fat, and hence high in carbohydrate (3). The low rates of CHD in the Far East have inspired many public health agencies in the United States to recommend that the general population adopt a low-fat diet (20,21). In contrast, most designers of controlled clinical trials chose to contrast saturated with polyunsaturated fatty acids. From the early 1960s, influential investigators had conceived of a definitive diet-heart trial, the purpose of which was to determine whether reduction of serum cholesterol levels with dietary modification will prevent CHD.

1. Diet–Heart Feasibility Study

In 1960, a group of investigators headed by Dr. Irvine H. Page established the Executive Committee on Diet and Heart Disease to explore the possibility of a "well controlled mass field trial to test the hypothesis that alternation of amount and type of fat and amount of cholesterol in the diet would decrease the incidence of clinical coronary heart disease in middle-aged American men." This committee surmised that the feasibility of a large study, which was estimated to require up to 100,000 men, could not be determined without the benefit of a smaller, pilot study. In 1962, therefore, the committee, including six principal investigators, obtained a research grant from the National Heart Institute (NHI) to carry out a collaborative feasibility study (22). This study was designed to address a variety of issues: Would participants be willing to volunteer for a prolonged dietary study? Could they adhere to the study diets? Would they be willing to be assigned to a control group? Would the costs of experimental foods be

Table 1 Summary of Results of the Diet-Heart Feasibility Study (National Diet-Heart Study)

Group	Diet	Serum cholesterol reduction (1 year)
B	30% total fat (P/S = 1.5) 275 mg cholesterol	10%
C	34% total fat (P/S = 2.0) 275 mg cholesterol	12%
D (control)	37% total fat (P/S = 0.4) 316 mg cholesterol	4%

prohibitively high? What would be the effects of seasonable variability of serum cholesterol levels in a long-term study? Would an adequate reduction in serum cholesterol concentration be achieved to effect a favorable outcome? Would a double-blind study be possible? One thousand men were recruited for this feasibility study. To assist the men to obtain the goals of therapy, they were given specially prepared meats, baked goods, oils, and dairy products. In addition, the participants underwent extensive dietary counseling throughout the study.

The composition of the diets and responses in serum cholesterol levels at 1 year of the Diet–Heart Feasibility Study (22) are shown in Table 1. The average serum cholesterol level at entry for all participants was 230 mg/dL. The groups receiving cholesterol-lowering diets had greater reductions in serum cholesterol than did the control group. However, the decline in cholesterol levels were about 25% less than predicted by standard diet equations. In spite of the relatively small decreases in serum cholesterol levels, the study investigators concluded that "a full-scale mass field trial can be carried out as a cooperative study with a similar approach to planning, design, protocol, structure, and organization [as the Diet–Heart Feasibility Study]." The investigators further recommended that "such studies should be planned and put into operation."

2. Diet–Heart Review Panel Report

In 1968, the National Heart Institute convened the Diet–Heart Review Panel under the chairmanship of Dr. Edward H. Ahrens Jr. to independently assess the results and recommendations of the Diet–Heart Feasibility Study. This panel aimed to evaluate the strength of the evidence relating diet to serum lipids and myocardial infarction, and to make recommendations as to what types of coronary prevention studies should be carried out. This panel presented its report in 1969

(23). Its authors emphasized that in the absence of proof of the diet–heart hypothesis a firm recommendation to the general public about dietary change is premature. The report concluded that the Diet–Heart Feasibility Study demonstrated that a large-scale trial can be double-blinded and could be carried out successfully. The panel went on to recommend the initiation of primary and secondary trials, which could be done simultaneously.

3.   National Heart and Lung Institute Task Force Review

In 1971, the National Heart and Lung Institute (NHLI) convened a task force to consider a major cholesterol-lowering trial. This task force was chaired by Dr. Elliott V. Newman. After extensive deliberations, the task force concluded that a national diet–heart trial is not feasible and should not be carried out (24). The recommendations of the Diet–Heart Review Panel (23) received special attention, but were not found to be fully convincing. In addition, the task force independently reviewed previous experience with clinical trials, and it held discussions with various governmental departments and industry representatives. Advice against a national diet–heart trial was based on three convictions: a definitive answer to the diet–heart question might not be forthcoming from such a trial; a well-controlled study in a large, free-living, mobile population probably is not practical; and the projected manpower and costs would be excessive. A variety of other practical and technical concerns were voiced. The task force instead recommended that priority be given to other trials that are more feasible, such as those employing cholesterol-lowering drugs. This report appears to have been the death knell for a national diet–heart trial, at least one sponsored by the National Institutes of Health.

B.   Earlier Diet Trials of Primary Prevention

Although a definitive diet–heart trial was never funded, a series of smaller dietary trials were carried out to examine the diet–heart question. The results of these studies are of interest both because they provided suggestive evidence of benefit from lowering of serum cholesterol levels by dietary alteration and because they revealed the limitations of small clinical trials. The results of these studies can be reviewed briefly.

1.   Los Angeles Study

This was a controlled trial of a diet high in polyunsaturated fatty acids for prevention of atherosclerotic complications (25). The study was carried out in 846 middle-aged and elderly male residents of a Veterans Administration domiciliary unit. Ages of patients ranged from 54 to 88 years (mean 65.5 years). Among these, 422 patients received a conventional control diet that was relatively high in saturated fatty acids; another 424 were fed the experimental diet, high in li-noleic acid. The latter diet reduced serum total cholesterol by 13% compared to the control diet. Patients continued on these diets for approximately 8 years. The primary endpoint of the study was sudden death or myocardial infarction. The difference in primary endpoint at the end of the study was not statistically sig-nificant. However, in the subgroup of patients who were < 65.5 years old, sta-tistical significance was obtained for total atherosclerotic events (sudden death + myocardial infarction + cerebral infarction). The authors reached the qualified conclusion that "lowering of serum-cholesterol levels by dietary means can lower the incidence of atherosclerotic complications in men aged 54–65, but not nec-essarily at old age levels." Even so, there appeared to be some excess non-atherosclerotic mortality in the experimental group in the late stages of the trial. The authors speculated that perhaps the experimental diet carried some "toxic effect." Subsequently, these authors reported that patients receiving the diet high in polyunsaturated fatty acids had an increased frequency of cholelithiasis (26).

2. Minnesota Study

This study was initiated in 1968 and was an outgrowth of the National Diet–Heart Feasibility Study (27). It was 4.5-year, open-enrollment, double-blind random-ized clinical trial of the effects of a lipid-lowering diet on cardiovascular risk. The trial was conducted in six Minnesota state mental hospitals and one nursing home. Subjects included 4393 institutionalized men and 4664 institutionalized women. The control diet contained 39% fat (18% saturated and 5% polyunsatu-rated fatty acids) and 446 mg/day of cholesterol. The intervention diet consisted of 38% fat (9% saturated and 15% polyunsaturated fatty acids) and 166 mg/day of cholesterol. Serum cholesterol levels were 28 mg/dL lower on the interven-tion diet. At the end of the study, no statistically significant differences between the two dietary regimens were observed for cardiovascular events, cardiovascu-lar deaths, or total mortality; still, a favorable trend for these endpoints was noted in younger groups. The relatively short duration of this study may well have less-ened its power to detect a significant difference between the two groups.

3. Finnish Study

In 1972, the results of a 12-year clinical trial were published from Finland (28). The trial was conducted in two mental hospitals; it included approximately 3000 patients. One hospital served a cholesterol-lowering diet, and the other, a con-trol diet. The cholesterol-lowering diet contained large amounts of linoleic acid. After 6 years, the two hospitals reversed their diets and continued for another 6 years. The two different dietary patterns resulted in significant differences in serum cholesterol levels. Overall, in the male patients, the cholesterol-lowering diet was associated with a significant reduction in CHD mortality; moreover, there was a strong trend toward a reduction in total mortality, although statistical significance was not achieved. In women, CHD mortality showed a trend toward a decrease, but the trend was not significant by statistics. Total mortality rates in women were not affected by the different diets. The finding of a highly significant reduction in CHD mortality among male patients in this primary prevention of diet modification study is remarkable. The results led the investigators to conclude that a cholesterol-lowering diet is efficacious in the prevention of CHD mortality.

4. Oslo Study

This study was initiated in 1972 and was a randomized, primary prevention trial of 1232 high-risk men in Oslo, Norway (29). During a 5-year period, the experimental group received dietary and antismoking advice. At entry about 80% of the men were daily cigarette smokers and their baseline serum cholesterol averaged 329 mg/dL. Patients in the treatment arm were counseled to eat less saturated fat and cholesterol and to consume a moderate increase in polyunsaturated fat. At the study's end, the treatment group experienced a 47% reduction in incidence of major coronary events compared to controls. Analysis of trial results indicated that a net difference of 10% in serum cholesterol between the groups accounted for most of the reduction in coronary events; antismoking efforts were relatively unsuccessful in the treatment arm.

5. Multiple Risk Factor Intervention Trial (MRFIT)

This large trial (30) included 12,866 high-risk men of ages 38 to 57 years. Subjects were randomized to either a special intervention (SI) regimen or usual care (UC). Special intervention consisted of intensified measures of blood pressure control, smoking cessation, and blood cholesterol lowering through dietary therapy. The three major risk factors (hypertension, smoking, and hypercholesterolemia) were mitigated in the SI group compared to the UC group. In spite of a reduction in risk factors, in the final analysis of the trial, there were no significant differences in CHD mortality between the two groups. Although the overall outcome was negative, the authors speculated that trends toward favorable benefit from reduction in smoking and blood cholesterol were confounded by an increase in fatal side effects from drug therapy for hypertension. MRFIT was not specifically a cholesterol-lowering trial, and it failed to document risk reduction from a cholesterol-lowering diet.

6. Unresolved Issues

Earlier dietary trials (25–30) of cholesterol lowering in primary prevention fell short of a definitive result to the diet-heart issue. Several trials nonetheless suggested benefit from cholesterol-lowering diets; and in fact, some showed statistically significant reductions in CHD events (or mortality) on a therapeutic diet. On the whole, however, these earlier dietary trials failed to convince a skeptical medical community that public health recommendations for dietary modification for the primary prevention of CHD are justified. In fact, in spite of their promising results, they generally have been considered to be insufficient to substitute for the definitive diet–heart study that was rejected by the NHLBI Task Force (24).

C. Earlier Drug Trials of Primary Prevention

Since drugs generally are more efficacious for lowering serum cholesterol levels than is dietary modification, the concept evolved that cholesterol-lowering trials using drugs should be more likely to prove benefit than would trials of dietary modification. Moreover, drug trials should be less expensive and less cumbersome than dietary trials. Even so, one drawback of these earlier trials was that highly efficacious cholesterol-lowering drugs were not available. Nonetheless, it was hoped that any positive result from drug trials would be a strong impetus for both dietary change and development of new drugs.

1. World Heath Organization Trial

This was a large, double-blind trial to test the efficacy of a fibric acid, clofibrate, to reduce the incidence of ischemic heart disease in middle-aged men (31). The WHO study was a multicenter trial carried out in the United Kingdom and Eastern Europe. It included 15,745 men, aged 30 to 59 at entry. The trial had two control groups; 5000 men were selected randomly from the bottom third of the serum cholesterol distribution and another 5000 were selected from the top third. These two control groups were compared to another 5000 men from the top third of the cholesterol distribution who were treated with clofibrate. Treatment with clofibrate reduced serum cholesterol levels by 9%. In comparison with the high-cholesterol group, clofibrate therapy lowered the incidence of ischemic heart disease by 20% ($P < .05$); nonfatal myocardial infarctions were reduced by 25%. In contrast, total mortality in the group treated with clofibrate was significantly higher than the high-cholesterol control group ($P < .05$). The excess of deaths in the clofibrate group appeared to be related mainly to diseases of the biliary and gastrointestinal tracts. The increase in total mortality by clofibrate therapy in the WHO trial created considerable alarm in the medical community. Several possible dangers of cholesterol lowering were visualized. First, clofibrate in particular, and fibric acids in general, could be toxic drugs. Indeed, other studies demonstrated that treatment with fibric acids raises the likelihood of cholesterol gallstones; and complications from gallstones could underlie increased mortality. Alternatively, a reduction in serum cholesterol levels per se might be dangerous. If so, introduction of cholesterol-lowering therapy for the purpose of reducing coronary risk could be counterproductive. Following the outcome of the WHO trial, Dr. Michael Oliver, its principal investigator, strongly emphasized the potential dangers of cholesterol-lowering therapy (32).

2. Lipid Research Clinics (LRC) Coronary Primary Prevention Trial (CPPT)

This trial was a multicenter study to test the efficacy of cholesterol lowering with cholestyramine for reducing risk for CHD in 3806 middle-aged men with hypercholesterolemia (33,34). The study lasted 7.4 years. LDL cholesterol levels during cholestyramine therapy were 12.6% lower than those in the placebo group. This lower LDL level was accompanied by a 19% reduction in major coronary events ($P < .05$). In addition, cholestyramine therapy reduced incidence rates for new positive exercise tests by 25%, new angina by 20%, and need for coronary bypass surgery by 21%. Moreover, patients having a fall in LDL cholesterol levels on therapy experienced an incidence of CHD half that of men who remained at pretreatment levels. In spite of the significant reduction in major coronary events and in other clinical indicators of coronary progression, cholestyramine therapy failed to reduce total mortality. In fact, noncardiovascular deaths were numerically greater in the treatment group than in the placebo group; this increment, was not, however, statistically significant. Although the LRC-CPPT was not powered to test the total mortality question, skeptical investigators pointed to the failure to reduce total mortality as a lack of overall efficacy of cholestyramine therapy.

3. Helsinki Heart Study

This trial randomized 4061 asymptomatic middle-aged men (ages 40 to 55 years) to either gemfibrozil therapy or placebo to test efficacy of drug therapy for reducing major coronary events (35). Gemfibrozil, like clofibrate, is a fibric acid. Agents of this class have their primary action on triglyceride metabolism; they are not effective LDL-lowering drugs. Recent research (36) indicates that fabric acids act primarily by activating nuclear receptors called PPARα. Activation of these receptors enhances or suppresses the synthesis of a series of proteins known to affect triglyceride and HDL metabolism; among these actions are an inhibition of synthesis of apolipoprotein CIII, stimulation synthesis of lipoprotein lipase, and enhancement of production of enzymes involved in fatty acid oxidation. These changes result primarily in a lowering of serum triglyceride levels. HDL cholesterol levels are raised somewhat by gemfibrozil therapy secondary to triglyceride lowering (35). Any reduction in LDL cholesterol, if it occurs at all, is small.

The Helsinki Heart Study (35) lasted for 5 years. In the group receiving the active drug, major coronary events were reduced by 34% compared to placebo ($P < .02$). Rates of fatal myocardial infarction, however, were not different between the placebo and treatment groups. Again, the study was not powered to test the total mortality issue; consequently, in spite of reducing major coronary events, treatment with gemfibrozil did not significantly reduce all-causes mortality. The decrease in major coronary events on gemfibrozil therapy was attributed mainly to a rise in HDL cholesterol levels (35) and, more recently, to a reduction in triglycerides (37).

D. Unresolved Issues

The three studies in this category (31,33–35) provided strong evidence that lipid lowering can reduce major coronary events. Statistically significant reductions in coronary events occurred in all of these studies. Decreases in serum cholesterol levels were nonetheless relatively small; further, the benefit in risk reduction was not great enough to convince skeptics that lipid-lowering drug therapy in primary prevention is justified. Still, the results of the LRC-CPPT(33,34) helped to convince the National Heart, Lung, and Blood Institute (NHLBI) to sponsor the National Cholesterol Education Program (NCEP) (38,39). Protagonists of the NCEP believed that the LRC-CPPT (33,34) adequately proved the cholesterol hypothesis (Fig. 1B). Although the hypothesis was proven with drug therapy, the emphasis of the NCEP was shifted to the dietary prevention of CHD (36). In essence, the LRC-CPPT (33,34) was meant to serve as a substitute for a definitive diet-heart trial, and its results were taken to support the diet-heart hypothesis.

On the other hand, the critics of the cholesterol hypothesis withheld their assent to the sufficiency of the LRC-CPPT (33,34) and other primary prevention trials. Foremost among their criticisms was the failure of cholesterol-lowering therapy to reduce total mortality (40). Although these primary prevention trials were not powered to test the total mortality question, their inability to show a reduction in total mortality was taken to mean that cholesterol-lowering therapy in toto is not beneficial. Worse, in the WHO trial (31), there was a statistically significant *increase* in total mortality in patients receiving clofibrate. This finding raised several concerns about the safety of cholesterol-lowering therapy. One concern was that fibrates as a class may carry significant toxicity; but even more ominous for the cholesterol hypothesis was the possibility that reducing cholesterol levels per se is dangerous. Concern about the latter was heightened by reports from epidemiological studies that the subgroups of populations with the lowest cholesterol levels suffer a higher incidence of total mortality than do those with more average levels. Several mechanisms were visualized whereby a lowering of serum cholesterol might be dangerous. Thus, the WHO trial cast a cloud on the cholesterol hypothesis for many years. To make matters worse, even in the LRC-CPPT (33,34) and the Helsinki Heart Study (35), the numbers of noncardiovascular deaths were numerically higher in the treatment group than in the placebo group; although these increments were not statistically higher in the treatment group, they nonetheless heightened skepticism in the minds of some investigators.

II. EARLIER SECONDARY PREVENTION TRIALS

It is of interest that earlier secondary prevention studies of cholesterol-lowering therapy were ever carried out. For many years pathologists had speculated that elevated serum cholesterol levels affect atherogenesis only in its early stages (41). In simplistic terms, the prevailing concept of atherogenesis was that an elevated serum cholesterol promotes the development of fatty streaks whereas other risk factors—cigarette smoking, hypertension, and diabetes—transformed fatty streaks into more advanced lesions (i.e., fibrous plaques and complicated plaques). If this sequence truly holds, cholesterol-lowering therapy should have little benefit in patients with advanced atherosclerotic disease and clinical CHD. In spite of this theoretical improbability of success of cholesterol-lowering therapy in secondary prevention, several trials of this type were mounted and carried out. They included both dietary and drug studies.

A. Earlier Dietary Studies of Secondary Prevention

1. MRC Low-Fat Trial

In 1965, investigators in the U.K. published the results of a secondary prevention trial sponsored by the Medical Research Council using a low-fat diet to lower serum cholesterol levels (42). The study included 264 men having a history of myocardial infarction. In the treatment group, fat intake was reduced to 40 g/day. The control group followed a "normal diet." Serum total cholesterol concentrations were significantly lower in the group receiving the low-fat diet. In spite of lower cholesterol levels in the treatment group, there was no significant difference in rates of recurrent myocardial infarction between the two groups. The authors therefore concluded that for postinfarction men under age 65, a low-fat diet does not improve the prognosis.

2. MRC Soya Study

In 1968, the Research Committee to the Medical Research Council (43) reported a trial designed to test the benefit of diet to lower serum cholesterol levels in men under age 60 who had previously suffered from a myocardial infarction. The treatment group received a diet containing 85 g of soya bean oil daily; the control group consumed their ordinary diet. The duration of the trial averaged 4 years. A 22% reduction in serum cholesterol was achieved with the soya diet, compared to the control group. In spite of the reduction in serum cholesterol levels on the soya diet, there was not a statistically significant reduction in recurrence rates of myocardial infarction between treatment and control groups. The number of recurrent events was lower in the treatment group, but the difference did not reach a level of statistical significance. The authors therefore concluded that the results of this trial "lend little support to the suggestion that a diet [rich in polyunsaturated fatty acids] should be recommended in the treatment of patients who have suffered from a myocardial infarction."

3. Oslo Diet–Heart Study

In 1970, Leren (44) published the result of a secondary prevention dietary study in 412 men, ages 30 to 64. The experimental group received a diet low in saturated fats and cholesterol, and high in polyunsaturates. The control group continued on a "typical Norwegian" diet. After 5 years of study, the treatment groups experienced significantly fewer fatal and nonfatal myocardial infarctions. After 11 years of follow-up, the group receiving the cholesterol-lowering diet had a significantly lower incidence of myocardial infarction mortality. Although the number of patients in this study was small, the results were impressive; they contributed to the evidence that effective cholesterol-lowering therapy is efficacious in secondary prevention.

B. Earlier Drug Studies of Secondary Prevention

1. Scottish Study

In 1971, the Scottish Society of Physicians reported a secondary trial using clofibrate therapy (45). In this trial, 350 patients with established CHD were treated with clofibrate, and another 367 received placebo. After 6 years of therapy, patients treated with clofibrate showed a strong trend towards beneficial outcomes. Total mortality in patients with preexisting angina pectoris was reduced significantly by 62%. In addition, angina patients treated with clofibrate showed a significant 53% reduction in all coronary events (fatal and nonfatal). Patients of all categories combined had a 44% decrease in nonfatal myocardial infarction. The authors concluded that clofibrate therapy reduced mortality and coronary morbidity in patients with angina pectoris, but this benefit was less evident in patients having a history of myocardial infarction.

2. Newcastle Study

In 1971, a group of physicians in the Newcastle-upon-Type Region reported a secondary prevention study in which clofibrate was compared to placebo (46). This 5-year, double-blind clinical trail randomized 497 patients with established CHD to drug and placebo regimens. At the end of the trial, the group treated with clofibrate had sustained significantly fewer deaths and nonfatal myocardial infarctions than did the group on placebo. Drug therapy was well tolerated, and side effects appeared minimal. The protection against recurrent coronary events seemingly was greater in smokers than in nonsmokers. Although many patients had elevations in serum triglycerides at baseline, the favorable effects of clofibrate therapy did not appear to be limited to those who had baseline hypertriglyceridemia.

3. Coronary Drug Project

The major objective of this study was to evaluate the efficacy and safety of several lipid-influencing drugs on recurrent coronary events in patients with previous myocardial infarction (47). The study included 8341 male patients, studied in 53 project centers. The design of the study included six regimens: conjugated estrogens 2.5 mg/day; conjugated estrogens 5.0 mg/day; clofibrate 1.8 g/day; dextrothyroxine (D-thyroxine) 6. 0 mg/day; niacin 3.0 g/day; and placebo. There were approximately five placebo patients for every two patients on any drug regimen. The primary endpoint was total mortality, and secondary endpoints included various major coronary events. During the course of the study, three of the study regimens—2.5 mg/day and 5.0 mg/day of estrogen and 6.0 mg/day of D-thyroxine—were discontinued because of trends indicative of adverse effects. The three other regimens—clofibrate, nicotinic acid, and placebo—were continued until the end of the study. Most patients in the latter three groups had at least 5 years of follow-up.

The patients treated with clofibrate showed no evidence of benefit, either in total mortality or in cause-specific mortality. Mortality rates were essentially the same for the two groups, and no identifiable subgroups had a reduction in total mortality on clofibrate therapy. Patients receiving clofibrate showed a trend toward a reduction in major coronary events (coronary death and myocardial infarction), but this trend was not statistically significant. Moreover, patients getting clofibrate had a significantly significant increase in thromboembolism, and cardiac arrhythmia and had a twofold increase in the rate of gallstone disease. The authors noted that the findings of the Scottish and Newcastle studies (45,46) that clofibrate therapy reduced coronary events in secondary prevention was not confirmed by the Coronary Drug Project (47). They therefore concluded that clofibrate therapy is probably not beneficial in secondary prevention of CHD.

Better results were obtained with nicotinic acid therapy. The latter did not yield a significant reduction in total mortality, nor was a subgroup identified in which a decrease in all-causes mortality was noted. On the other hand, the nicotinic acid group did experience a significantly lower incidence in major coronary events. Risk for recurrent CHD was decreased by about 25%. Further, long-term follow-up of patients who had been treated with nicotinic acid revealed a decrease in total mortality (48). Even so, nicotinic acid was accompanied by a relatively high frequency of side effects, which are known to be characteristic of this drug. The authors concluded that considering the lack of reduction in total mortality and the high incidence of side effects, the benefit of niacin therapy in secondary prevention were marginal at best.

4. Stockholm Study

Another secondary prevention trial was carried out later, in Stockholm (49). This trial combined clofibrate and nicotinic acid into a single-drug regimen. It recruited survivors of myocardial infarction and randomized them into a control group (n = 276) and a treatment group (n = 279). The study lasted 5 years. Combined drug therapy reduced serum cholesterol and triglycerides by 13% and 17%, respectively. Drug treatment was associated with a 26% reduction in total mortality ($P < .05$). Ischemic heart disease mortality was reduced by 36% ($P < .01$). Most of the benefit was observed for patients who had some elevation of serum triglycerides. In fact, about 50% of the patients in the trial carried a diagnosis of *hypertriglyceridemia* (serum triglyceride > 2.0 mmol/L). The authors postulated that the results are consistent with the mode of action of the two drugs. Both can be classified as triglyceride-lowering drugs. A logical conclusion of this study was that triglyceride-lowering drugs would be effective in reducing risk for CHD primarily for patients who are hypertriglyceridemic.

5. Program on the Surgical Control of Hyperlipidemia (POSCH)

This study was a clinical trial of 831 patients with a history of myocardial infarction who were divided between ileal bypass operation and control (50). The ileal bypass operation interrupts the enterohepatic circulation of bile acids; this action releases the feedback inhibition of bile acids on the conversion of cholesterol into bile acids in the liver. As a result, hepatic content of cholesterol is reduced and the synthesis of LDL receptors is enhanced. In the POSCH study, LDL cholesterol levels fell by 38% in patients receiving the ileal bypass. Patients who underwent the ileal bypass operation experienced a 35% reduction in major coronary events, a marked reduction in need for coronary procedures, and a 35% reduction in coronary deaths. Moreover, there was a strong trend toward a reduction in coronary mortality and total mortality. Among earlier secondary prevention trials, POSCH provided the strongest evidence for benefit from cholesterol-lowering therapy. The strength of the POSCH trial grew out of a marked and persistent lowering of LDL cholesterol levels; this study demonstrated clearly that cholesterol lowering reduces major CHD events.

B. Unresolved Issues

Earlier secondary prevention trials provided mixed results although several of them gave positive results. Generally, the drug trials were more positive than the dietary trials. The two MRC dietary studies (42,43) were taken to be negative; small trends toward benefit of cholesterol lowering did not reach statistical significance. The Oslo study (44) was positive, but weaknesses in experimental design left most investigators unconvinced. The drug trials were more suggestive of benefit. Two United Kingdom trials (45,46) of clofibrate therapy generally were positive for a decrease in recurrent coronary events in patients with established CHD. In the Coronary Drug Project (47), clofibrate therapy failed to produce a significant reduction in recurrent events, but a positive result was obtained with nicotinic acid. Also, in the Stockholm trial (49), the combination of clofibrate and nicotinic acid was accompanied by a reduction in total mortality. Nevertheless, these trials suggesting favorable outcomes with fibrate therapy were overshadowed by the negative results of the WHO primary prevention trial with clofibrate (31). The WHO trial (31), combined with the negative result for the clofibrate arm of the Coronary Drug Project (47), left little enthusiasm for fibrate therapy in secondary prevention. The positive result of the more recent POSCH study (50) provided strength to belief that cholesterol lowering will reduce the risk for recurrent coronary events; however, the ileal bypass operation did not offer a practical therapeutic option for many patients with established CHD. Finally, the lack of reduction in total mortality for most of the trials added to the lingering doubt that a true benefit for prolongation of life is achieved by reducing serum cholesterol levels.

III. META-ANALYSES OF EARLIER CHOLESTEROL-LOWERING TRIALS

Since earlier cholesterol-lowering trials taken individually did not satisfactorily confirm the descending limb of the cholesterol hypothesis (Fig. 1B), several investigators have attempted to obtain a more definitive result by pooling the data from multiple trials. This approach has gained wide acceptance in the medical community as constituting proof provided the outcome is robust. The limitations of meta-analysis nonetheless must be recognized. Validity of results depends foremost on a similarity in experimental design. When multiple therapies are used to achieve a single therapeutic response, as in cholesterol-lowering trials, definitive similarities exist, but so do differences. For example, a positive outcome may convincingly show that cholesterol lowering is efficacious; on the other hand, side effects accompanying one mode of therapy may obscure the safety of another. The heterogeneity of multiple studies must be kept in mind when evaluating the results of meta-analysis of cholesterol-lowering trials.

In 1990, Mouldoon et al. (51) reported a meta-analysis of earlier primary prevention trials. The report included results from 24,847 men of mean age 47.5 years. Their analysis indicated that cholesterol-lowering therapy in general produces a strong trend towards a reduction in CHD mortality ($P = .06$). Separate analysis of drug trials revealed a significant reduction in CHD mortality ($P = .04$). On the other hand, this benefit appeared to be negated by an increase in noncardiovascular deaths. In the pooled data, Mouldoon et al. (51) found that total mortality was increased in the aggregate treatment groups. This outcome was heavily influenced by the adverse outcome of the WHO trial (37), but other trials contributed some excess deaths in patients receiving cholesterol-lowering therapy. The authors were particularly impressed by excess deaths from violence—e.g., accidents, suicide, and homicide. They postulated that reducing serum cholesterol levels may produce changes in mental function that predisposes to violent behavior.

In 1990, a meta-analysis of earlier secondary prevention trials was published by Rossouw et al. (52). Their analytic exercise pooled data from available trials in which either diet or drugs were employed for cholesterol lowering therapy. This analysis was updated in 1993 (53). Serum cholesterol levels in these earlier trials were reduced by an average of about 15%. Compared to the control groups, patients receiving cholesterol-lowering therapy had a 26% reduction in nonfatal myocardial infarction, a 14% decrease in fatal myocardial infarction, a 11% fall in all cardiovascular deaths, and a 9% decrease in total mortality. The apparent reduction in total mortality approached statistical significance (confidence interval = 0.81 to 1.01). Noncardiovascular deaths showed no trend toward an increase; this finding spoke against the concept that reductions in serum cholesterol levels per se leads to serious or fatal side effects.

Also in 1990, Holme (54) published another meta-analysis of randomized trials evaluating effects of cholesterol reduction on CHD incidence and total mortality. This analysis combined both primary and secondary prevention trials, including both dietary and drug intervention. In this analysis, cholesterol-lowering therapy was found to reduce CHD by about 10%. Drug trials were found to be more effective for reducing CHD incidence presumably because of a greater cholesterol lowering. The overall results showed that for every 1% reduction in serum total cholesterol, the incidence of CHD was reduced by about 2.5%. On the other hand, total mortality was found to be increased by intervention by about 4% compared to control ($P > .10$). Effect of cholesterol intervention was not uniform in all categories. For example, a trend toward a decrease in total mortality appeared to be greater in secondary than in primary prevention trials.

Other meta-analyses were subsequently carried out. In 1992, Ravnskov (55) reported a combined analysis of all clinical trials of cholesterol-lowering therapy. He found that in 22 controlled, cholesterol-lowering trials the CHD mortality and total mortality were not significantly reduced by treatment. Moreover, in his view, other meta-analyses showing a decrease in CHD morbidity were biased in selection of studies and were based on preferential citation of supportive trials. This assessment led Ravnskov (55) to conclude that "lowering serum cholesterol concentrations does not reduce mortality and is unlikely to prevent CHD." Later, Gordon (56,57) provided a scholarly review of the strengths and weaknesses of meta-analysis of cholesterol-lowering trials. He emphasized that a strong possibility of bias in the selection of trials occurs when meta-analyses are constructed. Still, according to Gordon, some selection is warranted; the fact remains that trials differ in quality and mode of therapy. For example, inclusion of trials of hormones for lipid lowering is problematic; results obtained in these trials revealed that hormones unique toxic effects that could obscure any benefits from other modes of therapy (47). In addition, Gordon (57) suggested that fibric acids are suspect as carriers of toxicity. Regardless, he concluded that in spite of limitations, meta-analysis of earlier primary and secondary prevention trials yields two solid points. First, cholesterol-lowering therapy almost certainly reduces the incidence of CHD. But second, comprehensive meta-analysis of earlier cholesterol-lowering trials failed to demonstrate a significant overall reduction in all-causes mortality. This latter point was undoubtedly a major factor in the reluctance of the medical community before 1993 to accept the value of cholesterol-lowering therapy in either primary or secondary prevention.

In spite of the failure of meta-analysis to confirm a decrease in all-causes mortality by cholesterol-lowering therapy, earlier clinical trials largely proved the downward arm of the cholesterol hypothesis (Fig. 1B). They reveal that reducing serum cholesterol levels will in fact reduce the risk for CHD. In addition, Law et al. (58) attempted to assess "by how much and how quickly does reduction of serum cholesterol concentrations lower risk for ischemic heart disease?" This assessment was carried out by comparing data derived from epidemiology and clinical trials. They compared the results from 10 prospective (cohort) study, 3 international studies, and 28 randomized controlled trials. For the latter, mortality data were analyzed according to allocated treatment to ensure avoidance of bias. They assumed a close and elastic link in the relation between serum cholesterol levels and CHD incidence whether obtained from prospective studies or clinical trials. Evidence to support this linkage was presented. Law et al. found that results from cohort studies, international comparisons, and clinical trials were remarkably consistent. Cohort studies indicated that a 10% reduction in serum cholesterol levels reduces lifetime risk for CHD by 50% if achieved by age 40, falling to 20% at age 70. Further, the randomized trials, which have included 45,000 men and 4000 CHD events, suggest that the full effect of risk reduction is achieved after 5 years.

ANGIOGRAPHIC TRIALS

The availability of angiographic techniques for visualizing the lumen of coronary arteries naturally led cardiologists to consider the possibility that cholesterol-lowering therapy might delay progression of atherosclerosis or promote lesion regression. Although it was recognized that atherosclerotic lesions develop slowly, the introduction of quantitative angiography provided the opportunity to detect relatively small changes in lesion size. One of the potential advantages of use of angiographic endpoints in clinical trials is that the sample size required to detect significant changes should be much smaller than in trials relying on clinical endpoints. Another strong impetus to angiographic trials was the introduction of more efficacious cholesterol-lowering agents which should produce a robust effect. In the past decade, a sizable number of angiographic trials have been published. These trials have employed a variety of therapies for cholesterol lowering. They have been categorized and reviewed by Holmes et al. in the current volume (59). Three trials (60–62) employed intervention on life habits alone; two studies (60,62) used bile acid sequestrants; two others (64,65) employed fibrates alone; and nine trials (66–74) made use of statins alone. Five other trials (75–80) employed combined drug therapy, whereas one study (50) employed the ileal bypass to obtain cholesterol lowering. The details of these trials are presented in other chapters in this volume.

A.   Angiographic versus Clinical Endpoints

In most of the angiographic trials, clinical endpoints were recorded in addition to angiographic trials. This is fortunate because a marked discrepancy was present between changes in coronary lesion size and in clinical endpoints. Without question, cholesterol-lowering therapy led to a favorable reduction in lesion size (regression) or in rate of progression of lesion development. However, the difference in change in lesion size was small, being in the range of ~ 0.4 mm difference in minimum lesion diameter. Such small changes in lesion diameter might not be expected to produce significant changes in clinical outcome. And yet, major coronary events in patients on therapy were reduced by about 34% (59). This remarkable difference suggests that cholesterol-lowering therapy had benefits that were not recognized by changes in coronary plaque size.

B.   Clinical and Mechanistic Implications

The dissociation between angiographic changes and clinical endpoints requires an explanation of how cholesterol-lowering therapy can so effectively reduce incidence of acute coronary events. In particular, by what mechanism does cholesterol lowering markedly decrease coronary events without substantially changing lesion size? A critical clue may come from independent studies on how acute coronary events happen. These studies reveal that acute coronary events, i.e., unstable angina and myocardial infarction, occur mainly from coronary thrombosis secondary to plaque rupture (81,82). Thus, to prevent acute coronary events, it may not be necessary to reduce the gross size of coronary lesions; instead, the key may be to stabilize lesions so as to prevent plaque rupture. Cholesterol-lowering therapy may promote plaque stabilization. If so, angiographic studies might not be able to detect this stabilizing action. The angiographic trials thus could be revealing by their "negative" result on the impact of cholesterol lowering on coronary lesion size. In essence, they have contributed to a change in thinking about underlying causes of acute coronary events and about the mechanisms of benefit of cholesterol lowering therapy. As a result of this change, new strategies for prevention of acute coronary events are under consideration and investigation.

C.   Unresolved Issues

Angiographic trials have contributed importantly to the overall conclusion that decreasing serum cholesterol concentrations will reduce the risk for CHD. The demonstration that cholesterol lowering slows progression and sometimes causes regression of coronary lesions is reassuring. Nonetheless, angiographic studies cannot be substituted entirely for trials of clinical endpoints. For example, they do not reveal the quantitative benefit of cholesterol-lowering therapy in terms of clinical endpoints. Moreover, they provide little information about the side effects of therapy. If a particular regimen is grossly intolerable, or if major side effects of therapy occur, these may be uncovered. On the other hand, more subtle adverse effects of the treatment may well be overlooked. Finally, angiographic studies give us little information about cost-effectiveness of therapy. The angiographic approach thus probably will have limited usefulness for future investigation of new forms of cholesterol-lowering therapy. Although angiographic trials support the concept that efficacy of cholesterol lowering relates more to qualitative modifications than to quantitative changes in coronary plaques, they yield little insight in the nature of these modifications. Nonetheless, thanks in no small part to angiographic trials, study of vascular biology of the atherosclerotic plaque has shifted to a study of the unstable plaque. Many questions about the causes and structure of unstable plaques have emerged, and many vascular biologists believe that future advances in the prevention of CHD will reside mainly in clinical intervention to stabilize vulnerable coronary lesions.

V. STATIN TRIALS

The introduction of HMG CoA reductase inhibitors (statins) to treat elevated serum cholesterol levels constituted a true breakthrough in cholesterol management. The first drug in this class, compactin, was discovered in Japan by Endo (83). This agent inhibits HMG CoA reductase, a rate-limiting enzyme in the synthesis of cholesterol. Because of a high first-pass clearance, the action of these agents is limited almost exclusively to the liver. Inhibiting cholesterol synthesis reduces hepatic cholesterol content, a change that stimulates the expression of LDL receptors. Studies in both laboratory animals (84) and humans (85) showed that compactin markedly lowers LDL cholesterol concentrations. Subsequent investigations in humans demonstrated that agents of this class act mainly by enhancing receptor-mediated clearance of circulating LDL (86). The early success with compactin therapy initiated intensive investigation into the potential of statins for treatment of hypercholesterolemia and to efforts by the pharmaceutical industry to develop new agents of they compactin type. Further, the pharmaceutical industry has underwritten a series of clinical trials to evaluate the potential of statin therapy for reducing coronary risk. These have included both secondary prevention and primary prevention studies. Their findings can be summarized.

A. Secondary Prevention Trials

Three major secondary prevention trials have been carried out with statin therapy. These were the Scandinavian Simvastatin Survival Study (4S) (87), Cholesterol and Recurrent Events (CARE) study (88), and the Long-Term Intervention with Pravastatin in Ischaemic Disease (LIPID) study (89). In addition, another recent angiographic trial testing a statin, the Post Coronary Artery Bypass Graft (Post CABG) study, has clinical implications closely linked to those of other secondary prevention trials.

1. The 4S Trial

This study (87) was a multicenter study performed in Scandinavia to determine the efficacy and safety of simvastatin for preventing recurrent myocardial infarction and reducing mortality in hypercholesterolemic patients with established CHD. In this trial, 4444 patients with a history of angina pectoris or previous myocardial infarction were randomized to either simvastatin or placebo. The dose of simvastatin was adjusted to either 20 mg/day or 40 mg/day as required to achieve a total cholesterol of <200 mg/dL. The study duration averaged 5.4 years. Simvastatin therapy reduced total cholesterol levels by 25% (and LDL cholesterol levels by 35%). Concurrent with this reduction in cholesterol levels was a 35% decrease in recurrent major coronary syndromes. In addition, on simvastatin therapy coronary revascularization decreased by 37%, coronary mortality by 42%, and total mortality by 30%. Another important finding was a reduction in stroke. All of these reductions in clinical endpoints were highly statistically significant. Of equal importance, no major side effects occurred with simvastatin therapy, nor was noncardiovascular mortality increased. The results of this landmark study strikingly demonstrated that aggressive cholesterol-lowering therapy is highly efficacious and safe for reducing recurrent coronary heart disease in hypercholesterolemia patients with established CHD.

2. The CARE Trial

This study (88) was carried out in North America, included 4159 patients (14% women) with CHD, and lasted 5 years. In contrast to the 4S trial, which included hypercholesterolemic patients, CARE patients generally had serum cholesterol levels in the "average" range (mean = 209 mg/dL). Patients were randomized to receive either pravastatin, 40 mg/day, or placebo. Pravastatin therapy reduced LDL cholesterol levels from a baseline of 137 mg/dL to an average of 98 mg/dL. The group receiving pravastatin had a 25% reduction in major recurrent coronary events, a 24% reduction in coronary death, a 27% decrease in need for coronary bypass surgery or angioplasty, and a 31% reduction in stroke. CARE documented that CHD patients who have only average serum LDL cholesterol levels benefit from aggressive reduction in serum cholesterol levels.

3. The LIPID Trial

LIPID (89) was performed in 87 sites in Australia and New Zealand. It included 9014 patients with established CHD, defined as a history of either myocardial infarction or unstable angina. Among LIPID patients, 1511 were women, 3516 were ≥ 65 years, and 777 had diabetes. Serum cholesterol levels at entry ranged from 115 to 270 mg/dL. Patients received either placebo or pravastatin (40 mg/ day) for a period of 5 years. The group on pravastatin had a 29% reduction in major coronary events, a 24% decrease in coronary death, and a 24% reduction in need for coronary bypass surgery. On pravastatin therapy, stroke was decreased by 20% and all-cause mortality by 23%. The reduction in total mortality was statistically significant. LIPID richly confirmed the results of both 4S and CARE ; moreover, it provided strong evidence that the benefits of cholesterol-lowering therapy extend to almost all subgroups of patients having established CHD.

4. Subgroup Analysis of Secondary Prevention Trials with Statins

For each of these trials (87–89), subgroups analyses were performed and yield additional information about the benefits of statin therapy in various subgroups. Analyses showed that statins significantly reduce major coronary events in both men and women, in older and younger patients, in smokers and nonsmokers, in hypertensive and normotensive patients, and in diabetic and nondiabetic patients. Thus, statin therapy appears to be effective "across the board" in patients with CHD, regardless of subgroup.

5. The Post CABG Trial

This trial (69) stands in the boundary between angiographic trials and secondary prevention trials. Its major endpoints were angiographic changes, but the issues addressed relate closely to those of secondary prevention trials. Post CABG asked whether aggressive LDL-lowering therapy is superior to moderate LDL lowering for retarding the progression of atherosclerotic disease in coronary bypass graphs. The trial included 1351 patients with a history of coronary artery bypass graft (CABG). The combination of lovastatin and cholestyramine was adjusted as necessary to achieve the LDL targets for the two groups of patients. LDL cholesterol levels under moderately intensive therapy averaged ~135 mg/dL; and under intensive therapy, they were <100 mg/dL. The results demonstrated that more aggressive lowering of LDL cholesterol levels reduced progression of coronary atherosclerosis compared to less aggressive therapy. Also, there was a strong trend toward a greater reduction in major coronary events in patients who received more aggressive therapy.

6. Unresolved Issues

The results of the secondary prevention trials with statins have eliminated many of the uncertainties left by meta-analysis of earlier trials. In the recent trials, statin therapy markedly reduced risk for major CHD events. In addition, in 4S (87) and LIPID (89) trials, statin therapy significantly decreased total mortality. The latter finding ruled out the possibility that a reduction of serum cholesterol levels per se causes an increase in non-CHD mortality that offsets the reduction in CHD mortality. At the same time, these trials revealed that statins do not have significant toxicity, at least over the duration of the trials. Even if statins have some adverse events in the long term, their immediate and profound benefit in very high-risk patients, who have established CHD, justifies their use in patients of this type. Nonetheless, in spite of the risk reduction brought about by statin therapy for patients with CHD, there are several unresolved issues that are worthy of consideration.

One issue that is brought to the fore by these trials pertains to the optimal target for LDL cholesterol in secondary prevention. The NCEP (53) recommends a treatment goal for LDL cholesterol of ≤ 100 mg/dL. This recommendation derives from several observations, e.g. (1) the low rates of CHD in populations whose LDL-cholesterol levels are ≤ 100 mg/dL; (2) the relatively low reinfarction rates in patients with CHD who have very low cholesterol; and (3) the favorable outcome from angiographic trials in which a marked LDL lowering was achieved (53). Particularly strong support for more aggressive LDL-lowering therapy comes from the Post CABG trial (69).

Since the recent statin trials provide a wealth of information about the efficacy of serum cholesterol lowering, we can ask whether they reveal the quantitative relationship between LDL cholesterol levels and risk for recurrent coronary events in secondary prevention. Figure 2 shows four possibilities for this relationship. Figure 2A suggests a threshold relation, i.e., that a therapeutic low-

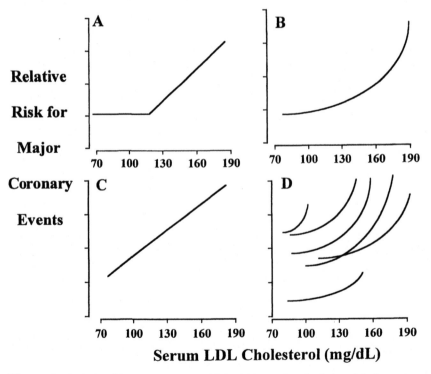

Figure 2 Relationship between serum LDL cholesterol and relative risk for major coronary events. (A) Threshold relationship. (B) Curvilinear relationship. (C) Linear relationship. (D) Variable relationships.

ering of LDL-cholesterol produces benefit down to a certain level, below which no further risk reduction occurs. In contrast Figure 2B denotes a curvilinear relationship such that continuing risk reduction occurs at progressively lower levels, but with diminishing returns. A third relationship is a linear decline in risk that continues to lower and lower levels (Fig. 3C). Finally, the possibility must be considered that reality conforms to a family of curves such that each individual has a unique relationship (Fig. 4D). If the latter holds, the occurrence of CHD in any individual signifies a variable susceptibility to coronary disease and different LDL cholesterol levels; if this pertains, risk reduction likely would be achieved by LDL lowering at any baseline LDL cholesterol concentration.

Recent posthoc analyses of secondary prevention trials may shed some light on the possible relationships outlined in Figure 2. For example, posthoc analysis of the CARE study (90) suggested that maximal benefit is achieved when LDL cholesterol levels are reduced to 125 mg/dL (see Fig. 2A). From a theoretical viewpoint, a threshold relation of this type is problematic. Earlier epidemiological studies suggested a threshold relation between serum cholesterol concentrations and CHD risk, but these gave way to a curvilinear association when population sizes were increased (91). Certainly, posthoc analysis in CARE (90) did not have the statistical power to differentiate between a threshold and a curvilinear association. If a threshold relation is prematurely judged to hold and the concept is implemented, some patients may fail to derive the maximal benefit that is possible from cholesterol-lowering therapy.

Indeed, in a posthoc analysis of the 4S trial (92), a curvilinear relation was identified between declining LDL cholesterol concentrations and recurrent coronary events (Fig. 2B). This finding is consistent with the shape of the association previously observed in a large epidemiological study. It should be noted that this shape is linear when plotted in a semi-log scale. Previous investigators have emphasized that the association between cholesterol levels and CHD events in general is log-linear. If so, the magnitude of the reduction in recurrent coronary events will be smaller when LDL cholesterol is lowered at levels < 130 mg/dL, compared to the magnitude at higher levels. Nonetheless, even if a 1% reduction in LDL cholesterol produces only an 0.5% decrease in CHD risk in the lower ranges, a 30 mg/dL reduction would still yield another 10% to 15% decrease in risk. This further decline is not trivial and would justify more aggressive LDL-lowering therapy. Thus, the consistency between large epidemiological studies and the posthoc results of the 4S trial supports the likelihood of continuation of benefit from reducing LDL cholesterol levels from 130 mg/dL to <100 mg/dL.

The recognition that LDL cholesterol concentrations can be safely driven to well below 100 mg/dL by high doses of statins has led to speculation that "the lower, the better." Some investigators continue to speculate that the relation between LDL cholesterol levels and CHD events is linear down to very low LDL concentrations (Fig. 2C). This possibility has led the pharmaceutical industry to initiate clinical trials of high-dose statins. These trials will eventually reveal whether the concept of "the lower, the better" is valid and if so, whether high doses of statins can be routinely in secondary prevention.

Preoccupation with the general shape of the relation between LDL cholesterol levels and coronary events (Fig. 2A–C) should not cause us to overlook the possibility that individuals vary in their susceptibility to a given level of LDL (Fig. 2D). Indeed, considering the fact of biological variability, such would be surprising if this were not the case. Recent trials demonstrate that statin therapy reduces recurrent events in coronary patients even when LDL cholesterol levels are in the "average" range for the general population (88,89). Thus, on the basis of these trials, a case can be made that all coronary patients should receive aggressive cholesterol-lowering therapy regardless of their baseline LDL cholesterol levels. The results of the recent LIPID trial (89), in particular, appears to support this concept.

Three practical questions emerge from these general considerations. First, should most coronary patients with LDL-cholesterol levels exceeding 130 mg/dL be started immediately on cholesterol-lowering drugs? Second, should coronary patients who have a baseline LDL cholesterol in the range of 100 to 129 mg/dL receive drug therapy? And third, should most coronary patients be treated with cholesterol-lowering drugs regardless of their baseline LDL cholesterol concentrations.

Regarding the first, there has been a tendency on the part of the medical community to delay use of drugs in CHD patients in favor of dietary and life-habit changes. This suggests a pervasive reluctance for using drug therapy for cholesterol lowering, perhaps due to previous concerns about the dangers of reducing serum cholesterol levels. Also, uncertainty about whether the responsibility for initiating drug therapy lies with the cardiologist or primary care physician seems to be a barrier to starting drugs. The American Heart Association has recently issued a recommendation that cholesterol-lowering drugs should be initiated immediately in CHD patients whose baseline LDL cholesterol is ≥ 130 mg/dL (93). In such patients, drug therapy almost certainly will be required to achieve a target goal of ≤ 100 mg/dL.

Next, let us ask about patients whose baseline LDL-cholesterol levels are in range of 100 to 129 mg/dL. Should they too be immediately started on cholesterol-lowering drugs? Unfortunately, the available data from controlled clinical trials do not directly address this issue. Nonetheless, subgroup analysis of the LIPID trial (89) revealed a strong trend toward benefit of statin therapy in coronary patients whose baseline LDL-cholesterol was in this range. Furthermore, a growing number of authorities hold the opinion that such patients deserve cholesterol-lowering therapy to achieve an LDL cholesterol level of ≤ 100 mg/dL. For many patients, this target can be achieved by combining low doses of cholesterol-lowering drugs with maximal nondrug therapy.

And finally, should almost all patients with CHD receive cholesterol-lowering drugs, regardless of baseline LDL cholesterol levels? If the relationship in Figure 2D holds, drugs to reduce cholesterol levels probably would be indicated in almost all patients. Certainly one way to interpret the statin trials is to take the comparison of statin versus placebo at face value and not take it be interpreted in terms of lipoprotein responses. An analogy is the comparison of aspirin versus placebo in which the findings are not directly linked to differential changes in coagulation parameters. The routine use of cholesterol-lowering drugs in coronary patients would have the advantage of simplicity, but the potential value of tritrating LDL cholesterol levels down to a low range (< 100 mg/dL) must not be overlooked.

A final issue that follows from the previous one is whether statins have unique properties to reduce CHD risk differently from other modalities of cholesterol-lowering therapy. This question naturally arises because of the marked reduction in CHD incidence observed in the statin trials. Perhaps it is pertinent to divide this last question into three parts: (1) Can all the CHD risk reduction observed with statin therapy be explained by LDL reduction? (2) Can some of the benefit of statin therapy be explained by favorable modification of other lipoproteins (e.g., reducing VLDL remnants and raising HDL levels)? and (3) Do statins have antiatherogenic effects that are independent of their actions on lipoprotein metabolism? Regarding the first, a recent analysis of available cholesterol-lowering trials suggests that CHD risk reduction parallels the degree of reduction of LDL-cholesterol, regardless of agent, including the statins (94). In other words, statins are more powerful in reducing LDL cholesterol levels than are other agents, but the greater effects of statins on CHD risk is quantitative, but not qualitative. Still, there is no doubt that statins lower VLDL remnants as well as LDL (95,96), and the strong possibility exists that reduction of remnant lipoproteins very well could contribute to the antiatherogenic action of statins. Finally, a growing body of data suggests that statins can produce a variety of other metabolic alterations (97); whether these modify risk independently of circulating lipoproteins is uncertain. If so, these effects probably are small compared to the greater benefit derived from the modification of serum lipoproteins (94).

B. Primary Prevention Trials

Two major primary prevention trials have been carried out with statins: the West of Scotland Coronary Prevention Study (WOSCOPS) (98), and the Air Force/ Texas Coronary Atherosclerosis Prevention Study (AFCAPS/TexCAPS) (99). These important trials almost certainly will have a major impact on thinking about the use of cholesterol-lowering drugs in primary prevention. Their main features and results can be reviewed.

1. WOSCOPS Trial

This study (98) was conducted out in 6595 men ranging in age from 45 to 64 years. Patients mostly were without clinical evidence of CHD, but all had hypercholesterolemia at entry. Many of the patients had other risk factors, and the group as a whole can be considered to be at high risk for CHD. Patients received either pravastatin (40 mg/day) or placebo, and were followed for 4.9 years. Pravastatin therapy lowered total cholesterol by 20%. Associated with this change were a 31% reduction in recurrent coronary events, a 33% decrease in coronary deaths, and a 22% reduction in total mortality. Pravastatin therapy was well tolerated; of particular importance, the treatment group did not experience an increase in noncardiovascular mortality. In this trial, older patients responded approximately as well to treatment as did younger patients.

2. AFCAPS/TexCAPS Trial

This study (99) included 6605 patients (15% women) in two centers in Texas. Participants received lovastatin (20 to 40 mg/day) or placebo. At entry most patients were at only moderate risk for future CHD; mean baseline LDL cholesterol was 150 mg/dL, and HDL cholesterol was 37 mg/dL. During treatment with lovastatin, the LDL cholesterol level fell to an average level of 115 mg/dL. After 5 years, patients on lovastatin therapy experienced a 37% reduction in first acute major coronary event (sudden death, fatal and nonfatal myocardial infarction, or unstable angina pectoris), a 44% reduction in fatal and nonfatal myocardial infarction, and a 33% reduction in need for coronary revascularization. Since most patients were at only moderate risk at entry, the study was not powered to test total mortality. Lovastatin therapy was well tolerated; no significant untoward effects occurred, nor was noncardiovasular mortality increased.

3. Unresolved Issues

WOSCOPS and AFCAPS/TexCAPS trials (98,99) provide overwhelming evidence that statin therapy effectively reduces the risk for new-onset CHD in primary prevention. In both trials, statin therapy reduced risk for new CHD events by about one-third. WOSCOPS (98) also revealed a reduction in total mortality during statin therapy. These trials thus amplify secondary prevention trials by confirming that cholesterol lowering with statins substantially reduces CHD risk.

The efficacy and safety of statins in primary prevention studies raises an important question: *Who is an appropriate patient for statin therapy?* This question is intimately bound with the definition of the high-risk state. Recent statin trials support the concept that aggressive cholesterol-lowering therapy is appropriate for patients whose short-term risk for CHD is high. In such patients, safety and cost concerns apparently are not sufficient to negate the benefit of cholesterol-lowering drugs. Most investigators agree that patients having severe hyper-cholesterolemia (LDL cholesterol > 220 mg/dL) are candidates for primary prevention with statin therapy (53). In addition, those who have moderate hyper-cholesterolemia (LDL cholesterol 160 to 219 mg/dL) along with multiple other risk factors warrant drug therapy. A provocative question is whether the patient who has multiple other risk factors but who has only a borderline-high LDL cho-lesterol level (130 to 159 mg/dL) is a candidate for statin therapy. AFCAPS/TexCAPS (99) results bring this question to the fore. Recent recommendations of the Second Joint Task Force of European and other Societies on Coronary Prevention (100) have recently advocated cholesterol-lowering therapy in patients judged to be at high risk, regardless of baseline LDL cholesterol levels.

Since statin therapy has been shown to reduce risk in primary prevention, a critical issue becomes how best to determine absolute CHD risk. The conventional approach is to sum CHD risk factors, either by adding of categorical risk factors (53) or by summing of graded risk factors (101). This approach has obvious limitations in risk prediction (102), and aggressive therapy may be instituted in many patients whose true risk is low. For this reason, the possibility of defining *total atherosclerotic plaque burden* has received increased attention. Patients with a heavy plaque burden are likely better candidates for aggressive risk reduction than are those with low plaque burden. Methods currently under evaluation to assess plaque burden are electron beam computerized tomography (EBCT) to assess coronary calcification, sonography to estimate carotid athero-sclerosis, and magnetic resonance imaging. Combining traditional risk factor analysis with noninvasive assessment of plaque burden may provide an improved approach for selecting patients for aggressive cholesterol-lowering therapy.

One subgroup of patients that deserves special attention in primary prevention are those with diabetes mellitus, particularly patients with type 2 diabetes. Prospective data indicate that patients with type 2 diabetes have a risk for acute coronary events that approaches that of nondiabetic patients with established CHD (103). Moreover, once diabetic patients develop CHD, their prognosis for survival is much worse that for nondiabetic patients with established CHD (104,105). It thus, seems critical to know whether diabetic patients without CHD deserve the aggressive risk reduction usually reserved for nondiabetic patients with CHD. Subgroup analysis of recent statin trials provide strong support for benefit from intensive cholesterol-lowering therapy in diabetic patients (89,104,105).

The decision to employ statin therapy in high-risk, primary prevention goes beyond absolute risk estimates. Both safety and costs of statin therapy must be taken into account. WOSCOPS and AFCAPS/TexCAPS (98,99) revealed a high level of safety of statins over the duration of the trials. This fact helps to justify their use in high-risk patients. Preliminary analysis of costs suggests that statin therapy falls in an acceptable range cost-effectiveness, e.g., <$50,000 per year of life saved, when a patient's absolute risk is > 2% risk per year (106). These estimates nonetheless are preliminary, and more detailed analyses are needed to provide more precise estimates of costs. Improved estimates could be one factor in the definition of the high-risk state, with respect to need for drug therapy to reduce serum cholesterol levels.

Because of the efficacy of statin therapy in primary prevention, enthusiasm for broader usage of these agents in the general population is growing. For example, might it be appropriate to use long-term statin therapy in persons whose absolute risk is in the range of 1% to 2% per year? Such patients can be considered to be at *intermediate risk*. One question however that has not been fully resolved is whether statins are safe to use for many years, e.g., 10 to 30 years. At present their use in high-risk patients, such as men of WOSCOPS, seems warranted because short-term benefit apparently outweighs short-term risk. Indeed, serious side effects of statin therapy over the 5 years of WOSCOPS and AFCAPS/TexCAPS were almost nonexistent. On the other hand, if use of statins were to be extended to intermediate-risk patients, they would have to be taken for many years to realize a significant population benefit. At present, 20-year or 30-year safety data are not available; in the long term, adverse effects could emerge that have not been detected so far. This uncertainty adds a note of caution about use of statins for intermediate-risk primary prevention.

Another factor to consider in use of statins in intermediate-risk primary prevention is their cost-effectiveness. In fact, the WOSCOPS and AFCAPS/TexCAPS trials provide a wealth of information for making cost-effectiveness analyses. Such analyses are currently being done. Preliminary estimates suggests that statin therapy in high-risk patients, such as those of WOSCOPS, are favorable compared to other standard medical therapies (106). Seemingly, the major factor influencing cost is the price of the medication. After statins go off patent, their cost will probably decline significantly. Still, drug costs are not the only cost consideration; patient identification, evaluation, and long-term monitoring all add to the cost of cholesterol-lowering therapy. Whether use of statins for intermediate-risk primary prevention is cost-effective remains to be determined. At some point on the declining spectrum of risk, long-term use of statins will incur costs that are unacceptably high. In the future, cost analysis will become an important component in clinical decisions about when to use statins in cholesterol-lowering therapy.

VI. CHOLESTEROL-LOWERING DRUGS VERSUS TRIGLYCERIDE-LOWERING DRUGS

Although each of the lipid-lowering drugs has some effect on all of the lipoprotein species, it is convenient to separate these agents into cholesterol-lowering drugs and triglyceride-lowering drugs. The former include statins and bile acid sequestrants; the latter are nicotinic acid and the fibric acids. The statins mainly lower LDL cholesterol, although they also reduce triglyceride-rich lipoproteins to some extent. Nicotinic acid mainly lowers triglyceride-rich lipoproteins, but has some LDL-lowering action as well. Fibric acids have a much lesser effect on LDL cholesterol, but are highly effective for reducing triglyceride-rich lipoproteins. Most of the serum cholesterol lowering that occurs with fibric acids is limited to triglyceride-rich lipoproteins and remnant lipoproteins. The NCEP has identified LDL as the major atherogenic lipoprotein and LDL cholesterol as the primary target of cholesterol-lowering therapy. This fact alone would favor statins over nicotinic acid and fabric acids, which have their major actions on triglyceride-rich lipoproteins. Indeed, many workers have questioned whether any benefit at all will derive from triglyceride-lowering therapy.

Certainly, the consistent benefit obtained from clinical trials with statin therapy supports the priority given to cholesterol-lowering drugs in prevention. Nonetheless, a review of existing trials with fibric acids suggests more benefit from triglyceride-lowering therapy than has generally been recognized. For example, a reduction in major coronary events with fibric acid treatment was reported in the following trials: Scottish physicians (45); Newcastle on Tyne Physicians (46); WHO clofibrate trial (31); Helsinki Heart Study (35); Stockholm Secondary Prevention Study (49); and the soon-to-be-published Veterans Affairs–HIT trial. Favorable angiographic outcomes with fibric acid therapy also have been published in the BECAIT trial (65) and in a Finnish study (64). Further, a marked risk reduction was noted in the subgroup of patients in the BIP trial who had hypertriglyceridemia (107). It is true that the clofibrate arm of the Coronary Drug Project (47) failed to show a significant reduction in recurrent CHD, but a trend in this direction was noted. Thus, taken as a whole, the trials of fibric acids suggest at least some benefit from triglyceride-lowering therapy.

The mechanisms whereby triglyceride-lowering drugs reduce CHD risk are uncertain. They have three different effects on lipoproteins metabolism: (1) they reduce triglyceride-rich lipoproteins including remnant lipoproteins; (2) they convert small LDL particles to normal size LDL; and (3) they raise HDL cholesterol concentrations (108). Any or all of these actions could retard atherogenesis and reduce risk. Regardless of the precise mechanism, triglyceride-lowering drugs offer the potential to augment or substitute for statin therapy in some patients, particularly those with elevated serum triglycerides. Several studies show that the combinations of statins and fibric acids (109,110) or statins and nicotinic acid (78) are highly effective for normalizing the whole lipoprotein profile. Therefore, future clinical trials would do well to examine whether combined drug therapy is more efficacious in reducing CHD risk than single-drug therapy.

VII. CURRENT AND FUTURE CLINICAL TRIALS

Some of the unresolved issues considered above will be addressed in current clinical trials. Certainly all of the important questions about primary and secondary prevention as they relate to cholesterol-lowering therapy have not been answered by previous trials. Subgroup analysis has addressed some of these questions and has provided tentative answers; even so, some of these questions will require new trials for definitive answers. In the discussion to follow, several current trials will be discussed. The results of some of these trials have very recently been reported, either in publication or at international meetings. The author is grateful to Ms. Jeanne Regnange of Merck and Co. for providing a compilation of ongoing clinical trials. They will be grouped according to category of trial.

A. Aggressive Statin Therapy

The Simvastatin SEARCH study (*Simvastatin-Study to Evaluate Additional Reduction of Cholesterol and Homocysteine*) is a 5-year trial being carried out in the United Kingdom in 12,000 high-risk patients; it compares simvastatin (20 mg/day), simvastatin (80 mg/day), folic acid, and placebo in 2×2 design. The primary aim of the study is to determine whether greater LDL cholesterol lowering produced by high dose simvastatin will decrease major coronary events more than will moderate doses. The study also will test the efficacy of supplemental folic acid in reducing CHD risk.

The *Treatment to New Target* (TNT) study has a similar aim. TNT is a secondary prevention study in 8600 patients who have established CHD; it is designed to determine whether high-dose atorvastatin (80 mg/day) will be more effective in reducing recurrent major coronary events than a more moderate dose (10 mg/day) TNT is a multicenter trial carried out North America, Africa, South Africa, and Australia. This study will be completed in 2004.

B. Statins + Antioxidants

The *Simvastatin Heart Protection Study* (Simvastatin HPS) is a 5-year trail in 20,000 high-risk patients (30% women), ages 40 to 80 years, to test 40 mg/day simvastatin plus and antioxidant cocktail (vitamin E, vitamin C, and beta-carotene) for prevention of major coronary events and total mortality. The primary endpoint of the trial is all-cause mortality, and the secondary endpoint, specific causes of mortality and stroke. The trial is expected to be completed in 2001.

C. Statin Comparisons

The IDEAL study is an open-label trial in 7500 men and women with established CHD to compare atorvastatin (80 mg/day) versus simvastatin (20 mg/day). IDEAL is a 7-year trial scheduled for completion in 2005. It is being carried out in 150 centers in Scandinavia and other parts of Europe. The primary endpoint of the study is major coronary events; secondary endpoints include total mortality, all coronary events, and hospitalizations.

D. Diabetic Patients

The three major secondary prevention trials using statins (89,104,105) included a sizable number of patients with diabetes. All suggested benefit for treatment of diabetic patients. If the data on diabetic patients from these three studies were to be pooled, this meta-analysis might help to confirm the efficacy of statin therapy in diabetic patients. Such an analysis has some urgency because of the need to define the appropriate targets for cholesterol management in diabetic patients even before completion of current clinical trials that specifically target such patients.

The *Fenofibrate Intervention and Event Lowering in Diabetes* (FIELD) trial was scheduled to begin in 1997 and to terminate in 2003. This trial will randomize 8000 patients with type 2 diabetes (50% men/50% women) to fenofibrate (200 mg/day) or placebo. Primary endpoints of this trial are major coronary events (total CHD, CHD deaths, and nonfatal myocardial infarction). The trial will be conducted in approximately 100 centers in Australia, New Zealand, and Finland. A second trial employing fenofibrate is the *Lipids in Diabetes Study* (LDS). This trial will compare fenofibrate, fenofibrate + cerivastatin, and placebo in a 2×2 factorial design in 4000 patients with type 2 diabetes in 21 to 25 centers in the U.K. It will be a primary prevention trial in which primary endpoints are cardiac death, myocardial infarction, PTCA, or CABG. The trial was scheduled to begin in 1998 and to end in 2003.

The *Atorvastatin Cardiovascular Outcomes Prevention Study in NIDDM* (ASPEN) is a 4-year trial to compare atorvastatin (10 mg/day) to placebo in 2250 patients with type 2 diabetes. Primary endpoints will include first acute cardiovascular events (cardiovascular death, nonfatal myocardial infarction, revascularization procedures or stroke). A portion of patients entering the trial will have established CHD, so that the study is a mixed primary/secondary prevention trial. It is being performed in the United States, Europe, South Africa, and Australia. The study is scheduled for completion in 2001.

The *Collaborative Atorvastatin Diabetes Study* (CARDS) has a similar design as the ASPEN study. It will compare atorvastatin (10 mg/day) to placebo in 2120 male and female patients (ages 40 to 70 years) with type 2 diabetes in the U.K. Primary endpoints are acute, major coronary events and revascularization. The study is projected for completion in 2001.

E. Secondary Prevention Trials with Fibric Acids

Bezafibrate Infarction Prevention Study (BIP) was a 6.25-year study in 3122 patients (8% women) with established CHD to examine whether bezafibrate (400 mg/day) will prevent recurrent coronary events. The study was recently completed and preliminary results were presented (107). The trial showed a trend toward reduction in clinical endpoints, but the decrease was not statistically significant. On the other hand, a subgroup of patients having hypertriglyceridemia revealed a striking reduction in recurrent coronary events. This finding suggests that fibric acids may be most effective for risk reduction in hypertriglyceridemic patients.

Veterans Affairs HDL Intervention Trial (VA-HIT) compared extended release gemfibrozil (1200 mg) with placebo in 2500 men with established CHD for their effects on major coronary events. The study lasted 7 years. Recruitment criteria included HDL cholesterol ≤ 40 mg/dL, LDL cholesterol ≤ 140 mg/dL, and triglyceride ≤ 300 mg/dL. Most of the patients had low LDL and low HDL levels, and had abdominal obesity, which is characteristic of the *insulin resistance syndrome*. The outcomes for this trial should be published in the near future. Oral preliminary presentation of the trial results indicated that gemfibozil therapy caused a significant reduction in major coronary events. Although the patients did not have categorical hypertriglyceridemia at entry, a reduction in triglyceride levels appeared to be the major action of the drug.

F. Hypertensive Patients

1. Anglo-Scandinavian Cardiac Outcomes Trial (ASCOT)

This is a 2-year study being carried out in U.K., Ireland, and Scandinavia. It includes 18,000 patients, age > 50 years who have hypertension and at least two other CHD risk factors. The trial will evaluate the effects of atorvastatin 10 mg/day in combination with various antihypertensive agents. Two antihypertensive regimens regimens will be compared: beta blocker + diuretic, vs. calcium blocker + ACE inhibitor. Half of each group will receive either placebo or atorvastatin. Primary endpoints include major coronary events, stroke, heart failure, and all-cause mortality.

2. Antihypertensive and Lipid Lowering Heart Attack Trial (ALLHAT)

This is a larger trial to compare four different first-line antihypertensive agents in 40,000 patients. Patient characteristics include 45% women, 55% African-American, 35% diabetic, and 37% > 70 years. Half the patients will be enrolled in cholesterol-lowering treatment. The trial is being done in 623 community-based clinics in the United States. The trial aims to determine the extent to which antihypertensive drugs + a statin (pravastatin 40 mg) will reduce risk for all-causes mortality in high-risk patients with mild to moderately raised cholesterol levels.

G. Postmenopausal Women

1. Heart and Estrogen-Progestin Replacement Study (HERS)

This trial was a secondary prevention trial conducted in 6000 postmenopausal women. Drug therapy included a combination of estrogen and progestin. A significant portion of the patients received lipid-lowering therapy, although 91% did not reach the NCEP goal for LDL cholesterol of <100 mg/dL. The results of this trial were recently reported (111). Hormone replacement therapy did not reduce the risk for recurrent coronary events. The data from this study raise serious doubts about the efficacy of hormone replacement therapy for secondary prevention in postmenopausal women. The positive outcome of cholesterol-lowering therapy with statins in secondary prevention in women thus shifts attention toward cholesterol-lowering drugs and away from hormone replacement therapy for postmenopausal women with established CHD.

2. Secondary Prevention of Ischemic Disease by Estrogen Replacement or Statins (SPIDERS)

This is 2×2 randomized trial being conducted in 4000 women with a history of myocardial infarction, angina, or transient ischemic attack. The study is based in the Netherlands, and will last 5 years. Three questions are being addressed: (1) whether hormone replacement therapy is more effective than placebo in preventing vascular events in postmenopausal women with established CHD; (2) whether simvastatin is more effective than placebo in reducing new vascular events; and (3) whether the combination of hormone replacement therapy + simvastatin is more effective than monotherapy. Primary endpoints will be major coronary events and coronary procedures.

H. Post Renal Transplant Patients

Assessment of Lescol in Renal Transplantation (ALERT) is a 4-year study in 2050 post renal transplant patients to assess efficacy of fluvastatin (40 mg/day) to reduce major coronary events. In addition, effects of therapy on all-causes mortality and chronic transplant dysfunction will be examined. All patients will receive at least 3 years of treatment.

I. Stroke

Studies are now being developed to evaluate the ability of statin therapy to prevent stoke in older patients. Clinical trials have already been started in Europe that include cerebrovascular endpoints. On of these trials, called PROSPER, will examine the efficacy of pravastatin to reduce risk of stroke and loss of cognitive function in elderly patients.

VIII. SUMMARY

The history of controlled clinical trials of cholesterol-lowering therapy records one of the great accomplishments in modern medicine. The dogged pursuit of the cholesterol hypothesis represents a five-decade effort on the part of an international group of committed researchers. The clinical trial effort was fueled by a large body of research that has documented a strong association between serum cholesterol and the development of CHD. Studies in experimental animals reach back almost a century to generate the cholesterol hypothesis. The discovery of familial hypercholesterolemia and the finding of its powerful association with premature atherosclerotic disease were other key links in the chain of evidence. Still, by the middle of the twentieth century, the connection between serum cholesterol and atherosclerotic disease was still viewed as tenuous. In retrospect, the years between 1900 and 1950 were a period of incubation for the cholesterol hypothesis. Thereafter, advances were made rapidly on many fronts; these included a wealth of revealing research in laboratory animals, investigations on the nature and metabolism of lipoproteins, the findings of the atherogenic potential imparted by various genetic forms of hypercholesterolemia, and the conformation of a relationship of elevated serum cholesterol to CHD risk through epidemiological studies. By 1975, there was little doubt that high serum cholesterol levels, particularly an elevated LDL cholesterol, contributed importantly to the development of CHD.

However, in spite of these advances, there was widespread skepticism that serum cholesterol levels could be effectively reduced or, even if reduced, would have a significant impact on the incidence of CHD. Nonetheless, a relatively small number of committed investigators believed that cholesterol-lowering therapy, whether by dietary or pharmaceutical means, would reduce the risk for CHD. These investigators mounted a series of clinical trials of increasing sophistication to test the hypothesis that lowering serum cholesterol would result in benefit. Even after almost three decades, from early 1960s to early 1990s, a conclusive answer was still not obtained; this proof remained elusive in spite of an enormous effort and multiple trials. Most of the earlier clinical trials had too few participants, were too short, or employed cholesterol-lowering regimens that were too weak to provide a definite result. In a word, their statistical power generally was inadequate. Gradually, these earlier trials increasingly revealed that reducing serum cholesterol yields some reduction in risk for CHD; this finding however was not entirely satisfying because a decrease in total mortality had not been proven unequivocally. The discovery of statins in the 1970s was a major step forward in cholesterol-lowering therapy. These agents were found to be both highly efficacious and relatively safe. The introduction of statins opened the door to more definitive clincal trials. In the past 5 years, a series of clinical trials have fully confirmed the major aspects of the cholesterol hypothesis and, importantly, have shown that effectively reducing serum cholesterol reduces total mortality. These trials thus have issued in a new era of CHD prevention with cholesterol-lowering therapy; they also have engendered many new trials that will more fully define the potential for cholesterol-lowering therapy to reduce risk for CHD, in both secondary and primary prevention. Thus, at the end of the twentieth century, we can say that the cholesterol hypothesis has been firmly proven as a fact. Clinical trials of cholesterol-lowering therapy have been a vital and necessary part of the effort to establish this fact.

REFERENCES

1. Anitschkow N, Chalatow S. Veber experimentelle cholesterinseatose und ihre bedeutung fur die entstehung einiger pathologisher prozesse. Zentralbl Allg Pathol Pathol Anat 1913; 24:1–9.
2. Katz LN, Stamler J. Experimental Atherosclerosis. Publication No. 124, American lectures series. Monograph in Bannerstone Division of American Lectures in Metabolism. Springfield, IL: Charles C. Thomas, 1953.
3. Keys A. Coronary heart disease in seven countries. Circulation 1970; 41:I1–I211.
4. Kannel WB. Contributions of the Framingham study to the conquest of artery disease. Am J Cardiol 1988; 62:1109–1112.
5. Goldstein JL, Hobbs HH, Brown MS. Familial hypercholesterolemia. In: Scriver CR, Beaudet AL, Sly WS, Valle D (eds). The Metabolic and Molecular Bases of Inherited Diseases. New York: McGraw-Hill, 1995:1981–2030.
6. Mahley RW, Rall SC Jr. Type III hyperlipoprotein (dysbetalipoproteinemia): the role of apolipoprotein E in normal and abnormal lipoprotein metabolism. In: Scriver CR, Beaudet AL, Sly WS, Valle D (eds). The Metabolic and Molecular Bases of Inherited Diseases. New York: McGraw-Hill, 1995:1953–1980.
7. Navab M, Berliner JA, Watson AD. The yin and yang of oxidation in the development of the fatty streak. Arterioscler Thromb Vasc Biol 1996; 16:831–842.
8. Dole VP, Gordis E, Bierman EL. Hyperlipemia and arteriosclerosis. N Engl J Med 1963; 269:686–689.
9. Mann GV. Diet-heart: end of an era. N Engl J Med 1977; 297:644–650.
10. McMichael J. Fats and athroma: an inquest. Br Med J 1979; 1:173–175.
11. Oliver MF. Might treatment of hypercholesterolaemia increase non-cardiac mortality? Lancet 1991; 1529–1531.
12. Oliver MF. Reducing cholesterol does not reduce mortality. J Am Coll Cardiol 1988; 12:814–817.

13. Stason WB. Costs and benefits of risk factor reduction for coronary heart disease: insights from screening and treatment of serum cholesterol. Am Heart J 1990; 119:718–724.

14. Field K, Thorogood M, Silagy C, Normand C, O'Neill C, Muir J. Strategies for reducing coronary risk factors in primary care: which is most cost effective? Br Med J 1995; 310:1109–1112.

15. Keys A, Anderson JT, Grande F. Serum cholesterol response to changes in the diet. II. The effect of cholesterol in the diet. Metabolism 1965; 14:759–765.

16. Grundy SM, Barrett-Connor E, Rudel LL, Miettinen T, Spector AA. Workshop on the impact of dietary cholesterol on plasma lipoproteins and atherogenesis. Arteriosclerosis 1988; 8:95–101.

17. Ahrens EH, Hirsch J, Insull W, Tsaltas TT, Blomstrand R, Peterson ML. The influence of dietary fats on serum-lipid levels in man. Lancet 1957; 1:943–953.

18. Keys A, Anderson JT, Grande F. Serum cholesterol response to changes in the diet. IV. Particular saturated fatty acids in the diet. Metabolism 1965; 14:776–787.

19. Hegsted DM, McGandy RB, Myers ML, Stare FJ. Quantitative effects of dietary fat on serum cholesterol in man. Am J Clin Nutr 1965; 17:281–295.

20. U.S. Department of Agriculture, Agriculture Research Service, Dietary Guidelines Advisory Committee. Report of the dietary guidelines advisory committee on the dietary guidelines for Americans. In: To the Secretary of Health and Human Services and the Secretary of Agriculture. Washington: Government Printing Office, 1995.

21. Krauss RM, Deckelbaum RJ, Ernst N, et al. Dietary guidelines for healthy American adults: a statement for health professionals from the Nutrition Committee, American Heart Association. Circulation 1996; 94:1795–1800.

22. National Diet-Heart Study Research Group. The National Diet–Heart Study Final Report. In: American Heart Association Monograph No. 18. New York: American Heart Association, 1968.

23. Diet-Heart Review Panel of the National Heart Institute. The MARS field trials of the diet–heart question. Their significance, timeliness, feasibility, and applicability. In: Ahrens EH Jr (ed). American Heart Association Monograph No. 28. New York: American Heart Association, 1969.

24. Arteriosclerosis. A Report by the National Heart and Lung Institute on Arteriosclerosis. In: National Institutes of Health, Vol. II. Bethesda, MD: DHEW Publication No. (NIH) 72-219, 1971.

25. Dayton S, Pearce ML, Goldman H. Controlled trial of a diet high in unsaturated fat for prevention of atherosclerotic complications. Lancet 1968; 1060–1062.

26. Sturdevant RAL, Pearce ML, Dayton S. Increased prevalence of cholelithiasis in men ingesting a serum cholesterol-lowering diet. N Engl J Med 1973; 288:24–27.

27. Frantz ID, Jr, Dawson EA, Ashman PI. Test of effect of lipid lowering by diet on cardiovascular risk. The Minnesota coronary survey. Arteriosclerosis 1989; 9:129–135.

28. Miettinen M, Karvonen MJ, Turpeinen O, Elauso R, Paavilainen E. Effect of cholesterol lowering diet on mortality from coronary heart diseases and other causes. A twelve year clinical trial in men and women. Lancet 1972; II:835–838.

29. Hjermann I, Holme L, Velve Byre K, Leren P. Effect of diet and smoking intervention on the incidence of coronary heart disease: report from the Oslo study group of a randomized trial in healthy men. Lancet 1981; 2:1303–1310.
30. Multiple Risk Factor Intervention Trial (MRFIT). A national study of primary prevention of coronary heart disease. JAMA 1976; 235:825–827.
31. Committee of Principal Investigators. A cooperative trial in the primary prevention of ischaemic heart disease using clofibrate. Br Heart J 1978; 40:1069–1103.
32. Oliver MF. Serum cholesterol—the knave of hearts and the joker. Lancet 1981; 2:1090–1095.
33. Lipid Research Clinics Program. The Lipid Research Clinics coronary primary prevention trial results. I. Reduction in the incidence of coronary heart disease. JAMA 1984; 251:351–364.
34. Lipid Research Clinics Program. The Lipid Research Clinics coronary primary prevention trial results. II. The relationship of reduction in incidence of coronary heart disease to cholesterol lowering. JAMA 1984; 251:365–374.
35. Frick MH, Elo O, Haapa K, et al. Helsinki Heart Study: primary prevention trial with gemfibrozil in middle-aged men with dyslipidemia: safety of treatment, changes in risk factors, and incidence of coronary heart disease. N Engl J Med 1987; 317:1237–1245.
36. Vu-Dac N, Schoonjans K, Kosykh V. Fibrates increase human apolipoprotein A-II expression through activation of the peroxisome proliferator-activated receptor. J Clin Invest 1995; 96:741–750.
37. Manninen V, Huttunen JK, Heinonen OP, Tenkanen L, Frick MH. Relation between baseline lipid and lipoprotein values and the incidence of coronary heart disease in the Helsinski Heart Study. Am J Cardiol 1989; 63:42H–47H.
38. Consensus Conference. Lowering blood cholesterol to prevent heart disease. JAMA 1985; 253:2080–2097.
39. Lenfant C. A new challenge for America: the National Cholesterol Education Program. Circulation 1986; 73:855–856.
40. Smith GD, Pekkanen J. Should there be a moratorium on the use of cholesterol lowering drugs? Br Med J 1992; 304:431–434.
41. McGill HC Jr: Persistent problems in the pathogenesis of atherosclerosis. Arteriosclerosis 1984; 4:443–451.
42. Ball KP, Hanington E, McAllen PM. Low-fat diet in myocardial infarction. A controlled trial. Lancet 1965; 501–504.
43. Research Committee to the Medical Research Council. Controlled trial of soyabean oil in myocardial infarction. Lancet 1968; 2:693–700.
44. Leren P. The effect of plasma cholesterol lowering diet in male survivors of myocardial infarction: a controlled clinical trial. Acta Med Scand 1966; 466:1–92.
45. Research Committee of the Scottish Society of Physicians. Ischaemic heart disease: a secondary prevention trial using clofibrate. Br Med J 1971; 4:775–784.
46. Group of Physicians of the Newcastle upon Tyne Region. Trial of clofibrate in the treatment of ischaemic heart disease: five year study by a group of physicians of the Newcastle upon Tyne region. Br Med J 1971; 4:767–755.

47. Coronary Drug Project Research Group. The coronary drug project: clofibrate and niacin in coronary heart disease. JAMA 1975; 231:360–381.
48. Canner PL, Berge KG, Wenger NK. Fifteen year mortality in coronary drug project patients: long-term benefit with niacin. J Am Coll Cardiol 1986; 8:1245–1255.
49. Carlson LA, Rosenhamer G. Reduction of mortality in the Stockholm ischaemic heart disease secondary prevention study to combined treatment with clofibrate and nicotinic acid. Acta Med Scan 1988; 223:405–418.
50. Buchwald H, Varco RL, Matts JP. Report of the Program on the Surgical Control of Hyperlipidemias (POSCH): effect of partial ileal bypass surgery on mortality and morbidity from coronary heart disease in patients with hypercholesterolemia. N Engl J Med 1990; 323:946–955.
51. Muldoon MF, Manuck SB, Matthews KA. Lowering cholesterol concentrations and mortality: a quantitative review of primary prevention trials. Br Med J 1990; 301:309–314.
52. Rossouw JE, Lewis B, Rifkind BM: The value of lowering cholesterol after myocardial infarction. N Engl J Med 1990; 232:1112–1119.
53. Expert Panel on Detection Evaluation and Treatment of High Blood Cholesterol in Adults. National cholesterol education program: second report of the expert panel on detection, evaluation, and treatment of high blood cholesterol. Circulation 1994; 89:1329–1445.
54. Holme I. An analysis of randomized trials evaluating the effect of cholesterol reduction on total mortality and coronary heart disease incidence. Circulation 1990; 82:1916–1924.
55. Ravnskov U. Cholesterol lowering trials in coronary heart disease: frequency of citation and outcome. Br Med J 1992; 305:15–19.
56. Gordon DJ. Cholesterol lowering and total mortality. In: Rifkind BM (ed). Lowering Cholesterol in High Risk Individuals and Populations. New York: Marcel Dekker, 1995:33–48.
57. Gordon DJ. Cholesterol and mortality: what can meta-analysis tell us. In: Gallo LL (ed). Cardiovascular Disease 2. New York: Plenum Press, 1995:333–340.
58. Law MR, Wald NJ, Thompson SG. By how much and how quickly does reduction in serum cholesterol concentration lower risk of ischaemic heart disease? Br Med J 1994; 308:351–352.
59. Holmes CL, Schulzer M, Mancini GBJ. Angiographic results of lipid- lowering trials: a systematic review and meta-analysis. (In current volume.).
60. Schuler G, Hambrecth R, Schlierf G, et al. Regular physical exercise and low-fat diet: effects on progression of coronary artery disease. Circulation 1992; 86:1–11.
61. Watts GF, Lewis B, Brunt JNH. Effects on coronary artery disease of lipid-lowering diet, or diet plus cholestyramine, in the St. Thomas' Atherosclerosis Regression Study (STARS). Lancet 1992; 339:563–569.
62. Ornish D, Brown SE, Scherwiz LW. Can lifestyle changes reverse coronary heart disease: the Lifestyle Heart Trial. Lancet 1990; 336:129–133.
63. Brensike JF, Levy RI, Kelsey SF, et al. Effects of therapy with cholestyramine on progression of coronary artery disease: results of the NHLBI type II coronary intervention study. Circulation 1984; 69:313–324.

64. Frick MH, Syvanne M, Nieminen MS, Kauma H, Taskinen MR. Prevention of the angiographic progression of coronary and vein-graft atherosclerosis by gemfibrozil after coronary bypass surgery in men with low levels of HDL cholesterol. Circulation 1997; 96:2137–2143.

65. Ericsson CG, Hamsten A, Nilsson J, Grip L, Svane B, de Faire U. Angiographic assessment of effects of bezafibrate on progression of coronary artery disease in young male postinfarction patients. Lancet 1996; 347:849–853.

66. Herd JA, Ballantyne CM, Farmer JA. Effects of fluvastatin on coronary atherosclerosis in patients with mild to moderate cholesterol evaluations (lipoprotein and coronary atherosclerosis study [LCAS]). Am J Cardiol 1997; 80:278–280.

67. Bestehorn HP, Rensing UFE, Rosamm H, Betz P, Benesch L, Schemeitat K. The effect of simvastatin on progression of coronary artery disease; the multicenter coronary intervention study (CIS). Eur Heart J 1990; 335:1109–1113.

68. Tamura A, Mikuriya Y, Nasu M, Coronary Artery Regression Study (CARS) Group. Effect of pravastatin on progression of coronary atherosclerosis in patients with serum total cholesterol levels from 160 to 220 mg/dl and angiographically documented coronary artery disease. Am J Cardiol 1997; 79:893–896.

69. Post Coronary Artery Bypass Graft Trial Investigators. The effect of aggressive lowering of low-density lipoprotein cholesterol levels and low-dose anticoagulation on obstructive changes in saphenous-vein coronary-artery bypass grafts. N Engl J Med 1997; 336:153–162.

70. Jukema JW, Bruschke AVG, van Boven AJ, et al. Effects of lipid lowering by pravastatin on progression and regression of coronary artery disease in symptomatic men with normal to moderately elevated serum cholesterol levels. The Regression Growth Evaluation Statin Study (REGRESS). Circulation 1995; 91:2528–2540.

71. Pitt B, Ellis SG, Mancini GBJ, Rosman HS, McGovern MF. Design and recruitment in the United States of a multicenter quantitative angiographic trial of pravastatin to limit atherosclerosis in the coronary arteries (PLAC I). Am J Cardiol 1993; 72:31–35.

72. Waters D, Higginson L, Gladstone P. Effect of monotherapy with an HMG-CoA reductase inhibitor on the progression of coronary atherosclerosis as assessed by serial quantitative arteriography: the Canadian Coronary Atherosclerosis Intervention Trial. Circulation 1994; 89:959–968.

73. MAAS Investigators. Effect of simvastatin on coronary atheroma: the multicentre anti-atheroma study (MAAS). Lancet 1994; 344:633–638.

74. Blankenhorn DH, Azen SP, Kramsch DM. Coronary angiographic changes with lovastatin therapy: the Monitored Atherosclerosis Regression Study (MARS). Ann Intern Med 1993; 119:969–976.

75. Sacks FM, Pasternak RC, Gibson CM, Rosner B, Stone PH. Effect on coronary atherosclerosis of decrease in plasma cholesterol concentrations in normocholesterolaemic patients. Lancet 1994; 344:1182–1186.

76. Haskell WL, Alderman EL, Fair JM. Effects of intensive multiple risk factor reduction on coronary atherosclerosis and clinical cardiac events in men and women

with coronary artery disease: the Stanford Coronary Risk Intervention Project (SCRIP). Circulation 1994; 89:975–990.

77. Kane JP, Malloy MJ, Ports TA, Phillips NR, Diehl JC, Havel RJ. Regression of coronary atherosclerosis during treatment of familial hypercholesterolemia with combined drug regimens. JAMA 1990; 264:3007–3012.

78. Brown G, Albers JJ, Fisher LD. Regression of coronary artery disease as a result of intensive lipid-lowering therapy in men with high levels of apolipoprotein B. N Engl J Med 1990; 323:1289–1298.

79. Brown BG, Zhao X-Q, Bardsley J, Albers JJ. Secondary prevention of heart disease amongst patients with lipid abnormalities: practice and trends in the United States. J Intern Med 1997; 241:283–294.

80. Blankenhorn DH, Nessim SA, Johnson RD, Sanmarco MER, Azen SP, Cashin-Hemphill L. Beneficial effects of combined colestipol-niacin therapy on coronary atherosclerosis and coronary venous bypass grafts. JAMA 1987; 257:3233–3240.

81. Constantinides P. Plaque hemorrhages, their genesis and their role in supraplaque thrombosis and atherogenesis. In: Glagov S, Newman WP, Schaffer SA (eds). Pathobiology of the Human Atherosclerotic Plaque. New York: Springer-Verlag, 1990:393–411.

82. Davies MJ. A macro and micro view of coronary vascular insult in ischemic heart disease. Circulation 1990;82(suppl II):38–46.

83. Endo A. The discovery and development of HMG CoA reductase inhibitors. J Lipid Res 1992; 33:1569–1582.

84. Kuroda M, Tsujita Y, Tanzawa K, Endo A. Hypolipidemic effects in monkeys of ML-236B, a competitive inhibitor of 3-hydroxy-3-methylglutaryl coenzyme A reductase. Lipids 1979; 14:585–589.

85. Yamamoto A, Sudo H, Endo A. Therapeutic effects of ML-236B in primary hypercholesterolemia. Atherosclerosis 1980; 35:259–266.

86. Grundy SM, Bilheimer DW. Influence of inhibition of 3-hydroxy-3-methyl-glutaryl-CoA reductase by mevinolin in familial hypercholesterolemia heterozygotes: effects on cholesterol balance. Proc Natl Acad Sci USA 1984; 81:2538–2542.

87. Scandinavian Simvastatin Survival Study Group. Randomised trial of cholesterol lowering in 4444 patients with coronary heart disease: the Scandinavian Simvastatin Survival Study (4S). Lancet 1994; 344:1383–1389.

88. Sacks FM, Pfeffer MA, Moye LA. The effect of pravastatin on coronary events after myocardial infarction in patients with average cholesterol levels. N Engl J Med 1996; 335:1001–1009.

89. Long-Term Intervention with Pravastatin in Ischaemic Disease (LIPID) Study Group. Prevention of cardiovascular events and death with pravastatin in patients with coronary heart disease and a broad range of initial cholesterol levels. N Engl J Med 1998; 339:1349–1357.

90. Sacks FM, Moye LA, Davis BR, et al. Relationship between plasma LDL concentrations during treatment with pravastatin and recurrent coronary events in the Cholesterol and Recurrent Events trial. Circulation 1998.

91. Grundy SM. Cholesterol and coronary heart disease. A new era. JAMA 1986; 256:2849–2858.

92. Pedersen TR, Olsson AG, Faergeman O, et al. Lipoprotein changes and reduction in the incidence of major coronary heart disease events in the Scandinavian Simvastatin Survival Study (4S). Circulation 1998; 97:1453–1460.
93. Grundy SM, Balady GJ, Criqui MH, et al. When to start cholesterol-lowering therapy in patients with coronary heart disease. A statement for health professionals from the American Heart Association task force on risk reduction. Circulation 1997; 95:1683–1685.
94. Gould AL, Rossouw JE, Santanello NC, Heyse JF, Furberg CD. Cholesterol reduction yields clinical benefit: impact of statin trials. Circulation 1998; 97:946–952.
95. East CA, Grundy SM, Bilheimer DW. Preliminary report: treatment of type 3 hyperlipoproteinemia with mevinolin. Metabolism 1986; 35:97–98.
96. Mostaza JM, Schulz I, Vega GL, Grundy SM. Comparison of pravastatin with crystalline nicotinic acid monotherapy in treatment of combined hyperlipidemia. Am J Cardiol 1997; 79:1298–1301.
97. Rosenson RS, Tangney CC. Antiatherogenic properties of statins: implications for cardiovascular event reduction. JAMA 1988; 279:1643–1650.
98. Shepherd J, Cobbe SM, Ford I. Prevention of coronary heart disease with pravastatin in men with hypercholesterolemia. N Engl J Med 1995; 333:1301–1307.
99. Downs JR, Clearfield M, Whitney E, Shapiro D, Beere PA, Gotto AM. Primary prevention of acute coronary events with lovastatin in men and with women average cholesterol levels. Results of AFCAPS/TexCAPS. JAMA 1998; 279:1615–1622.
100. Wood D, De Backer G, Faergemann O, Graham I, Mancia G, Pyorala K. Prevention of coronary heart disease in clinical practice. Recommendations of the second joint task force of European and other societies on coronary prevention. Eur Heart J 1998; 19:1434–1503.
101. Wilson PWF, D'Agostino RB, Levy D, Belanger AM, Silbershatz H, Kannel WB. Prediction of coronary heart disease using risk factor categories. Circulation 1998; 97.
102. Grundy SM, Balady GJ, Criqui MH, et al. Primary prevention of coronary heart disease: guidance from Framingham. A statement for healthcare professionals from the American Heart Association's Task Force on Risk Reduction. Circulation 1998; 97.
103. Haffner SM, Lehto S, Ronnemaa T, Pyorala K, Laakso M. Mortality from coronary heart disease in subjects with type 2 diabetes and in nondiabetic subjects with and without prior myocardial infarction. N Engl J Med 1998; 339:229–234.
104. Pyorala K, Pederson TR, Kjekshus J, et al. Cholesterol lowering with simvastatin improves prognosis of diabetic patients with coronary heart disease: a subgroup analysis of the Scandinavian Simvastatin Survival Study (4S). Diabetes Care 1997; 20:614–620.
105. Goldberg RB, Mellies MJ, Sacks FM, et al. Cardiovascular events and their reduction with pravastatin in diabetic and glucose-intolerant myocardial infarction survivors with average cholesterol levels. Subgroup analyses in the Cholesterol and Recurrent Events (CARE) trial. Circulation 1998; 98:2513–2519.

106. Jacobson TA. Improving health outcomes without increasing costs: maximizing the full potential of lipid reduction therapy in the primary and secondary prevention of coronary heart disease. Curr Opin Lipidol 1997; 8:369–374.

107. Kaplinsky E. Bezafibrate Infarction Prevention Study. Preliminary results presented at the annual meeting of the European Society of Cardiology, Vienna, August 1998.

108. Grundy SM, Vega GL. Fibric acids: effects on lipids and lipoprotein metabolism. Am J Med 1987; 83:9–20.

109. East C, Bilheimer DW, Grundy SM. Combination drug therapy for familial combined hyperlipidemia. Ann Intern Med 1988; 109:25–32.

110. Vega GL, Grundy SM. Management of primary mixed hyperlipidemia with lovastatin. Arch Intern Med 1990; 150:1313–1319.

111. Hulley S, Grady D, Bush T. Randomized trial of estrogen plus progestin for secondary prevention of coronary heart disease in postmenopausal women. Heart and Estrogen/progestin Replacement Study (HERS) Research Group. JAMA 1998; 280:605–613.

2
The Scandinavian Simvastatin Survival Study (4S)

Terje R. Pedersen
Aker University Hospital, Oslo, Norway

I. BACKGROUND

In February 1987, the author was informed by the Scandinavian Director of the pharmaceutical company Merck & Co. Ltd. that marketing of new, very potent cholesterol-lowering drugs was being planned for the near future. These drugs seemed to be very well tolerated and capable of reducing LDL cholesterol by approximately 40% on average. As a consequence of this information, a first-draft protocol was presented to Merck beginning of March the same year of a multicenter clinical trial that was later to be named "4S": the Scandinavian Simvastatin Survival Study (1,2). At that time, the cholesterol controversy was debated intensely within the medical community in northern Europe. Although many clinicians in Scandinavia felt that a low-fat, low-cholesterol diet was beneficial for patients at high risk of coronary heart disease, the vast majority had adopted a very restrictive policy in the prescription of cholesterol-lowering drugs. Such drugs were not even prescribed regularly to patients with familial hypercholesterolemia. Since cardiologists in Scandinavia had turned skeptical to treatment guidelines that were not based on hard endpoints from randomized clinical trials, the environment was ideal for a new trial with all-cause mortality as the primary endpoint. In August the same year, Merck agreed to support the study, and a study organization with a steering committee was founded. For this committee, the chairmen of the national cardiac societies and leading representatives of atherosclerosis research in each of the countries Denmark, Finland, Iceland, Norway, and Sweden were invited. The committee was reinforced by a statistician and joined by the author as principal investigator and scientific coordinator. The Scandinavian data coordinator within Merck was a nonvoting member of the steering committee. After having recruited centers, investigators, and study nurses, and having made revisions to the protocol, gained approval from national ethical review committees, and set up a study coordinating center in each of the countries, the first patient was recruited in February 1988.

II. STUDY DESIGN

A. Organization

The 94 clinical centers selected for the study were mostly university centers and major county hospitals, but several smaller community hospitals and medical centers also participated. At each center there was a minimum of two investigators, mostly specialists in cardiology or internal medicine, but also fellows, to recruit and follow up patients. In addition, each center had at least one study nurse to coordinate follow-up locally. In each country, the local Merck subsidiary set up a study coordinating center manned with data management personnel and study monitors that made regular visits to each center. The scientific coordinator worked in close cooperation with the subsidiaries and was at the disposition to all investigators concerning operational matters. The steering committee had semiannual meetings, receiving reports from the scientific coordinator and the Merck coordinators.

A Data Safety Monitoring Committee (DSMC) was organized with five members, including one statistician. This committee received reports from two sources. One was the statistician at Merck Research Laboratories, N.J., preparing reports, based on the case report forms sent from the clinical centers to the Merck coordinating center in each country. The study endpoints were first classified by an independent endpoint classification committee, consisting of two experienced cardiologists who did not have access to the study allocation code before processing data for the DSMC. The other source of information was the investigators who sent copies of death reports directly to the DSMC, at the same time as sending copies to the national coordinating center. When reviewing preliminary results, the DSMC considered only the direct reports from the investigators as a basis for stopping rules to be applied.

Lipids and lipoproteins were analyzed at a central laboratory located at the National Hospital in Oslo, Norway. All blood samples were drawn after at least 12 hours fasting. After coagulation and centrifugation one-third of the serum was sent unfrozen to the central laboratory the same day. The rest was frozen at $-20°C$ and shipped in batches. The enzymatic method of Boehringer Mannheim was used to determine total cholesterol and triglycerides. HDL cholesterol was determined after precipitating apo B–containing lipoproteins using MnCl. LDL cholesterol was calculated with the Friedewald formula (3). The other analyses, like liver enzymes and routine hematological and urine examinations, were performed in the local laboratories.

Electrocardiograms (ECG) from the baseline examination and each annual visit were analyzed at the Minnesota ECG Coding Center, University of Minnesota, USA, for the purpose of identifying definite myocardial infarctions for which there were no such "hospital-verified" event.

ECGs recorded in association with hospital-verified major coronary events were classified at the Clinical Events ECG Coding Center at the University of Kuopio, Finland.

B. Patient Recruitment and Randomization

A systematic screening of patients' records was performed at each center to identify potential candidates that might fulfill the eligibility criteria (see Table 1). Patients that appeared to be eligible were invited to the clinic and informed about the study. After a preliminary consent, fasting total cholesterol and triglycerides were determined locally. Patients with cholesterol >5.5 mmol/L (212 mg/dL) were invited to the study and given dietary advice on how to lower their cholesterol. The advice was consistent with recommendations from the Study Group of the European Atherosclerosis Society (4) which are similar to the Step I diet of the NECP guidelines (5) (<30% of energy from fat).

Eight weeks later serum samples were obtained and analyzed at the central laboratory. The same day the patients were given single-blind placebo for 2 weeks. When the analysis revealed that the serum lipids were within the entry "window," a final eligibility check was performed, fasting serum lipids and lipoproteins were analyzed at the central laboratory, and the patients were randomly allocated to double-blind therapy with simvastatin or placebo. Randomization was made in blocks of unknown size, separate for each center. The centers received prepackaged study medication and assigned patients with previous myocardial infarction the lowest available allocation number and patients with angina only the highest number.

C. Follow-up

During the first 18 months of follow-up, patients returned to the centers every 6 weeks for serum sampling and analysis of CK and the liver enzymes ALAT and ASAT, thereafter semiannually. Lipid measurements were performed at the central laboratory at 6, 18, and 24 weeks, thereafter semiannually. Queries for adverse experiences were done after 6, 12, and 24 weeks, thereafter semiannually. A clinical examination was performed by the investigator and recordings of ECG were made before randomization and annually thereafter.

The goal of simvastatin therapy was to reduce serum total cholesterol to <5.2 mmol/L (200 mg/dL) but not to <3.0 mmol/L (116 mg/dL). The starting dose

Table 1 Eligibility Criteria for the 4S

Inclusion criteria

Age	35–69 years
Sex	Both
Diagnosis	Myocardial infarction > 6 months ago

based on at least two of the following:
    a. Typical chest pain for >30 min
    b. Two enzymes > normal range
    c. Serial ECGs with pathological Q-waves
       or
       ST elevation followed by T inversion
       In ≥ two leads
or
Angina pectoris, stable for > 3 months
based on a history of exertional angina and
at least one of the following:
    a. Coronary angiogram showing >70% stenosis
    b. ECG showing definite signs of ischemia during
       1. An episode of pain
       2. An exercise test

Total cholesterol 5.5–8.0 mmol/L (212 - 309 mg/dL) 2 months following
dietary advice

Exclusion criteria
 1. Congestive heart failure requiring therapy
 2. Chest X-ray demonstrating cardiomegaly or pulmonary congestion
 3. Persistent atrial fibrillation or flutter
 4. Antiarrhythmic therapy
 5. Premenopausal women, unless sterilized
 6. Secondary hypercholesterolemia
 7. Fasting triglycerides > 2.5 mmol/L (221 mg/dL)
 8 Unstable angina
 9. Hemodynamically important valvular heart disease
10. History of stroke
11. Impaired hepatic function (liver function tests >50% above normal)
12. Partial ilial bypass
13. Drug or alcohol abuse
14. Poor mental function
15. Reduced life expectancy
16. Current treatment with investigational drug
17. Hypersensitivity to HMG-CoA reductase inhibitors
18. Tendon xanthomas or serum total cholesterol > 8 mmol/L (309 mg/dL) after dietary
    advice

of simvastatin was 20 mg daily, to be taken in the evening. If total cholesterol was still >5.2 mmol/L after 6 or 18 weeks, the investigator received a computer-generated message to double the dose to 40 mg daily. To avoid unblinding of the investigator, an algorithm was built into the central laboratory computer to issue similar messages for randomly selected patients in the placebo group. In the case that total cholesterol was reduced to a value < 3.0 mmol/L the computer generated a message to cut the dose to 10 mg per day (this occurred in only two cases). If repeated samples indicated a serum total cholesterol > 9.0 mmol/L (347 mg/dL), the investigator received a message to reinforce dietary advice. If at the next scheduled measurement the level was still > 9 mmol/L, study medication was withdrawn and the patient was treated according to the physician's discretion.

D. Endpoint Definition

The steering committee, in collaboration with the endpoint classification committee, worked out a manual with definition criteria for all hospital-verified cardiovascular events, based on the WHO MONICA method (6). The primary endpoint was mortality. The secondary endpoint was "major coronary events," defined as coronary deaths, definite or probable hospital-verified nonfatal acute myocardial infarction, resuscitated cardiac arrest, and ECG-verified definite silent myocardial infarction. The first of the tertiary endpoint was "CHD-related event-free survival," including all events of the secondary endpoint, myocardial revascularization procedures, and hospitalizations for acute CHD events. Other tertiary endpoints were "atherosclerosis-related event-free survival," "myocardial revascularisation procedures," "hospitalization for acute CHD without confirmed myocardial infarction," and "number of hospitalizations for acute atherosclerosis-related events." In addition, the manual for endpoint definitions included criteria for classification of subcategories of non-CHD cardiovascular diseases, such as cerebrovascular diseases and definitions of noncardiac deaths.

The two members of the endpoint classification committee classified each endpoint separately throughout the duration of the study. They used all available information from hospitals, practicing physicians, and relatives to establish the cause of death. In case of disagreement about any type of endpoint category, they met to discuss the event and reached consensus in each case.

E. Statistical Considerations

The design of the study included a 1-year recruitment period and a follow-up period from the last patient randomization of a minimum of 3 years. The power calculations for the study assumed a uniform entry rate and a total mortality rate in the placebo group of 11%. With a two-sided $\alpha = 0.05$ and a $1-\beta = 0.95$ for detecting an overall ~30% reduction in the mortality rate by simvastatin, the total sample size needed was 3600. Shortly prior to the randomization period, a study simulation was performed, applying the in- and exclusion criteria on the population in the Göteborg Post-Myocardial Infarction Clinic. This simulation suggested a mortality rate that was lower than originally estimated. To achieve the above power the simple size was revised to approximately 4000. Toward the end of the recruitment period, the actual total mortality in the two treatment arms combined turned out to be lower than the last estimate and the sample size was revised again, to approximately 4400. This number was estimated to be achieved if recruitment for the dietary period was stopped end of May 1989. When the last eligible patient was randomized August 16, 1989, 4444 patients had been included in the double-blind phase of the trial since May 19, 1988. The three interim analyses to be performed by the DSMC had been predetermined to take place at 100, 200, and 350 deaths (Table 2). As the study progressed, the interim analyses could not be done on the exact number of deaths specified in the protocol for practical reasons, but the method used allowed for adjustment of the stopping rule according to the actual number of events. Also, the lag time from a death occurred until the report reached the DSMC varied greatly, but the average time decreased from approximately 9 months during the beginning of the study to less than 6 months toward the end. Following the third interim analysis of reported deaths, the DSMC on May 27, 1994, advised the stopping of the study, as the α-value crossed the boundary of the statistical stopping rule (Table 2). For practical reasons the cutoff date was set to August 1, 1994. At this date the actual number of deaths reached 438, very close to the protocol-specified study target of 440 deaths.

The statistical assessments of baseline and efficacy data were performed according to a predefined data analysis plan, incorporated into the study protocol. All analyses were performed according to the intention-to-treat principle. The log-rank test was used to assess treatment group differences, and the Cox regression model, including relevant baseline factors, was used for a secondary analysis of the main endpoint data.

Table 2 Planned vs. Actual Interim Analyses and
Corresponding Stopping Guideline

Interim analyses and stopping guideline, study protocol			
Deaths	100	200	350	440
Cumulative α	.0014	.0049	.0219	.05
Interim analyses and stopping guideline at third analysis			
Deaths	100	221	384	440
Cumulative α	.0014	.0060	.0195	.05

III. RESULTS

Of all those invited to the study, a total of 7027 patients gave preliminary consent and received dietary advice. At the second visit, 2 months later or at the day of randomization, 2583 patients (36.8%) had one or more exclusion criteria. The main reasons for exclusion were a total serum cholesterol value outside the "entry window" in 1300 patients (18.5%), a serum triglyceride value >2.5 mmol/L in 864 (12.3%), or unwillingness to participate among 396 (5.6%). There were no main differences in baseline characteristics between the two treatment groups (Table 3). The double-blind period ranged from 4.9 to 6.3 years among those who survived (median 5.4 years). Following the cutoff date, all surviving patients were called in for a final physical examination in the next 6 weeks and query of endpoints and adverse experiences. Although some failed to turn up at this visit, vital status was confirmed for every participant in the study. On August 8, when the principal investigator had completed the final physical examination of "his" study patients, he and the chairman of the steering committee were unblinded of the main study results. The principal investigator immediately started to prepare the draft of the main publication. This was presented to the other members of the steering committee 3 weeks later for discussion and editing. The draft was updated with the final results as they continued to be confirmed based on continuous data quality control. The results were presented for the first time at the American Heart Association Scientific Sessions in Dallas, Texas, November 19, 1994, and were published in the *Lancet* 2 days later.

From an administrative aspect, the study was successfully conducted with very high adherence rate. Only 13% of the patients in the placebo group and 10% of the simvastatin group stopped taking the study medication; for 5% of the patients in both study groups adverse experiences were reported to be the reason.

A. Lipids and Lipoproteins

After 6 weeks therapy LDL cholesterol was reduced by 38% and total cholesterol by 28% in the simvastatin group. Figure 1 shows the frequency distribution of total and LDL cholesterol in the two treatment groups after 1 year. At this point the serum mean total cholesterol was 4.85 (SD 0.81) mmol/L in the simvastatin group and 6.74 (SD 0.87) mmol/L in the placebo group (7). The mean LDL cholesterol was 3.06 (SD 0.75) mmol/L in the simvastatin group and 4.91 (SD 0.83) mmol/L in the placebo group. Over the remaining course of the study the difference between the simvastatin and the placebo group were only slightly diminished, mostly due to a rising number of patients discontinuing therapy and open-label cholesterol-lowering therapy being started in 35 placebo patients. The mean reductions over 5.4 years were in the placebo group and the simvastatin group, respectively: serum total cholesterol, +1%, –25%; triglycerides, +7%,

Table 3   Baseline Characteristics of 4444 Patients Randomized

	Placebo n = 2223 (%)	Simvastatin n = 2221 (%)
Male	1803 (100)	1814 (100)
Mean age	58.1 yr	58.2 yr
Age ≥ 65 yr	388 (22)	391 (22)
Qualifying diagnosis:		
Myocardial infarction	1490 (83)	1515 (84)
Angina only	313 (17)	299 (16)
Smoking status:		
Smokers	499 (28)	459 (25)
Ex-smokers	939 (52)	995 (55)
Nonsmokers	365 (20)	360 (20)
Female	420 (100)	407 (100)
Mean age, female	60.5 yr	60.5 yr
Age ≥ 65 yr	115 (27)	127 (31)
Qualifying diagnosis:		
Myocardial infarction	278 (66)	247 (61)
Angina only	142 (34)	160 (39)
Smoking status:		
Smokers	97 (23)	83 (20)
Ex-smokers	127 (30)	126 (31)
Nonsmokers	196 (47)	198 (49)
Both genders		
Previous history:		
Hypertension	584 (26)	570 (26)
Diabetes mellitus	97 (4)	105 (5)
CABG or PTCA	151 (7)	189 (9)
Years since first diagnosis of MI or angina:		
≤1 yr	589 (26)	602 (27)
2–5 yrs	961 (43)	929 (42)
≥5 yrs	673 (30)	690 (31)
Other therapy at baseline:		
Beta blockers	1266 (57)	1258 (57)
Aspirin	815 (37)	822 (37)
Calcium antagonists	668 (30)	712 (32)
Baseline lipids, mmol/L (mg/dL):		
Total cholesterol	6.75 (260.6)	6.74 (260.2)
HDL cholesterol	1.19 (45.9)	1.18 (45.5)
LDL cholesterol	4.87 (187.5)	4.87 (187.5)
Triglycerides	1.51 (133.6)	1.49 (131.9)

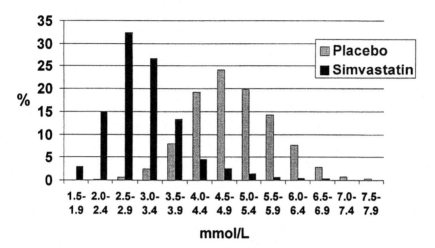

Figure 1 Frequency distribution of serum LDL cholesterol after 1 year treatment in the placebo and simvastatin groups.

−9%; HDL, cholesterol, +1%, +8%; LDL cholesterol, +1%, −34%; apo A-I, −3%; and apo B, −3%, −27%.

B. Fatal and Nonfatal Study Endpoints

The principal finding of the study was the significant difference in all-cause mortality. In the placebo group 256 patients (11.5%) died as against 182 (7.8%) in the simvastatin group, a relative risk reduction of 30% (95% confidence interval 15–42%) (P = .0003) (Fig. 2) (8). This was entirely explained by a 42% relative risk reduction in CHD mortality (95% confidence interval 17 - 54%) (P < .00001). Annual all-cause mortality increased during the study in the placebo group, reflecting the growing age of the population. The low mortality at the beginning of the study is also explained by the selection procedure, excluding patients with signs or symptoms of fatal disease. Table 4 shows the development of coronary and total mortality rate during the study. Toward the completion of the study, the annual coronary mortality rate in the simvastatin group was only half that in the placebo group.

Table 5 gives the causes of death by study group. There was no noncoronary category that had a significantly higher rate in the simvastatin group than in the placebo group. There were too few deaths from cerebrovascular or other cardiovascular causes to demonstrate any significant differences between the treatment groups for these categories. "Major coronary events" was the secondary endpoint and comprised coronary deaths plus nonfatal definite or probable myocardial infarction, resuscitated cardiac arrest, and silent myocardial infarction as diag-

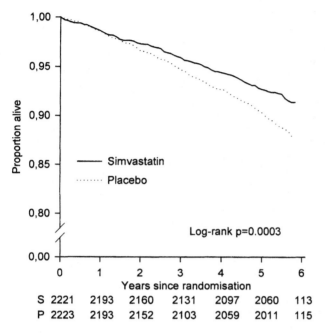

Figure 2 Kaplan-Meier curves for all-cause mortality. Number of patients at risk at the beginning of each year is shown below the horizontal axis. (From Ref. 2.)

nosed by new Q-waves in annual ECG recordings. This composite endpoint occurred in 622 (28.0%) in the placebo group and in 431 (19.4%) in the simvastatin group, a relative risk reduction of 34%. For comparison with the CARE study results, analyses have been performed, excluding resuscitated cardiac arrest and silent myocardial infarction, from the composite endpoint. In 4S there were 562 patients (25.3%) in the placebo group and 369 patients (16.6%) in the simvastatin group with the CARE-type primary endpoint, a relative risk reduction of 37% (95% confidence interval 28–45%) ($P < .00001$). The nonfatal endpoints are

Table 4 Annual Coronary and Total Mortality Rate During the Study Based on Kaplan-Meier Estimate

Year	Coronary mortality (%)		Total mortality (%)	
	Placebo	Simvastatin	Placebo	Simvastatin
1	1.2	1.0	1.4	1.4
2	1.3	0.8	2.1	1.4
3	1.5	1.0	1.9	1.3
4	1.4	0.8	2.0	1.6
5	1.8	0.9	2.4	1.7

Table 5 Mortality and Causes of Death

	No. (%) of patients	
Cause of death	Placebo (n = 2223)	Simvastatin (n = 2221)
Definite acute MI	63	30
Probable acute MI	5	5
Acute MI not confirmed:		
Instantaneous death	39	29
Death within 1h[a]	24	8
Death within 1–24 h	15	9
Death >24 h after onset of event	11	10
Nonwitnessed death[b]	23	13
Intervention-associated[c]	9	7
All coronary:	189 (8.5)	111 (5.0)
Cerebrovascular	12	14
Other cardiovascular	6	11
All cardiovascular:	207 (9.3)	136 (6.1)
Cancer	35	33
Suicide	4	5
Trauma	3	1
Other	7	7
All noncardiovascular	49 (2.2)	46 (2.1)
All deaths	256 (11.5)	182 (8.2)

Relative risk, calculated by Cox regression analysis.
[a]Following acute chest pain, syncope, pulmonary edema, or cardiogenic shock.
[b]With no likely noncoronary cause.
[c]Coronary death within 28 days of any invasive procedure.
Source: Ref. 2.

shown in Table 6. In addition to reducing the risk of needing coronary revascularization procedures by 37% ($P = .00001$), the risk reduction of fatal plus nonfatal cerebrovascular disease event was 27% ($P = .03$).

C. Subgroup Analyses

The study results have been analyzed in a large number of subgroups defined by various baseline characteristics (9–12). The results are remarkably consistent across all major groups defined by demographic and other variables. A selection of subgroup results is shown in Table 7. Most noteworthy is that the relative risk among subgroups of baseline LDL-cholesterol or HDL-cholesterol levels is remarkably similar. This means, e.g., that patients with CHD and LDL cholesterol levels in the range of 3.0–7.0 mmol/L have a detrimental level, regardless of what it may be. The results also shows that CHD patients are not protected from fur-

Table 6 Patients with Nonfatal Cardiovascular Events During Follow-up

	No. (%) of Patients*	
Event	Placebo (n = 2223)	Simvastatin (n = 2221)
Major coronary:		
Definite acute MI	270 (12.1)	164 (7.4)
Definite or probable acute MI	418 (18.8)	279 (12.6)
Silent MI	110 (4.9)	88 (4.0)
Resuscitated cardiac arrest	0	1
Acute MI, intervention-associated	25	12
Any major coronary[a]	502 (22.6)	353 (15.9)
Coronary surgery or angioplasty	383 (17.2)	252 (11.3)
Non-MI acute CHD	331 (14.9)	295 (13.3)
Acute non-CHD cardiac	109 (4.9)	109 (4.9)
Cerebrovascular:		
Stroke, nonembolic	33	16
Stroke, embolic	16	13
Stroke, hemorrhagic	2	0
Stroke, unclassified	13	15
Stroke, intervention-associated	10	3
Transient ischemic attack	29	19
Any cerebrovascular[a]	95 (4.3)	61 (2.7)
Other cardiovascular	33 (1.5)	24 (1.1)

[a]A patient with two or more events of different types will appear more than once in a column but only once in a row.
Source: Ref. 2.

ther events when they have high HDL cholesterol levels and such patients have the same relative risk reduction as those with the lowest levels. With baseline triglycerides, the results were different (Fig. 3). In patients with the highest triglyceride levels at baseline, there was a significantly greater benefit than in patients with the lowest levels. In the placebo group baseline triglycerides was significantly related to the risk, but this was not the case in the simvastatin group. This can be interpreted as evidence that statins are most effective in patients with a propensity for small, dense LDL particles, a situation which is typical in patients with high triglyceride levels.

There was no significant interaction between other concomitant medication and simvastatin. Table 7 shows examples with beta blockers, calcium antagonists, and aspirin. This means that additional protective effect of lipid lowering is seen in patients that are also treated with other drugs used for secondary prevention.

Table 7 Major Coronary Events in Subgroups of 4S

Category	No. (%) of patients		Relative risk (95% CI)	
	Placebo	Simvastatin		
Men	531 (29.4)	371 (20.5)	0.66	(0.58–0.75)
Women	91 (21.7)	60 (14.7)	0.66	(0.48–0.91)
Age < 60 yr	303 (27.6)	188 (17.6)	0.61	(0.51–0.73)
Age ≥ 60 yr	319 (28.3)	243 (21.0)	0.71	(0.60–0.86)
Age ≥ 65 yr	168 (33.4)	122 (23.6)	0.66	(0.52–0.84)
Hypertension				
Yes	190 (32.5)	122 (21.4)	0.63	(0.50–0.79)
No	432 (26.4)	309 (18.7)	0.68	(0.59–0.79)
Diabetes				
Yes	44 (45.4)	24 (22.9)	0.45	(0.27–0.74)
No	578 (27.2)	407 (19.2)	0.68	(0.60–0.77)
History of MI				
Yes	543 (30.4)	370 (20.8)	0.65	(0.57–0.74)
No	83 (18.2)	63 (13.9)	0.74	(0.53–1.03)
Smoking status:				
Nonsmoker	147 (26.2)	91 (16.3)	0.59	(0.46–0.77)
Ex-smoker	282 (26.5)	213 (19.0)	0.69	(0.58–0.82)
Smoking	193 (32.4)	127 (23.4)	0.69	(0.55–0.86)
Concomitant drugs on baseline:				
Beta blocker				
Yes	352 (27.8)	236 (18.8)	0.65	(0.55–0.76)
No	270 (28.2)	195 (20.2)	0.68	(0.57–0.82)
Calsium antagonist				
Yes	221 (33.1)	154 (21.6)	0.62	(0.50–0.76)
No	401 (25.8)	277 (18.4)	0.68	(0.59–0.80)
Aspirin				
Yes	230 (28.2)	161 (19.5)	0.66	(0.54–0.80)
No	392 (27.8)	270 (19.3)	0.67	(0.57–0.78)
Baseline lipids in quartiles:				
LDL cholesterol (mmol/L)				
≤4.39	135 (24.4)	89 (17.2)	0.65	(0.50–0.85)
4.4 – 4.84	158 (29.6)	112 (20.7)	0.67	(0.52–0.85)
4.85–5.34	153 (25.9)	108 (18.4)	0.68	(0.53–0.87)
≥5.35	175 (31.1)	118 (20.9)	0.70	(0.54–0.90)
HDL cholesterol (mmol/L)				
≤0.99	177 (32.1)	121 (22.5)	0.67	(0.53–0.84)
1.00–1.14	143 (28.0)	110 (20.4)	0.71	(0.55–0.91)
1.15–1.34	154 (27.4)	97 (16.6)	0.57	(0.44–0.74)
≥1.35	144 (24.4)	96 (17.7)	0.70	(0.54–0.90)

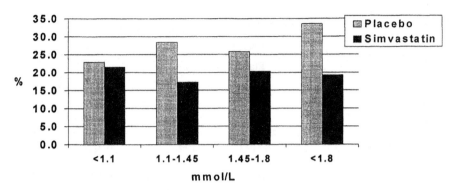

Figure 3   Major coronary events by baseline serum triglyceride quartiles.

D. Safety

The proportion of patients that withdrew from study therapy because of adverse reactions was 6% in both the placebo and simvastatin groups. For no single category by body system was there a significantly greater adverse experience rate in the simvastatin group than among the placebo group (13). There was one single case of myopathy in the simvastatin group; the patient recovered fully upon withdrawal of the study drug. Table 8 gives the most frequently reported noncardiovascular adverse experiences in the two treatment groups. There is reason to question whether simvastatin is causing myalgia in patients that have no sign of myopathy or any digestive symptoms. Such symptoms will occur in a large proportion of patients for reasons that are unrelated to their medication. In 4S, myalgia was reported in 3.2% of placebo-treated patients and 3.7% in the simvastatin group. On average, simvastatin-treated patients had significantly higher levels of serum enzymes related to the liver, but in the majority of cases the elevations were small and without clinical significance. As Table 8 shows, there were no significant differences between placebo and simvastatin patients with regard to consistent elevations to more than three times the upper normal level of aspartate aminotransferase or alanine aminotransferase. Hepatitis of unexplained cause occurred in 2 and 1 patients in the placebo and simvastatin groups, respectively.

  The frequency of cancer was similar in the two treatment groups, with 96 cases in the placebo and 89 in the simvastatin group.

E. Health-Economic Considerations

Application of the 4S study results may affect large groups of the population in industrialized countries, and health care providers are increasingly focusing on

Table 8 Noncardiovascular Adverse Experiences in 4S

	No. (%) of patients	
Adverse event	Placebo n = 2223	Simvastatin n = 2221
Digestive	521 (23.4)	559 (25.2)
Musculosceletal	592 (26.6)	613 (27.6)
Nervous system/ psychiatric	440 (19.8)	463 (20.8)
Cutaneous	355 (16.9)	374 (16.8)
Elevation to more than 3× upper normal limit of:		
ASAT[a] at least once	24 (1.1)	20 (1.0)
ASAT more than once	7 (0.3)	5 (0.2)
ALAT[b] at least once	32 (1.6)	46 (2.2)
ALAT more than once	12 (0.6)	14 (0.7)

[a]Aspartate aminotransferase.
[b]Alanine aminotransferase.

how to allocate limited resources. Therefore, health-economic analyses become more important in the assessment of how to use drugs in the prevention of common diseases. In 4S it was for the first time possible to assess cost against benefit of stain therapy in a prospectively designed trial where data had been recorded for this purpose. All relevant information about hospitalizations and use of health care resources for cardiovascular diseases had been recorded, as well as the dosage of simvastatin in each case.

The health-economic analyses were done using two different methods. First, a cost minimization analysis was performed (14). This method calculates the direct cost of simvastatin therapy during the trial and compares this with the direct costs of health care in the two treatment groups. For this analysis U.S. costs were applied. Average discounted cost of the drug Zocor (Merck) was obtained for the average dose used in the study. In May 1995 this would amount to $4679 per patient ($4400 discounted at 5%) over the mean 5.25 years of treatment. In addition, one could argue that the cost of laboratory measurements of lipids and liver enzymes should be added to treatment cost; this was calculated to be $250 per patient (discounted), making the total cost $4650. Then the cost of utilizing health care resources in the placebo and simvastatin groups were compared. In the placebo group 937 patients had been hospitalized 1905 times because of cardiovascular diseases and spent a total of 15,089 days in the hospital. The numbers for the simvastatin group were 720 patients hospitalized 1403 times and spending 9951 days in the hospital, a relative reduction of days in hospital of 34%.

The most frequency reasons for cardiovascular hospitalizations were acute myocardial infarction and coronary revascularization procedures. The cost of these hospitalizations were calculated using diagnose-related group (DRG)-based cost per case in the United States from MEDSTAT (15). The average reduction in cost for all cardiovascular disease per randomized patient was 31%, or $3872, over 5.4 years with the use of simvastatin. Thus, the effective cost of therapy would be reduced by 88%, from $2.30 per day on average to $0.28 per day when laboratory costs are not included and to $0.41 per day when such costs are included.

Many health economists prefer to use another method to demonstrate health-economic implications, namely the cost-effectiveness calculation. With this method the cost of the clinical benefits is assessed, mostly as cost per life-year saved. In this analysis the effective cost of therapy is calculated as shown above in the cost minimization analysis, taking into account savings from less use of health care resources. As a denominator is used the difference in life-years following randomization of the two treatment groups. This analysis has been done by the health-economic group at the Stockholm School of Economics (16,17). In their calculation it was assumed that simvastatin therapy was stopped at the completion of the double-blind study period and that there were no further benefits of the 5.4-year simvastatin therapy. The remaining life expectancy for the survivors in the two groups was modeled based on the average life expectancy for CHD patients of the same age and sex in Sweden. The gain in life expectancy with this model is probably quite conservative, since it is not reflecting the real situation, namely that simvastatin therapy would normally not be terminated after 5.4 years with the knowledge that the treatment yielded clinical benefit. On the other hand, the model assumed no further treatment costs either and therefore appear balanced. With this method it was calculated that the direct cost per life-year gained was on average for the entire study population SEK 56 400 (In 1995 ≈ $7700). The cost per life-year gained varied greatly among subgroups of patients, according to their level of risk (17). For example, in a sensitivity analysis it was shown that the cost per life-year gained was $3800 for men aged 70 years with the highest study total cholesterol level of 8.0 mmol/L (309 mg/dL) but $27,400 for 35-year-old females with the lowest entry cholesterol levels of 5.5 mmol/L (212 mg/dL). In these calculations, only direct costs are assessed. When including indirect costs, such as loss of labor production, the calculations give different results. For both men and women aged 35 years the treatment would be cost saving, regardless of the baseline cholesterol level. For men aged 59 years the cost per life-year gained would range from $1200 for those with total cholesterol levels of 8.0 mmol/L to $2100 for those with the lowest levels of the study.

The health-economic analyses performed from 4S indicate that this therapy is more cost-effective than many of the therapeutic regimens that are applied in the field of cardiovascular medicine today. For example, percutaneous transluminal coronary angioplasty has been shown to have a cost-effectiveness ratio ranging from \$5400–7400 per life-year gained in patients with severe angina to \$24,000–110,000 per life year saved for mild angina (18,19). The incremental direct cost per life-year gained of choosing tissue-type plasminogen activator instead of streptokinase for the treatment of acute myocardial infarction was calculated to be \$32,678, or more than four times the cost of simvastatin in the 4S population, and was still considered to be favorable (20).

IV. CONCLUSION

4S shows that long-term treatment with simvastatin in patients with CHD and moderately elevated cholesterol levels is safe and produces substantial benefit in terms of survival and cardiovascular morbidity with favorable cost-effectiveness ratios. The study results terminated the long-lasting skepticism about cholesterol lowering to prevent CHD that were based on misinterpretation of spurious results from previous trials that had been too small or had other design deficits. The impact of simvastatin seems to begin shortly after treatment is started, as the two Kaplan-Meier curves for major coronary events started to separate immediately, but the major benefits are evident only after 1 to 2 years. These findings are probably best explained by a retardation of the progression of atherosclerosis through the profound reduction in LDL cholesterol induced by simvastatin.

REFERENCES

1. Scandinavian Simvastatin Survival Study Group. Design and baseline results of the Scandinavian Simvastatin Survival Study of patients with stable angina and/or previous myocardiala infarction. Am J Cardiol 1993; 71:393–400.
2. Scandinavian Simvastatin Survial Study Group. Randomised trial of cholesterol lowering in 444 patients with coronary heart disease: the Scandinavian Simvastatin Survival Study (4S). Lancet 1994; 344:1383–1389.
3. Friedewald WT, Levy RI, Frederickson DS. Estimation of the concentration of low-density lipoprotein in plasma, without use of the preparative ultracentrifuge. Clin Chem 1972; 18:499–502.
4. Study Group European Atherosclerosis Society. Strategies for the prevention of coronary heart disease: a policy statement of the European Atherosclerosis Society. Eur Heart J 1987; 8:77–88.
5. Expert Panel of the National Cholesterol Education Program (NECP). Report of the National Cholesterol Education Program Expert Panel on detection, evaluation, and treatment of high blood cholesterol in adults. Arch Intern Med 1988; 148:36–69.
6. WHO MONICA Project. MONICA Manual, Rev ed. Geneva: Cardiovascular Diseases Unit, 1990.

7. Pedersen TR, Olson AG, Færgeman O, Kjekshus J, Wedel H, Berg K, Wilhelmsen L, Haghfelt T, Thorgeirsson G, Pyörälä K, Miettinen T, Christophersen B, Torbert JA, Musliner TA, Cook TJ. Lipoprotein changes and reduction in the incidence of major coronary heart disease events in the Scandinavian Simvastatin Survival Study (4S). Circulation 1998; 97:1453–1460.
8. Scandinavian Simvastatin Survival Study Group. Randomised trial of cholesterol lowering in 4444 patients with coronary heart disease: the Scandinavian Simvastatin Survival Study (4S). Lancet 1994; 344:1383–89.
9. Scandinavian Simvastatin Survival Study Group. Baseline serum cholesterol and treatment effect in the Scandinavian Simvastatin Survival Study (4S). Lancet 1995; 345:1274–1275.
10. Kjekshus J, Pedersen TR, Tobert JA. Lipid-lowering therapy for patients with or at risk of coronary artery disease. Curr Opin Cardiol 1996; 11:418–427.
11. Pyörälä K, Pedersen TR, Kjekshus J, et al. Cholesterol lowering with simvastatin improves prognosis of diabetic patients with coronary heart disease. Diabetes Care 1997; 20(4):614–620.
12. Miettinen TA, Pyörälä K, Olsson AG, et al. Cholesterol-lowering therapy in women and elderly patients with myocardial infarction or angina pectoris: findings from the Scandinavian Simvastatin Survival Study (4S). Circulation 1997; 96:4211–4218.
13. Pedersen TR, Berg K, Cook TJ, et al. Safety and tolerability of cholesterol lowering with simvastatin during 5 years in the Scandinavian Simvastatin Survival Study. Arch Intern Med 1996; 156:2085–2092.
14. Pedersen TR, Kjekshuh J, Berg K, Olsson Anders G, Wilhelmsen L, Wedel H, Pyorala K, Miettinen T, Haghfelt T, Faergeman O, Thorgeirsson G, Jonsson B, Schwartz J. Group for the Scandinavian Simvastatin Survival Study. Cholesterol lowering and the use of healthcare resources. Results of the Scandinavian Simvastatin Survival Study. Circulation 1996; 93:1796–1802.
15. MEDSTAT Groups Inc. DRG Guide: Descriptions and Normative Values. Ann Arbor, MI: 1994.
16. Jonsson B, Johannesson M, Kjekshus J, Olsson AG, Pedersen TR, Wedel H, Group for the Scandinavian Simvastatin Survival Study, et al. Cost-effectiveness of cholesterol lowering. Results from the Scandinavian Simvastatin Survival Study (4S). Eur Heart J 1996; 17:1001–1007.
17. Johannesson M, Jönsson B, Kjekshus J, Olsson AG, Pedersen TR, Wedel H. Cost effectiveness of simvastatin treatment to lower cholesterol levels in patients with coronary heart disease. N Engl J Med 1997; 336:332–336.
18. Tengs TO, Adams ME, Plsikin JS, G Safran D, Siegel JE, Weinstein MC, Graham JD. Five-hundred life-saving interventions and their cost-effectiveness. Risk Anal 1995; 15:369–390.
19. Wong JB, Sonnenberg FA, Salem D, Pauker S. Myocardial revascularization for chronic stable angina. Ann Intern Med 1990; 113:852–857.
20. Mark DB, Hlatky MA, Califf RM, Naylor CD, Lee KL, Armstrong PW, Barbash G, White H, Simoons ML, Nelson CL, Clapp-Channing N, Knight D, Harrell FE, Simes J, Topol EJ. Cost effectiveness of thrombolytic therapy with tissue plasminogen activator as compared with streptokinase for acute myocardial infarction. N Engl J Med 1995; 332:1418–1424.

3

Cholesterol and Recurrent Events Trial (CARE)

The Effect of Pravastatin on Coronary Events After Myocardial Infarction in Patients with Average Cholesterol Levels

Frank M. Sacks,[1] Marc A. Pfeffer,[2] Lemuel A. Moyé,[3] C. Morton Hawkins,[3] Barry R. Davis,[3] Jean L. Rouleau,[4] Thomas G. Cole,[5] and Eugene Braunwald[2] for the CARE Trial Investigators
[1]*Harvard School of Public Health, Brigham and Women's Hospital, and Harvard Medical School, Boston, Massachusetts;* [2]*Brigham and Women's Hospital and Harvard Medical School, Boston, Massachusetts;* [3]*University of Texas School of Public Health, Houston, Texas;* [4]*University Health Network and Mount Sinai Hospital, Toronto, Ontario, Canada;* [5]*Washington University, St. Louis, Missouri*

I. INTRODUCTION

Plasma total and LDL cholesterol are predictors of initial or recurrent coronary heart disease (1–4), and lowering LDL cholesterol from elevated levels prevents recurrent coronary events (5–8). However, epidemiological studies show that the relationship between plasma cholesterol and coronary events is stronger in elevated than in average ranges (1–4). Prior to the CARE trial, it had not been clear whether coronary events can be prevented by cholesterol-lowering therapy in patients who did not have hypercholesterolemia. This is an important public health issue because the large majority of patients with coronary disease have total cholesterol and LDL levels in the average, not elevated, range (9–12), similar to that of the general population (13) (Fig. 1). The CARE trial specifically studied whether pravastatin, an inhibitor of HMG CoA reductase, could prevent recurrent coronary events and stroke in a typical postmyocardial infarction population with plasma total cholesterol <240 mg/dL (6.2 mmol/L), and LDL cholesterol levels 115–174 mg/dL (3.0–4.5 mmol/L).

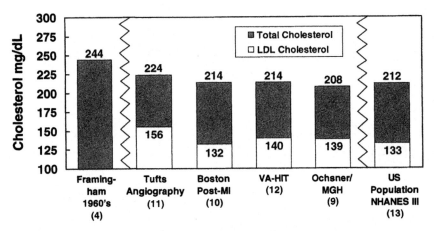

Figure 1   Cholesterol and LDL cholesterol in CHD patients.

The CARE trial addressed important subsidiary issues in coronary prevention. Several groups of patients had not been represented in most trials of cholesterol treatment. These include women, older persons (>65 years), diabetic patients, or those with impaired left ventricular function. By enrolling these persons, the trial determined whether they could benefit from pravastatin.

Information from clinical trials had not been available on whether the efficacy of cholesterol treatment is greater in patients with hypercholesterolemia than in those with pretreatment LDL concentrations in the average or below average range for North American and European populations. Coronary event reduction was evaluated according to the pretreatment LDL concentration in the CARE trial. Finally, little is known about the influence of the degree of LDL lowering or the achieved concentration of LDL on reduction of coronary events, and particularly in the below-average range easily achievable by standard doses of the HMG CoA reductase inhibitors. This information would be important in developing evidence-based guidelines for cholesterol treatment. This was also evaluated in CARE.

II. METHODS

The design of the CARE trial has previously been described in detail (14–16). The major eligibility criteria were acute myocardial infarction between 3 and 20 months prior to randomization, age 21 to 75 years, plasma total cholesterol <240 mg/dL (6.2 mmol/L), LDL cholesterol 115–174 mg/dL (3.0–4.5 mmol/L), fasting triglycerides <350 mg/dL (4.0 mmol/L), fasting glucose <220 mg/dL (12 mmol/L), left ventricular ejection fraction >25%, and no symptomatic congestive heart failure. Men and postmenopausal women were eligible. The diagnosis of myocardial infarction was confirmed centrally from information provided by the clinical centers (14). A stroke adjudication committee reviewed the medical records of all patients for whom a diagnosis was reported of stroke or TIA by the investigators at the clinical centers (17). The lipid levels from two or three qualifying visits at least 8 weeks after hospital discharge for myocardial infarction, and after 4 weeks of Step 1 diet treatment, were averaged for eligibility. The eligible patients were stratified by clinical center and randomly assigned to either pravastatin 40 mg once daily or a matching placebo by telephone call to the Data Coordinating Center. After randomization, clinic visits were conducted quarterly. Patients continued on prescribed cardiac and other medications. Patients were recruited from 80 participating centers—13 in Canada and 67 in the United States.

Plasma total cholesterol, HDL cholesterol, and triglyceride levels were measured by the core laboratory at baseline, 6, and 12 weeks after randomization, quarterly for the first year, and semiannually thereafter. LDL cholesterol was calculated (18). The primary endpoint of the trial was time to the occurrence of death from coronary heart disease (fatal myocardial infarction definite or probable, sudden death, death during a coronary interventional procedure, other coronary death) or nonfatal myocardial infarction (symptomatic unless during noncardiac surgery), each confirmed centrally using serum creatine kinase measurements (14). Deaths were reviewed by the Endpoints Committee without knowledge of treatment assignment or plasma lipid levels. The protocol was approved by the trial's Safety and Data Monitoring Committee and the Institutional Review Boards of all participating centers.

All analyses were performed on an intention-to-treat basis, and P values were two-sided. The effect of therapy on the primary endpoint of the trial was assessed using the log-rank test (19). All other hypothesis tests and all risk reductions were assessed during the Cox proportional hazard model (20). The size of the trial did not provide adequate power for the assessment of therapeutic efficacy for the primary endpoint within subgroups. Therefore, analysis of treatment effects was studied in several prespecified subgroups utilizing the endpoint major coronary events, defined as fatal coronary heart disease, nonfatal myocardial infarction, bypass surgery, or angioplasty. Cox proportional hazard analysis was used to assess the relationship between baseline characteristics and the occurrence of endpoints. Patients were then tracked at 3-month intervals.

The assessment of a potential difference between the therapeutic effect between subgroups of patients defined by differences in baseline characteristics, e.g., women and men, was addressed by adding an interaction term. If the effect of pravastatin on the endpoint was different in the subgroups, this interaction term would be statistically significant. Since a possible explanation for the statistically significant interaction might be the influence of other baseline risk factors (e.g., prior MI, time since the index MI, baseline LDL cholesterol, or diabetes), these, as well as other baseline risk factors, were added to this interaction model.

The influence of LDL cholesterol, HDL cholesterol, and triglycerides measured during the treatment period on coronary events was investigated in time-dependent Cox analyses separately and then together (21). Lipid concentrations during treatment, percentage reduction from baseline, and absolute reduction in concentration from baseline were studied. In these Cox regression analyses, the follow-up lipid measurement was the time-dependent covariate. The lipid concentrations were updated for each patient as the analysis progressed through patients' follow-up periods. The primary analysis of lipid levels during treatment combined patients in the pravastatin and placebo groups together into a single group called the "total cohort." Additional analyses examined the pravastatin and placebo groups separately. Unless otherwise stated, multivariate models adjusted for the baseline variables of age, sex, smoking history, diabetes, hypertension, and left ventricular ejection fraction.

The CARE trial was an investigator-initiated study proposed to and funded by Bristol-Myers Squibb Co. The data were collected and analyzed by and are now maintained at the Coordinating Center, University of Texas School of Public Health. As regards publication, the sponsor is entitled to provide comments to manuscripts before submission, which the authors may consider; the rights to publication reside contractually with the investigators. The sponsor was contractually required to fund the study until its conclusion. The sponsor was a member of the executive committee of the trial. The sponsor maintained information on adverse events and other trial data as required by federal regulation.

III. RESULTS AND DISCUSSION

A. Main Trial Results

1. Baseline Characteristics, Lipid Changes, and Adherence

The results of the trial have been reported for the total group (16,17) and for various patient groups (16,17,21–26). Baseline characteristics were similar in the placebo and pravastatin groups (Table 1). The median duration of follow-up was 5.0 years (range 4.0 to 6.2 years). Vital status was ascertained for the first 4 years on all patients and at the end on all but one patient. Mean plasma lipid concentrations were total cholesterol 209 mg/dL, LDL cholesterol 139 mg/dL, HDL cholesterol 39 mg/dL, and triglycerides 155 mg/dL. In comparison to the changes in the placebo group, pravastatin, during the 5-year treatment period, lowered total cholesterol by 20%, LDL cholesterol by 28%, and triglycerides by 14%, and raised HDL cholesterol by 5% (all $P <.001$). In the last year of follow-up, 86% of the placebo and 94% of the pravastatin group were taking their study medication.

Table 1 Baseline Characteristics

	Placebo (n = 2078)	Pravastatin (n = 2081)
Age (years)	59 ± 9	59 ± 9
Women/men	14%/86%	14%/86%
Race (white/other)	92%/8%	93%/7%
Country of residence		
USA/Canada	66%/34%	66%/34%
Hypertension	43%	42%
Current smoker	21%	21%
Diabetes	15%	14%
Body mass index (kg/m²)	28 ± 4	28 ± 4
Blood pressure (mm Hg)		
Systolic	129 ± 18	129 ± 18
Diastolic	79 ± 10	79 ± 10
Cardiovascular status		
Months from MI to randomization	10 ± 5	10 ± 5
Type of MI		
Q-wave/other	61%/38%	61%/38%
Angina	20%	21%
Congestive heart failure	4%	4%
CABG	28%	26%
PTCA	32%	34%
CABG or PTCA	54%	54%
Thrombolysis	40%	42%
Ejection fraction	53 ± 12%	53 ± 12%
Medication use		
Aspirin	83%	83%
Beta blocker	39%	41%
Nitrates	33%	32%
Calcium channel blocker	38%	40%
ACE inhibitor	14%	15%
Diuretic	11%	11%
Insulin	2.6%	2.4%
Oral hypoglycemic	7%	5%*
Estrogen (% of women)	10.3%	8.4%
Plasma lipids (mg/dL)		
Cholesterol		
Total	209 ± 17	209 ± 17
VLDL	27 ± 16	27 ± 16
LDL	139 ± 15	139 ± 15
HDL	39 ± 9	39 ± 9
Triglycerides	155 ± 61	156 ± 61

Percentage of patients or mean ±SD.
All differences between the groups not significant except as indicated: $*P < .05$.
(From Ref. 35.) © Massachusetts Medical Society

2. Coronary Endpoints

Pravastatin treatment significantly lowered the rate of the primary endpoint, fatal coronary heart disease, and confirmed nonfatal myocardial infarction, from 13.2% to 10.2%, representing a 24% relative risk reduction (95% confidence interval 9 to 36; P <.003) (Table 2; Fig. 2). Reductions were similar in the incidences of the components of the primary outcome variable, coronary death, and nonfatal myocardial infarction. Pravastatin decreased the rate of coronary bypass surgery by 26% (P =.005), and angioplasty by 23% (P =.01) (Table 2; Fig. 2).

3. Stroke

Stroke was a prespecified outcome variable because several previous lipid trials reported that rates of stroke and/or transient ischemic attack (TIA) were lower in treatment than control groups, although of borderline statistical significance (27,28). However, no previous trial found a significant reduction in stroke or any cerebrovascular diagnosis as an outcome. The incidence of confirmed stroke was 3.7% in the placebo group compared to 2.5% in the pravastatin group, representing a significant 32% reduction (Table 3) (17). This is very similar to the reduction in all strokes reported by the clinical centers, 31% (16). There were fewer strokes of specific types in the pravastatin than the placebo group, although none of the differences was statistically significant, probably owing to the small numbers of specific types of strokes (17) (Table 3). There was no increase in intracerebral hemorrhage, 6 vs. 2, in the placebo vs. pravastatin groups, respectively. The combined cerebrovascular endpoint of confirmed stroke or TIA occurred in 6.0% of the placebo group compared with 4.4% of the pravastatin group (risk reduction 27%; P = .02) (Table 3).

4. Mortality

The trial was not designed to detect significant changes in either total or cause-specific mortality. In all, 196 patients in the control group died compared to 180 in the pravastatin group (9% reduction in risk, 95% confidence interval -12 to 26; P = .37). Deaths from noncoronary causes occurred in 75 and 84 patients in the placebo and pravastatin groups, respectively—11 and 16 deaths due to cardiovascular but noncoronary causes, 45 and 49 due to cancer, 4 and 8 violent deaths, and 15 and 11 due to other causes, respectively, with no significant differences. The cause of death could not be determined for two patients in the placebo group.

   Total mortality was significantly reduced by pravastatin in older patients (>65 yr) (29), patients who had pretreatment LDL concentrations >125 mg/dL, and patients who had a CABG before enrollment (26) (Fig.3).

Table 2 Cardiovascular Events in the Prevention and Placebo Treated Patients[a]

Event	Placebo (N = 2078)		Pravastatin (N = 2081)		Risk Reduction with pravastatin (95% CI)
	No. of patients	Incidence (%)	No. of patients	Incidence (%)	
Death from CHD or nonfatal MI[b]	274	13.2	212	10.2	24 (9 to 36)
Death from CHD	119	5.7	96	4.6	20 (−5 to 39)
Nonfatal MI	173	8.3	135	6.5	23 (4 to 39)
Fatal MI	38	1.8	24	1.2	37 (−5 to 62)
Fatal MI or confirmed nonfatal MI	207	10.0	157	7.5	25 (8 to 39)
Clinical nonfatal MI[c]	231	11.1	182	8.7	23 (6 to 36)
CABG	207	10.0	156	7.5	26 (8 to 40)
PTCA	219	10.5	172	8.3	23 (6 to 37)
CABG or PTCA	391	18.8	294	14.1	27 (15 to 37)
Unstable angina	359	17.3	317	15.2	13 (−1 to 25)
Stroke	78	3.8	54	2.6	31 (3 to 52)

[a]Risk reduction and P values were based on Cox proportional-hazards analysis; P values are identical to those derived by log-rank analysis. Intention-to-treat analysis.
[b]This combined variable was the specified primary endpoint. Nonfatal myocardial infarctions were confirmed by the core laboratory.
[c]This variable comprises all nonfatal myocardial infarctions reported by investigators.

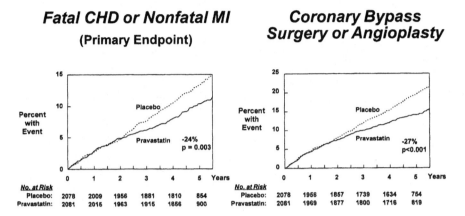

Fatal CHD or Nonfatal MI
(Primary Endpoint)

Coronary Bypass
Surgery or Angioplasty

Figure 2 Coronary events in the pravastatin and placebo groups: total cohort. (From Ref. 35.) © Massachusetts Medical Society

Patient Group

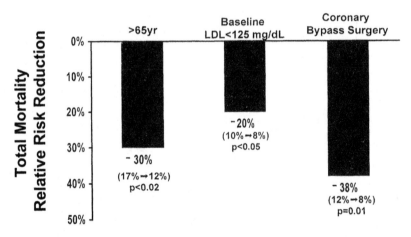

Figure 3 Reduction in total mortality in older patients (N=1283), those with pretreatment LDL ≥ 125 mg/dL (N = 3308), and after coronary artery bypass surgery (N = 1091). Relative risk reduction in the pravastatin group compared to the placebo group shown under the bars. Numbers in parenthesis under the bars are total mortality rates in the placebo group (left) connected by an arrow to the mortality rate in the pravastatin group (right), showing the reduction in mortality rate from the placebo to pravastatin treated patients.

Table 3 Stroke Category

	Placebo	Pravastatin	
Atherosclerotic	36	30	
Lacunar	8	6	
Embolic	11	9	
Tandem	3	1	
Procedural	5	2	
Undetermined	6	2	
Other/unusual	1	0	
Intra-axial	5	2	
Subarachnoid	1	0	
Total stroke	76 (3.7%)	52 (2.5%)	Risk reduction 32% (4–52%)$P = .03$
Stroke or TIA	124 (6.0%)	92 (4.4%)	Risk reduction 27% (4–44%) $P = .02$

B. Event Reduction in Subgroups of Patients

1. Women (22)

Woman had usually been excluded from lipid trials because the perception of trialists was that the major problem of cardiovascular disease was in men, and that a homogeneous study population is very important to achieve valid findings. It is now well established that coronary disease and stroke are the predominant causes of death in women. The baseline characteristics, lipid changes, medication use, and event reduction in women as well as men are published in detail (22). There were 576 women and 3583 men enrolled. All women were postmenopausal by protocol, and 10% were receiving estrogen replacement at entry. Women were older than men at randomization (average 61 compared to 58 years), and the prevalence at baseline of the following risk factors was significantly greater in women than men: hypertension 54% vs. 41%, diabetes 20% vs. 13%, current smoking 30% vs. 20%, and family history of CHD 45% vs. 40%.

Women were more likely than men to have multiple risk factors for CHD. Fewer women in CARE received thrombolytic therapy for their MI than men, 38% vs. 43%, and fewer underwent coronary artery bypass grafting between the time of their qualifying MI and randomization, 15% vs. 23%. The use of other coronary procedures including angiography and PTCA was not significantly different. Women had a higher mean total cholesterol concentration than men, 215 vs. 208 mg/dL, resulting from a higher HDL cholesterol, 45 vs. 38 mg/dL. LDL cholesterol, VLDL cholesterol, and triglycerides were similar in women and men. Pravastatin had similar effects on plasma lipids in women and men.

The risk of a recurrent event in the placebo-treated patients was similar in women and men. Pravastatin therapy caused a substantial reduction in all major cardiovascular events in women (Table 4) (25). The primary endpoint of CHD death or nonfatal MI was decreased by 43%, ($P = .035$), fatal or nonfatal MI by 56% ($P = .009$), PTCA 48% ($P = .025$), CABG 40% ($P = 0.13$), and stroke 56% ($P = .07$). The men in the pravastatin group also showed significant reduction in risk of the primary and expanded coronary endpoints and in revascularization procedures, but the magnitudes of the reductions were lower than in women, and similar to the results in the trial overall (22). The difference in event reduction between the women and men was nominally significant, $P = .048$ for combined coronary events. However, since it was not a prior hypothesis of the research group that women would benefit more than men, and since there is little precedence in the literature for such a result, the observed differences are difficult to interpret and could have resulted from the play of chance. Considering all hospitalizations for cardiovascular disease, including multiple events per person, 228 would be prevented for every 1000 women with MI treated with pravastatin for 5 years. One such hospitalization would be prevented for every four women treated during a 5-year period. The substantial reduction in cardiovascular events in the women taking pravastatin demonstrates that treatment is essential.

2. Older Patients (23)

Older patients were excluded from most previous trials of lipid therapy (23). Although their high event rate provides a strong motivation for including them to improve statistical power, there were theoretical reasons to think that blood cholesterol is less important in the etiology of cardiovascular events in older persons. Epidemiological studies generally found that cholesterol was a weak or insignificant predictor of coronary events in patients age 60 and older, even after multivariate adjustment for other risk factors and patient characteristics (30–33). However, there was long-standing suspicion that serious confounding was present in the patients who had low cholesterol, in the form of other chronic disease or debilitation, that both lowered cholesterol and caused death. This was only recently demonstrated in a cohort study of older patients by using serum iron and albumin as markers of debility, and excluding deaths in the first follow-up year (34). In this multivariate analysis, serum cholesterol was a significant predictor of coronary death, while it was not when the markers were not included. In the CARE trial, an upper age limit of 75 years at randomization (80 years at the end of the trial) was chosen.

The trial enrolled a total of 1283 patients who were > 65 years ("older patients"), and 2876 patients who were <65 years ("younger patients") (23). The mean age of the older patients was 69 years at randomization and 74 years at the end of the trial. Significant differences in baseline coronary risk factors were observed in the two age groups. Older patients as compared to younger patients more frequently were women (18% vs. 12%), hypertensive (48% vs. 40%), and

Table 4 Cardiovascular Events in Women

	Placebo (N = 290)		Pravastatin (N = 286)		% Risk reduction (95% CI)	P
	Patients (no.)	Incidence (%)	Patients (no.)	Incidence (%)		
Death from CHD or nonfatal MI	39	13.4	23	8.0	43 (4 to 66)	0.035
Death from CHD	14	4.8	11	3.8	21 (-74 to 64)	0.56
Nonfatal MI	28	9.7	14	4.9	51 (8 to 74)	0.028
Fatal MI	6	2.1	1	0.3	83 (-40 to 98)	0.099
Fatal MI or nonfatal MI	33	11.4	15	5.2	56 (18 to 76)	0.009
Clinical nonfatal MI[a]	38	13.1	18	6.3	54 (20 to 74)	0.006
CABG	23	7.9	14	4.9	40 (-17 to 69)	0.13
PTCA	34	11.7	18	6.3	48 (8 to 71)	0.025
CABG or PTCA	57	19.7	28	9.8	53 (26 to 70)	0.001
Unstable angina	65	22.4	56	19.6	14 (-23 to 40)	0.41
Stroke	16	5.5	7	2.4	56 (-7 to 82)	0.071
Combined coronary events[b]	80	27.6	46	16.1	46 (22 to 62)	0.001

[a]MI reported by the clinical center.
[b]CHD death, nonfatal MI, PTCA, and CABG.
(From Ref. 36.) © The American College of Cardiology

diabetic (19% vs. 12%), and were more likely to have had a second previous myocardial infarction (21% vs. 15%) but less commonly were current smokers (12% vs. 24%) or had a family history of coronary heart disease (33% vs. 44%) (all P <.05). The baseline total, HDL, and LDL cholesterol levels were similar in the two age groups, but triglycerides were slightly lower in the older patients, 149 vs. 159 mg/dL (P <.05). Pravastatin therapy compared to placebo had similar effects on plasma lipids in older patients compared to younger patients (23).

Older patients treated with placebo had a higher frequency of coronary events than the younger patients in the placebo group: 17% vs. 11% for coronary death or nonfatal MI, 10% vs. 4% for coronary death; 11.0% vs. 9.5% for CABG; 14% vs. 5% for CHF; and 7% vs. 2% for stroke (23). In contrast, PTCA was performed in 7% of the older vs. 12% of the younger patients in the placebo group. Older patients treated with pravastatin had a 39% lower incidence of the primary endpoint, fatal coronary heart disease or confirmed myocardial infarction, than the placebo group (95%) confidence interval, 18–55%; $P = .001$) (Table 5). Older patients also experienced a significant reduction in risk for stroke (40%), coronary artery bypass grafting (43%), and for major coronary endpoints (32%). The younger patients also experienced significantly reduced rate of coronary events in the pravastatin compared to the placebo group, although not as much reduction in either relative or absolute terms as the older group. Considering all cardiovascular events resulting in hospitalization, including multiple events in patients, a total of 225 would be prevented for every 1000 older patients (65–74 years) treated with pravastatin for 5 years compared to 121 in 1000 younger patients (<65 years). The number of older patients needed to treat for 5 years to prevent a hospitalization for cardiovascular disease is 4, a major coronary event 11, a CHD death 22, or a stroke 34 (23). The number of younger patients needed to treat to prevent a cardiovascular event is greater than for the older patients— 8 for hospitalizations, 20 for major coronary events, and 250 for stroke. Younger patients did not have reduction in CHD death. Absolute reduction in risk is greater in older than younger patients owing to their higher underlying event rate.

It is now clear that *relative* risk reduction from pravastatin treatment is not less for older than younger patients. Moreover, since the underlying risk of a cardiovascular event rises with age, the absolute risk reduction and the number of events prevented by treatment are greater in older than younger patients. Since there is no lessening of benefit, relative or absolute, as the age of patients increases, it is reasonable to question whether there is a biological upper limit of age for efficacy of lipid treatment. From a clinical perspective, the CARE trial findings suggest that pravastatin could be considered for any patient according to the level of risk and overall clinical condition, rather than the age per se.

Table 5 Cardiovascular Events, Age 65-75 Years at Entry

Event	Placebo (n = 643) (No.) (%)	Pravastatin (n = 640) No (%)	Relative risk reduction (95% CI)	P value
CHD death or nonfatal MI[a]	111 (17.3)	69 (10.8)	39% (18, 55)	0.001
CHD death	66 (10.3)	37 (5.8)	45% (18, 63)	0.004
Nonfatal MI	57 (8.9)	41 (6.4)	30% (-5, 53)	(0.09)
Fatal or nonfatal MI	73 (11.4)	50 (7.8)	33% (4, 53)	0.03
Clinical nonfatal MI[b]	78 (12.1)	54 (8.4)	33% (5, 53)	0.02
CABG	71 (11.0)	42 (6.6)	43% (16, 61)	0.004
PTCA	43 (6.7)	39 (6.1)	11% (-38, 42)	0.61
CABG or PTCA	104 (16.2)	73 (11.4)	32% (8, 49)	0.01
Unstable angina	96 (14.9)	103 (16.1)	-8% (-43, 18)	0.59
Congestive heart failure	89 (13.8)	70 (10.9)	23% (-6, 44)	0.10
Stroke	47 (7.3)	29 (4.5)	40% (4, 62)	0.03
Major coronary events[c]	181 (28.1)	126 (19.7)	32% (15,46)	0.001

Risk reductions computed by Cox proportional hazards analysis.
[a]Primary endpoint of the trial.
[b]MI reported by the clinical centers.
[c]CHD death, nonfatal MI, CABG, PTCA.

3. Diabetic Patients (24)

The CARE trial included 586 (14.1%) patients with a clinical diagnosis of diabetes (24). Of these, 45% were receiving sulfonylurea treatment and 19% were treated with insulin; the remainder were on diet therapy. Compared to patients without diabetes, those with diabetes were older (61 vs. 58 years), more overweight (mean BMI 29 vs. 27), and had a higher incidence of hypertension (52% vs. 41%) and CHF (13% vs. 6%). The group with diabetes had slightly but significantly lower LDL-C (average 136 vs. 139 mg/dL), lower HDL-C (38 vs. 39), and higher triglyceride (164 vs. 154) than the nondiabetes group. Baseline fasting glucose values in those with diabetes was higher than in the nondiabetes group (149 vs. 97 mg/dL). Pravastatin had similar effects on plasma lipid concentrations in the diabetes and the nondiabetes groups.

The diabetic patients experienced significantly higher rates of cardiovascular events than the nondiabetic patients with the exception of angioplasty (24). The primary endpoint occurred in 20% of diabetic and 12% of nondiabetic patients ($P < .001$); major coronary events (coronary death, myocardial infarction, CABG, PTCA) occurred in 37% of diabetic and 25% of nondiabetic patients ($P < .001$), and stroke occurred in 8% of diabetic and 3% of nondiabetic patients ($P < .001$). Pravastatin treatment reduced major coronary events by 25% ($P = .053$), a *relative* reduction which was similar to nondiabetic patients (Table 6) (24). There were similar reductions in *relative* risk in each of the nonfatal endpoints among patients with and without diabetes. Morever, because of their higher event rate, the diabetes group experienced greater *absolute* risk reduction than nondiabetic. For every 1000 diabetics treated for 5 years with pravastatin, 81 of them would be spared from having one or more major coronary endpoints compared with 52 in 1000 nondiabetics.

Risk stratification according to fasting glucose was considered in the CARE population (24). The new American Diabetes Association definitions for abnormal fasting glucose were applied to the CARE patients: normal <110, impaired fasting glucose 110–125 mg/dL, and diabetes >125 mg/dL. Cutpoint analysis demonstrated increased risk of coronary events in those with glucose >110 mg/dL, or even 110–125 mg/dL, compared with those with normal glucose (24). Thus the increased coronary risk supported the definition of abnormal fasting glucose. In the nondiabetic patients who had impaired fasting glucose, the pravastatin group experienced trends toward reduced event rates, although the group was too small for the differences to be significant (24). Since fasting glucose intolerance in those not diagnosed with diabetes is a risk factor for diabetes as well as for coronary events, these patients should be prioritized for lipid treatment.

Table 6 Cardiovascular Events in Diabetic Patients

	Placebo (N = 304)		Pravastatin (No. = 282)		% Risk reduction (95% CI)
	Patients	Incidence (%)	Patients	Incidence(%)	
Death from CHD or nonfatal MI	62	20.4	50	17.7	13% (4 to −26)
CHD Death	30	9.9	27	9.6	3% (42 to −63)
Nonfatal MI	37	12.2	28	9.9	18% (50 to −33)
Total MI	49	16.1	35	12.4	23% (50 to −19)
CABG	44	14.5	29	10.3	30% (56 to −12)
PTCA	33	10.9	24	8.5	23% (55 to −30)
CABG or PTCA	72	23.7	47	16.7	32% (53 to 2)
Unstable angina	63	20.7	52	18.4	14% (40 to −25)
Stroke	24	7.9	19	6.7	14% (53 to −57)
Major coronary events[a]	112	36.8	81	28.7	25% (43 to 0)

[a]Death from CHD, nonfatal MI, CABG, or PTCA.

4. Other High-Risk Patients

The risk of recurrent events was greater in smokers than nonsmokers, hypertensive than normotensive patients, and in those with impaired left ventricular ejection fraction (25–40%) (16). Relative risk reduction was similar among these higher vs. lower risk groups (16). Because of their higher event rates, absolute reductions for major coronary events were greater in smokers, hypertensives, and those with impaired ejection fraction than in the lower-risk groups, and the number of these patients needed to treat to prevent an event were relatively few (Fig. 4).

5. Lipid Concentrations

Pretreatment LDL.   The pretreatment LDL level influenced the reduction in coronary events during treatment (16,25). When the LDL concentrations were divided at the median, 137.5 mg/dL, cardiovascular risk reduction tended to be greater in those with LDL above than below the median, 137.5 mg/dL (Fig. 5). The same pattern was present for greater total cholesterol concentrations above compared to below the median, 209 mg/dL. When the pretreatment LDL range was futher subdivided, the differences in event reduction became greater. Risk reduction was attenuated progressively in baseline LDL strata below the median. The rate of major coronary events was 23% lower in the pravastatin than the placebo group in patients with baseline LDL below the median (137.5 mg/dL

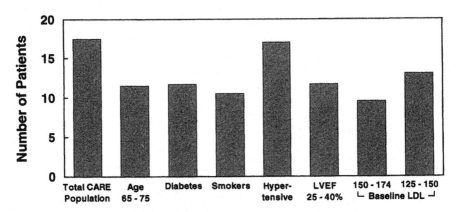

Figure 4   Number of patients needed to treat to prevent one from having a major coronary event during 5 years. CHD death, nonfatal MI, CABG, PTCA.

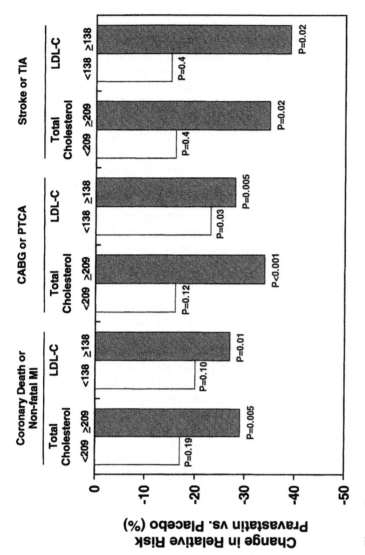

Figure 5   Risk reduction for coronary or cerebrovascular events according to pretreatment total or LDL cholesterol concentrations. Median values used for cutpoints (approximately 2080 patients per lipid group).

[3.55 mmol/L]), but only 15% lower in patients in the lowest tertile (≤130 mg/dL [3.36 mmol/L]), 10% lower in the lowest quartile (<127 mg/dL [3.28 mmol/L]), and 3% higher in the lowest quintile (<125 mg/dL [3.2 mmol/L]). Those patients with baseline LDL>150 mg/dL (3.9 mmol/L) (N = 953) experienced a 35% reduction in major coronary events, compared to a 26% reduction in those with LDL 125 to 150 mg/dL (3.2 to 3.9 mmol/L) (N = 2355), and a 3% increase for LDL <125 mg/dL (3.2 mmol/L) (N = 851) (P = .03 for interaction between baseline LDL and risk reduction) (Fig. 6). In the placebo group, baseline LDL predicted recurrent coronary events. For every 25 mg/dL increment in baseline LDL, the placebo group experienced 28% increased risk (95% CI 5% to 56%) (25) (Fig. 7). In the pravastatin group, event rates were similar for all baseline LDL categories. This suggests that treatment of LDL reaches a limit in reducing the 5-year event rate in this population to 20% for major coronary events. The same relationship was found for the primary endpoint, CHD death or nonfatal MI, which was reduced to 10% in these three groups of patients with high, average, or low pretreatment LDL concentrations. These findings also suggest that LDL did not materially influence coronary event rates at concentrations below 125 mg/dL.

Achieved LDL. In the total cohort, placebo and pravastatin groups taken together, the average LDL concentrations during treatment (LDL achieved) correlated significantly with the risk of coronary events (Fig. 8). The LDL concentration during follow-up, rather than the absolute *change* in LDL concentration, or the percentage change, was significant (21). Therefore, the analyses focused on the LDL concentrations achieved during treatment rather than the *change* in LDL during treatment. The relationship between follow-up average LDL and coronary events in the total cohort was determined to be nonlinear using time-dependent decile analysis. The relative risks for both the primary endpoint and expanded endpoint declined progressively from the 10th decile of achieved LDL (median LDL 162 mg/dL) to the sixth decile (median 121 mg/dL), after which there was no further reduction. The optimum cutpoint for an LDL concentration achieved during treatment that distinguished a higher from lower relative risk was determined to be 125 mg/dL using maximum likelihood criteria. The risk of a primary endpoint for patients who had follow-up LDL ≥ 125 mg/dL (mean 145 mg/dL), whether in the pravastatin or placebo group, was 43% greater than for those with followup LDL<125 mg/dL (P < .01).

In the placebo group, the relationship between LDL achieved and coronary events was similar to the relationship in the upper half of the LDL range of the total cohort, which was mostly composed of placebo patients (21). The relationship between LDL achieved during treatment and coronary events in the pravastatin group was also significant overall (P = .02) but nonlinear (Fig. 9). In the pravastatin group, the highest event rate was in the highest decile, median

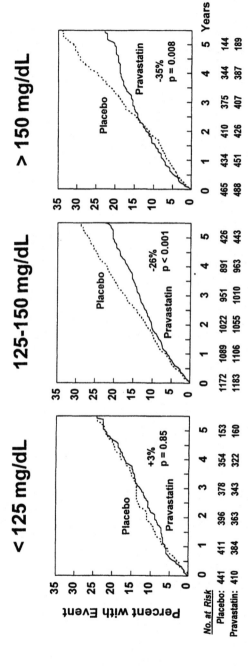

Figure 6 Coronary events in the pravastatin and placebo treated patients according to baseline LDL cholesterol concentration. CHD death, nonfatal MI, CABG, PTCA. Percentages indicate change in risk in pravastatin compared to placebo group. N = 851 for LDL<125; 2355 for LDL 125–150; 953 for LDL>150 mg/dL. (From Ref. 34.) © Massachusetts Medical Society

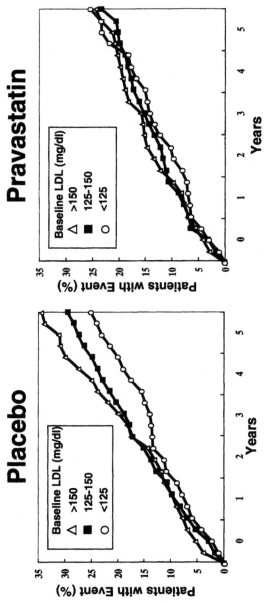

Figure 7 Coronary events according to baseline LDL cholesterol concentration in the placebo (left panel) and pravastatin (right panel) groups. CHD death, nonfatal MI, CABG, PTCA. Data are the same as in Figure 5. (From Ref. 37.)

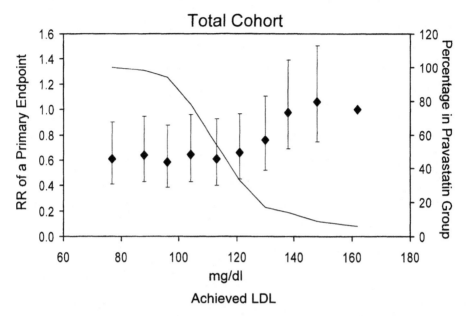

Figure 8   Coronary events and LDL cholesterol concentration achieved during follow-up. Placebo and pravastatin groups combined. Primary endpoint: fatal CHD or nonfatal MI. N = 486 patients with endpoint, 55 in 10th decile. Relative risk determined by Cox proportional-hazards analysis with time-dependent covariates. Data points show relative risks with 95% CIs for coronary events for deciles of average follow-up LDL concentrations. Percentages of patients in each decile of LDL concentration who are in pravastatin group are indicated by the solid line, corresponding to the right vertical axis. (From Ref. 21.) © American Heart Association

LDL 136 mg/dL, and the event rates were similar across all the other nine deciles that had medians between 71 and 117 mg/dL. Adherence, defined as taking study medication for at least 3 of the 5 years of follow-up, was 92% in the highest decile of LDL during treatment, 94% to 95% in the 8th to 9th deciles, and 98% to 99% in the remaining deciles. The average decrease in LDL concentration that resulted in the median LDL concentrations during follow-up of 125 mg/dL was 25 mg/dL, or 17% of the pretreatment concentration. Lower LDL concentrations that were produced by larger decreases in LDL by up to 53 mg/dL, or by 43%, from baseline were not associated with reductions in the coronary event rate below that associated with an LDL concentration of 125 mg/dL. Furthermore, the relative risk of an expanded coronary event in patients in the pravastatin group whose average follow-up LDL was ≤100 mg/dL was 0.97 compared to those with LDL 101 to 125 mg/dL, thereby demonstrating that the risks were nearly identical in both ranges of follow-up LDL. We used confidence intervals around this relative risk to calculate that the probability was 10% that a 15% reduction in end-

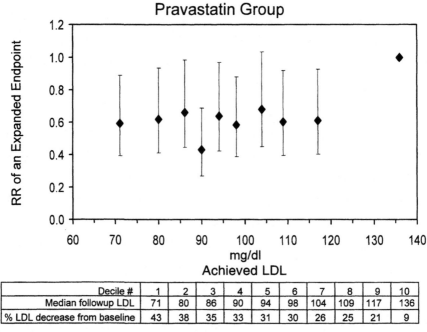

Decile #	1	2	3	4	5	6	7	8	9	10
Median followup LDL	71	80	86	90	94	98	104	109	117	136
% LDL decrease from baseline	43	38	35	33	31	30	26	25	21	9

Figure 9 Coronary events and LDL cholesterol concentration achieved during follow-up. Pravastatin group, n = 2081 patients. Coronary death, nonfatal MI, CABG, PTCA (430 patients with endpoint, 52 in 10th decile. Relative risk determined by Cox proportional-hazards analysis with time-dependent covariates. Data points show relative risks with 95% CIs for coronary events for deciles of average follow-up LDL concentration. Adherence, defined as taking study medication for at least 3 of the 5 years of follow-up; 92% in highest decile of follow-up LDL; 94–95% in 8th and 9th deciles; and 98–99% in remaining deciles. (From Ref. 21.) © American Heart Association

points in the LDL range ≤ 100 mg/dL compared to the range 101 to 125 mg/dL, and 15% for only a 10% reduction (21). This suggests that a clinically important reduction in events was unlikely to have been missed.

These findings regarding LDL during followup are consistent with those for pretreatment LDL, and are summarized in Table 7. They suggest an interpretation that LDL concentrations >125 mg/dL are an important and treatable cause of cardiovascular events, whereas those in a range <125 mg/dL may have less or no such relationship. For the 80% of patients who had a pretreatment LDL>125 mg/dL, coronary events (and all-cause mortality) were reduced by pravastatin therapy. Over 90% of the patients achieved an LDL <125 mg/dL during treatment—the range associated with benefit. The patients who did not

Table 7 Pretreatment and On-treatment LDL Cholesterol Concentrations and Coronary Risk Reduction

Pretreatment LDL concentration			
LDL range	115–125	125–150	150–174
N (% of study population)	851 (20%)	2355 (57%)	953 (23%)
Relative risk reduction[a]	-3%	26%	35%
On-treatment (achieved) LDL concentration			
LDL range, pravastatin group	52–100	100–125	>125
N (% of study population)[b]	1249 (60%)	687 (33%)	146 (7%)
Relative Risk Reduction	40%	37%	—(referent)

[a] Major coronary events: CHD death, MI, CABG, PTCA.
[b]Pravastatin group.
Units: LDL-cholesterol mg/dL.

achieve an LDL<125 mg/dL during treatment had a higher event rate than those who did. Finally, the small percentage of patients, 20%, who were enrolled with an LDL already <125 mg/dL did not appear to receive benefit from LDL lowering to a mean of 85 mg/dL.

HDL and Triglycerides.   Pretreatment HDL concentration was associated inversely with the coronary event rate in the placebo as well as pravastatin groups, and relative risk reductions were similar across HDL categories (25). The HDL concentration achieved during treatment was a signficant inverse predictor of coronary events in univariate analysis but not after adjustment for baseline nonlipid risk factors (21). Pretreatment triglyceride concentration was not a significant predictor of coronary events (25). The triglyceride level achieved during treatment was a weak but significant predictor of coronary events (21).

6. Revascularized Patients (26)

Before enrollment into the trial, 2245 of the 4159 patients had had a revascularization procedure: 1154 with PTCA, 876 with CABG, and 215 with both procedures. For analysis of risk reduction with pravastatin, the 215 patients who had CABG and PTCA were included in the analysis of CABG (26). Excluding the patients who had both procedures had little effect on the results. In patients who had had PTCA, the rate of CHD death or recurrent MI was reduced by 39% (11.9% in the placebo group vs. 7.5% in the pravastatin group, p = 0.01). The risk of another revascularization procedure was reduced significantly by 22% (23% vs 18%, $P = 0.05$). Risk of stroke was also reduced in those with PTCA, by 72% (3.6% vs 1.0%, $P = .006$). Patients who had had CABG were also benefited substantially by pravastatin. Risk of CHD death or nonfatal MI was reduced by 33% (from 13% to 9%, $P = .03$). Mortality was reduced in patients with CABG treated with pravastatin: CHD deaths by 44% (from 7.8% to 4.6% $P = .02$), and all-cause mortality by 38% (from 12% to 8%, $P = .01$).

Cholesterol and Recurrent Events (CARE) Trial: Main Trial Results

1. Pravastatin significantly lowered cardiovascular events in typical patients with myocardial infarction who have total cholesterol <240 mg/dL and LDL cholesterol 115–174 mg/dL
2. Reduction in coronary endpoints for the total group
   a. Coronary death or nonfatal myocardial infarction
   b. Myocardial infarction, fatal or nonfatal
   c. CABG
   d. PTCA
3. Reduction in stroke, as a prespecified outcome
4. Reduction in coronary endpoints for important patient groups
   a. Women, and men
   b. Older patients, 65–75 years, and younger patients
   c. Diabetic and nondiabetic patients
   d. Hypertensive and nonhypertensive patients
   e. Smokers and nonsmokers
   f. Patients with impaired left ventricular ejection fraction, 25–40%, or normal
5. No increase in adverse events or discontinuation of study medication in the pravastatin compared to placebo group

The implications of the CARE trial are that the vast majority of patients with coronary heart disease should receive lipid-lowering treatment. The average period of treatment in the trial was 5 years, but that is certainly not recommended to be the end of therapy for patients. It could be expected that the benefit would continue if treatment were extended beyond 5 years. Several groups of patients were included in CARE for whom the importance of cholesterol in coronary disease had not been established. Pravastatin reduced recurrent cardiovascular events in women, older patients age 65 to 75 years, diabetic persons, and those with impaired left ventricular function.

The broader public health implications are that cholesterol concentrations are very likely to be too high in the general population. Total and LDL cholesterol concentrations worldwide are converging near the means in the CARE population, since the levels are increasing in populations such as Japan, where cholesterol historically has been very low, and levels are decreasing in western

Europe and Scandinavia, where they have been higher than in North America. Since the pathobiology of disease progression in those who have or have not manifested clinical expression of the disease is likely to be similar, it can be inferred that lipid treatment would benefit this lower risk population worldwide.

REFERENCES

1. Rose G, Reid DD, Hamilton PJ, McCartney P, Keen H, Jarrett RJ. Myocardial ischemia, risk factors and death from coronary heart disease. Lancet 1977; 1:105–109.
2. Martin MJ, Hulley SB, Browner WS, Kuller LH, Wentworth D. Serum cholesterol, blood pressure, and mortality: implications from a cohort of 361,662 men. Lancet 1986; 2:933–936.
3. Pekkanen J, Linn S, Heiss G, et al. Ten-year mortality from cardiovascular disease in relation to cholesterol level among men with and without preexisting cardiovascular disease. N Engl J Med 1990; 322:1700–1707.
4. Kannel WB. Range of serum cholesterol values in the population developing coronary artery disease. Am J Cardiol 1995; 76:69C–77C.
5. Scandinavian Simvastatin Survival Study Group (4S). Randomized trial of cholesterol lowering in 4444 patients with coronary heart disease: the Scandinavian Simvastatin Survival Study (4S). Lancet 1994; 344:1383–1389.
6. Shepherd J, Cobbe SM, Ford I, et al. Prevention of coronary heart disease with pravastatin in men with hypercholesterolemia. N Engl J Med 1995; 333:1301–1307.
7. Byington RP, Jukema JW, Salonen JT, et al. Reduction in cardiovascular events during pravastatin therapy. Pooled analysis of clinical events of the pravastatin atherosclerosis intervention program. Circulation 1995; 92:2419–25.
8. Holme I. Cholesterol reduction and its impact on coronary artery disease and total mortality. Am J Cardiol 1995; 76:10C–17C.
9. Lavie CJ, Milani RV. National Cholesterol Education Program's recommendations and implications of "missing" high-density lipoprotein cholesterol in cardiac rehabilitation programs. Am J Cardiol 1991; 68:1087–1088.
10. Buring JE, O'Connor GT, Goldhaber SZ, et al. Decreased HDL2 and HDL3 cholesterol, apo A-I and apo A-II, and increased risk of myocardial infarction. Circulation 1992; 85:22–29.
11. Genest J Jr, McNamara JR, Ordovas JM, et al. Lipoprotein cholesterol, apolipoprotein A-I and B and lipoprotein(a) abnormalities in men and women with premature coronary heart disease. J Am Coll Cardiol 1992; 19:792–802.
12. Rubins HB, Robins SJ, Collins C, et al. Distribution of lipids in 8,500 men with coronary artery disease. Am J Cardiol 1995; 75:1196–1201.
13. Johnson CL, Rifkind BM, Sempos CT, et al. Declining serum total cholesterol levels among US adults. The National Health and Nutrition Examination Surveys. JAMA 1993; 269:3002–3008.

14. Sacks FM, Pfeffer MA, Moye L, et al. Rationale and design of a secondary prevention trial of lowering normal plasma cholesterol levels after acute myocardial infarction: the Cholesterol and Recurrent Events trial (CARE). Am J Cardiol 1991; 68:1436–1446.

15. Pfeffer MA, Sacks F, Moye LA, et al. Cholesterol and recurrent events: a secondary prevention trial for normolipidemic patients. Am J Cardiol 1995; 76:98C–112C.

16. Sacks FM, Pfeffer MA, Moye LA, et al. The effect of pravastatin on coronary events after myocardial infarction in patients with average cholesterol levels. N Engl J Med 1996; 335:1001–1009.

17. Plehn JF, Davis BR, Sacks FM, et al. Reduction in stroke incidence following myocardial infarction with pravastatin: the Cholesterol and Recurrent Events Study. Circulation 1999; 99:216–223.

18. Friedewald WT, Levy RI, Frederickson DS. Estimation of the concentration of low-density lipoprotein cholesterol in plasma, without the use of the preparative ultracentrifuge. Clin Chem 1972; 18:499–502.

19. Kalbfeisch JD, Prentice RL. *The Statistical Analysis of Failure Time Data.* New York: Oxford University Press, 1980.

20. Cox DR. Regression models and life-tables (with discussion). J R Stat Soc, series B 1972; 34:187–220.

21. Sacks FM, Moye LA, Davis BR, et al. Relationship between plasma LDL concentrations during treatment with pravastatin and recurrent coronary events in the Cholesterol and Recurrent Events trial. Circulation 1998; 97:1446–1452.

22. Lewis SJ, Sacks FM, Mitchell JS, et al. Effect of pravastatin on cardiovascular events in women after myocardial infarction: the Cholesterol and Recurrent Events trial. J Am Coll Cardiol 1998; 32:140–146.

23. Lewis SJ, Moye LA, Sacks FM, et al. Effect of pravastatin on cardiovascular events in 65–75 year old patients who have had myocardial infarction and have cholesterol levels in the average range. Ann Intern Med 1998; 129:681–689.

24. Goldberg RB, Mellies MJ, Sacks FM, et al. Cardiovascular events and their reduction with pravastatin in diabetic and glucose intolerant myocardial infarction survivors with average cholesterol levels: subgroup analysis in the CARE trial. Circulation 1998; 98:2513–2519.

25. Pfeffer MA, Sacks FM, Moye LA, et al. Influence of baseline lipids on effectiveness of pravastatin in the CARE trial. J Am Coll Cardiol 1999; 33:125–130.

26. Flaker GC, Warnica JW, Sacks FM, et al. Pravastatin prevents clinical events in revascularized patients with average cholesterol concentrations. J Am Coll Cardio 1999; 34:106–112.

27. Dayton S, Pearce ML, Hashimoto S, Dixon WJ, Tomiyasu U. A controlled clinical trial of a diet high in unsaturated fat in preventing complications of atherosclerosis. Circulation 1969; 40(suppl II):1–63.

28. Coronary Drug Project Research Group. Clofibrate and niacin in coronary heart disease. JAMA 1975; 231:360–381.

29. Lewis SJ, Sacks F, Braunwald E. Cholesterol lowering in older patients. Letter to the editor. Ann Intern Med 1999.

30. Krumholz HM, Seeman TE, Merrill SS, et al. Lack of association between choles-

terol and coronary heart disease mortality and morbidity and all-cause mortality in persons older than 70 years. JAMA 1994; 272:1335–1340.

31. Anderson KM, Castelli WP, Levy D. Cholesterol and mortality: 30 years of follow-up from the Framingham study. JAMA 1987; 257:2176–2180.
32. Kronmal RA, Cain KC, Ye Z, Omenn GS. Total serum cholesterol levels and mortality risk as a function of age. Arch Intern Med 1993; 153:1065–1073.
33. Manolio TA, Pearson TA, Wenger NK, Barrett-Connor E, Payne GH, Harlan WR. Cholesterol and heart disease in older persons and women: review of an NHLBI workshop. Ann Epidemiol 1992; 2:161–176.
34. Corti MC, Guralnik JM, Salive ME, et al. Clarifying the direct relation between total cholesterol levels and death from coronary heart disease in older persons. Ann Intern Med 1997; 126:753–760.
35. Sacks FM, et al. N Engl J Med 1996; 335:1001.
36. Sacks FM, et al. J Am Coll Cardiol 1998; 32:140.

4

Lessons from the West of Scotland Coronary Prevention Study (WOSCOPS)

James Shepherd and Allan Gaw
Royal Infirmary, Glasgow, Scotland

I. INTRODUCTION

Atheroma was recognized in ancient Egypt, but only recently has its impact on population mortality been fully appreciated. In industrialized countries, coronary atherosclerosis and its clinical manifestation, ischemic heart disease, is responsible for the majority of deaths among all adults age 40 years and over. Although the lesion is multifactorial in origin, its frequent enrichment in cholesterol has led to the growing conviction that it results from a steady and progressive lifetime uptake of plasma cholesterol into the arterial subintimal space. This concept, which accords with epidemiological evidence associating raised circulating cholesterol levels with increased risk of coronary mortality, has led to the evolution of the cholesterol hypothesis that since hypercholesterolemia is associated with coronary heart disease (CHD), cholesterol-lowering strategies should be instrumental in its avoidance. The West of Scotland Coronary Prevention Study (WOSCOPS) examined that hypothesis (1).

II. BACKGROUND

The results of three major primary prevention trials designed to test the potential of lipid-lowering agents for preventing CHD appeared in the literature prior to WOSCOPS. The first, the World Health Organization (WHO) clofibrate trial (2), demonstrated a reduction in the rate of nonfatal myocardial infarction in the drug-treated group but also indicated that clofibrate therapy was associated with a rise in total mortality. This finding raised doubts over the benefits of widespread use of lipid-lowering agents. The publication of the Lipid Research Clinics Coronary Primary Prevention Trial (LRC-CPPT) in 1984 (3,4) and the Helsinki Heart Study (HHS) in 1987 (5) reversed this attitude to some extent, but many still remained skeptical (6–8) because of the inability of these later trials to show a significant effect on coronary or total mortality. In the LRC-CPPT, an 11% decrease in the level of low-density lipoprotein (LDL) cholesterol was associated with a significant decrease (19%) in cardiac events (fatal plus nonfatal myocardial infarction). The performance of this study was less than predicted due to compliance problems. In HHS, gemfibrozil was less potent but was palatable, and compliance among subjects at trial visits was high. The drug reduced LDL cholesterol and triglyceride levels by 8% and 35%, respectively, and increased levels of high-density lipoprotein (HDL) cholesterol by about 10%. These lipid changes were associated with a significant reduction (34%) in the incidence of CAD as measured by the combined fatal plus nonfatal myocardial infarction endpoint.

Neither the LRC-CPPT nor HHS had the statistical power to address the question of the benefits of lipid-lowering agents in preventing coronary death. To do this, it was necessary to: (a) study a population with a higher event rate; (b) increase the sample size; and (c) use a more effective lipid-lowering agent. The WOSCOPS study, using pravastatin, addressed each of these issues. Pravastatin, one of the newly developed 3-hydroxy-3-methylglutaryl coenzyme A (HMG-CoA) reductase inhibitors, is a powerful cholesterol-reducing agent that is easily tolerated and uniformly effective. The increase in sample size and the recruitment of older men in a high-risk area (Scotland in 1988 had the world's highest incidence of CAD mortality) with associated higher event rates strengthened the power of the study to assess the impact of an improved lipid profile on CAD events.

III. WOSCOPS DESIGN AND FINDINGS

The WOSCOPS study was, in its day, the largest statin-based cholesterol-lowering trial ever conducted in subjects without established coronary heart disease, and the first of its kind to use a member of the 3-hydroxy-3-methylglutaryl coenzyme A reductase inhibitor or statin class of lipid-lowering drugs in this population. A group of 6595 men aged 45 to 64 years were recruited in the West of Scotland on the basis of a plasma cholesterol of 6.5 to 7.8 mmol/L (250–300 mg/dL) and an LDL cholesterol between 4.0 and 6.0 mmol/L (155–232 mg/dL). None of the subjects had had a previous myocardial infarction or coronary revascularization surgery although 5% had evidence of angina pectoris as determined by a positive self-reported Rose questionnaire (9), and individuals were admitted into the trial with minor ECG abnormalities, the latter being defined as Minnesota Codes 4-2, 4-3 or 5-2, 5-3. This effectively is equivalent to ST depression ≤ 1 mm with a horizontal or downward sloping ST segment, or T-wave inversion < 5 mm. It should be noted, however, that in WOSCOPS, which was a primary prevention study, individuals with more significant ST-T changes were excluded. Thus, none of the participants had a left bundle branch block on entry or ST depression > 1 mm or T-wave inversion > 5 mm. The men were randomly assigned to treatment with pravastatin (40 mg/day) or to a matching placebo. Smoking and cholesterol-lowering dietary advice was provided throughout the study which lasted, on average, 4.9 years. Treatment with pravastatin lowered plasma cholesterol by 20%, LDL cholesterol by 26%, and triglyceride by 12%. HDL cholesterol rose 5%. These lipid changes resulted (Table 1) in a one-third reduction in fatal and nonfatal myocardial infarctions (174 on pravastatin versus 248 on placebo). Cardiovascular deaths fell by the same percentage (50 deaths on pravastatin, 73 on placebo), and, since there was no increase in noncardiovascular mortality as a result of pravastatin therapy, deaths from any cause were reduced by about a fifth (105 deaths in the pravastatin group versus 135 on placebo ($P = .051$ by the log-rank test and $P = .037$ using the Cox model). Unexpectedly, the benefits of therapy began to emerge within about 6 months of initiating therapy and were maintained as long as it continued.

Poor compliance is always a problem in the conduct of primary prevention trials. Patients who do not have overt symptoms of the condition for which they are being treated are less motivated to comply with long-term therapy and may be less willing to tolerate minor side effects that could be associated with medication. There is a greater tendency for adverse events that develop after the initiation of therapy to be linked, rightly or wrongly, with the medication being taken. Hence, although it is appropriate that the primary analysis of any randomized trial is based on the intention-to-treat principle, some form of on treatment analysis is of great interest in the context of primary prevention both to provide a true picture of the benefits of therapy in subjects who are good compliers and to motivate future patients to optimize their compliance. At the same time, it should be acknowledged that analyses, which adjust for on-treatment measures such as compliance, are no longer based on truly randomized comparisons and hence should be interpreted cautiously.

The level of noncompliance in the WOSCOPS study, with approximately 30% of subjects withdrawn from randomized therapy at 5 years, was similar to that of previous primary prevention trials of cholesterol-lowering drugs (3–5). This was, in part, due to the causes outlined above and was considerably exacerbated by the cholesterol controversy (6–8) which received a high media profile throughout the study, providing a very difficulty climate within which to

Table 1 Primary Findings in the West of Scotland Coronary Prevention Study

	Placebo group (n = 3293)	Pravastatin group (n = 3302)	Percentage change	P value[a]
Plasma lipids (mg/dl)				
Total cholesterol	272	217	20↓	<.001
Total triglyceride	164	144	12↓	<.001
LDL cholesterol	192	142	26↓	<.001
HDL cholesterol	44	46	5↑	<.001
Endpoints (number of events)				
Nonfatal MI/CHD death	248	174	31↓	<.001
Nonfatal MI	204	143	31↓	<.001
Coronary angiography	128	90	31↓	.007
Revascularization procedures	80	51	37↓	.009
Cardiovascular deaths	73	50	32↓	.033
All-cause mortality	135	106	22↓	.051

[a]Log-rank test.

conduct a trail of this nature. There was no evidence in the trial of any treatment-related patterns of withdrawal from medication. The drug was very well tolerated. Posthoc analysis of the benefits to compliant subjects (10) indicated that they gained more than the entire cohort, analyzed on an intention-to-treat basis (Table 2). While subject to the usual subgroup analysis caveats, these findings support the view that physicians ought to take steps to improve patient motivation to adhere to prescription advice. The data also carry an important message with regard to assessment of the cost-effectiveness of treatment (see later).

The primary findings of WOSCOPS established the safety and efficacy of pravastatin and opened new perspectives on the selection and management of hypercholesterolemic patients at risk of a first myocardial infarction. Paramount in the mind of the practising clinician are the following issues:

1. Who should be treated?
2. What is the treatment strategy?
3. How much will it cost?

The WOSCOPS findings have helped answer these questions.

IV. LESSONS FROM WOSCOPS

A. Who Should Be Treated?

Interpretation of the results of the WOSCOPS study in its broadest sense would lead to the conclusion that since cholesterol reduction with pravastatin is virtu-

Table 2 Benefits of Good Compliance in WOSCOPS

Endpoint	Entire cohort	Percent risk reduction in	
		>75% compliers[a] (unadjusted)	>75% compliers[b] (adjusted)
Definite CHD death or	31	37	38
nonfatal MI	($P < .001$)	($P < .001$)	($P < .001$)
Cardiovascular death	32	35	37
	($P < .033$)	($P < .035$)	($P < .029$)
All deaths	22	32	32
	($P < .051$)	($P < .017$)	($P < .015$)
Revascularisation procedures	37	46	46
(CABG and PTCA)	($P < .009$)	($P < .0026$)	($P < .031$)
Incident cancers	−8	4	6
	(NS)	(NS)	(NS)

[a]Log-rank test.
[b]Cox model.

ally devoid of side effects[1] all patients eligible for recruitment to the trial would benefit from intervention and consequently merit treatment. However, implementation of such a liberal strategy is clearly economically impractical. Instead, it is more rational to identify and intervene only in those individuals whose risk of an event exceeds an accepted specified value. In 1994 a joint commission from the European Atherosclerosis Society, the European Society of Cardiology, and the European Society for Hypertension ruled (11) that a 20% risk of a vascular event over the next 10 years (or, broadly, a 2% risk per annum) warranted consideration for intervention, fully aware that such an arbitrary decision required enough flexibility in its interpretation to meet the circumstances of populations with differing financial and healthcare constraints.

With that caveat in mind, how can one target individuals for treatment on the basis of their prospective risk of an event? Consideration of the frequency and distribution of cardiovascular events in the placebo group of the WOSCOPS cohort permits such a selection to be made. A number of categorical and continuous coronary heart disease risk predictors was identified and the impact of each risk predictor quantified both individually and in aggregate (12). As is evident from Table 3, age is an important univariate predictor of risk of fatal or nonfatal myocardial infarction as are systolic blood pressure and plasma HDL cholesterol and triglyceride. Total plasma cholesterol or even LDL cholesterol, on the other hand, is much less predictive of the likelihood of disease. Categorical predictors based on preexisting disease (like diabetes, angina pectoris, or prescribed nitrate therapy), an unhealthy life-style (e.g., cigarette smoking), or psychosocial factors (e.g., widowhood, unemployment, or stultified education) again weigh heavily on risk of CHD in their own right (Table 3). In aggregate, they permit the constitution of a risk scoring system (Table 4) which locates only 9% of the fatal and nonfatal coronary events in the lowest risk quartile of the placebo group and 45% in the highest quartile. Similarly, only 5% of coronary revascularisations and 9% of deaths from any cause occurred in the lowest risk quartile in distinction to 54% and 56% respectively in the highest quartile.

Figure 1 translates these arbitrarily designated risk scores into the terms of the European Society Joint Guidelines (11). Individuals with a 5-year risk of a fatal or nonfatal myocardial infarction of 10% expressed an arbitrary risk score of 6.2. Forty-five percent of all coronary events occurred in individuals with a score above this value, and these events were confined to individuals in the highest quartile of risk in the placebo group. In other words, adoption of the European Joint Guidelines will focus on that quarter of hypercholesterolemic coronary

---

[1]Following publication of the WOSCOPS results the U.S. Federal Drug Administration on the strength of the study safety data, eliminated the need for repeated liver function testing for pravastatin and recommencded only that testing was required prior to and 12 weeks following initiation of effective treatment.

Table 3 Predictors of Risk in the WOSCOPS Placebo Group

		Definite CHD death or nonfatal MI	
Variate		Univariate risk ratio (95% CI)	Multivariate risk ratio (95% CI)
Continuous	Median value		
Age (5 yrs)	55.2	1.35 (1.23, 1.48)	1.35 (1.23, 1.48)
Systolic BP (20 mm Hg)	134	1.29 (1.17, 1.44	
Diastolic BP (10 mm Hg)	84	1.17 (1.07, 1.28)	1.19 (1.09, 1.31)
Total cholesterol (20 mg/dL)	269	1.04 (0.96, 1.13)	
LDL cholesterol (20 mg/dL)	189	1.09 (0.98, 1.21)	
HDL cholesterol (10 mg/dL)	43	0.73 (0.65, 0.82)	
Log triglyceride (0.5 log [mg/dL])	5.00	1.24 (1.10, 1.40)	
Total/HDL cholesterol ratio (0.5)	6.28	1.10 (1.07, 1.14)	1.08 (1.05, 1.12)
Categorical	Prevalence (%)		
Current smoker	44	1.78 (1.47, 2.16)	1.82 (1.49, 2.21)
Diabetes mellitus	1.2	2.63 (1.48, 4.67)	2.10 (1.18, 3.73)
Nitrate consumption	2.1	3.56 (2.43, 5.23)	1.90 (1.26, 3.02)
Angina pectoris	5.1	2.41 (1.78, 3.26)	1.54 (1.07, 2.22)
Family history of CHD	5.7	1.58 (1.12, 2.22)	1.71 (1.22, 2.42)
Widowhood	2.5	2.35 (1.54, 3.57)	1.66 (1.09, 2.54)
Unemployment	29	1.62 (1.33, 1.96)	
No school certificate	56	1.26 (1.03, 1.56)	

Table 4 CHD Risk Distribution in the WOSCOPS Placebo Group

		Quartiles of risk score		
Endpoint	1	2	3	4
Definite CHD death or nonfatal MI				
5-yr probability of event (%)	<4.48	≥4.48, <6.53	≥6.53, <9.62	≥9.62
Observed events (% of total)	9	19	27	45
Definite or suspect CHD death or nonfatal MI				
5-yr probability of event (%)	<5.27	≥5.27, <7.68	≥7.68, <11.3	≥11.3
Observed events (% of total)	9	18	29	44
Coronary revascularisation (PTCA or CABG)				
5-yr probability of event (%)	<1.27	≥1.27, <1.79	≥1.79, <2.73	≥2.73
Observed events (% of total)	5	15	26	54
Definite or suspect CHD death				
5-yr probability of event (%)	<0.763	≥0.763, <1.27	≥1.27, <2.20	≥2.20
Observed events (% of total)	11	13	12	64
Cardiovascular mortality				
5-yr probability of event (%)	<0.960	≥0.960, <1.58	≥1.58, <2.71	≥2.71
Observed events (% of total)	11	13	17	59
All-cause mortality				
5-yr probability of event (%)	<1.82	≥1.82, <3.05	≥3.05, <5.21	≥5.21
Observed events (% of total)	9	16	19	56

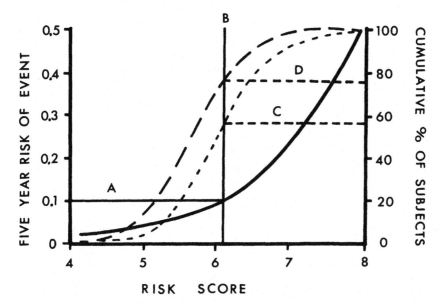

Figure 1 Risks and events in the WOSCOPS placebo group. This figure describes the risk of CHD death or nonfatal MI in the WOSCOPS placebo group. The 5-year risk of an event (right vertical axis) is plotted as a continuous line against subjects' risk score, calculated from their baseline risk factor profile. The cumulative percentage of randomized subjects (left vertical axis) who did (dotted line) or did not (dashed line) suffer an event throughout the trial is also indicated in relation to their risk score. Horizontal line A, representing a 5-year risk of 10%, the value targeted by the joint EAS, ESC, ESH guidelines (reference), is associated with a risk score of ≥ 6.2, as depicted by the vertical line B. Horizontal lines C and D indicate that 45% of all coronary events occurred in subjects with a risk score of >6.2 (C) and that 25% of the population had a risk score that exceeded this value (D).

event-free individuals at highest risk, will successfully identify 45% of patients in this category who will eventually have a myocardial infarction, and, by inference, will disregard 55% of all first myocardial infarct victims. Such patients can be located on the basis of their non-lipid-risk-factor profile, cogent argument that cholesterol screening of asymptomatic individuals should be limited to middle-aged men at high short-term risk of a coronary event, with treatment restricted to this group and to those who present with atherosclerotic disease. Compared with a strategy of untargetted screening and treatment, this approach would ensure the maximum absolute benefit and cost effectiveness of therapy. Treatment of subjects in the top quartile of risk distribution in WOSCOPS would address 45% of all coronary events and 64% of coronary fatalities. However, the majority of events (55%), which occur in the lower three quartiles of the risk distribution, would be missed, even although it is clear that pravastatin induced the same relative reduction in coronary risk across the entire spectrum of the WOSCOPS cohort. But, the number needed to be treated to prevent one coronary event in the lower three quartiles would be about 140, suggesting the case for aggressive lifestyle intervention in these subjects before any consideration is given to the introduction of drug therapy. Economic constraints (see later) wield an ever growing influence on clinical decision making.

Another way to arrive at a decision on which patients should receive lipid-lowering drug therapy is to compare the benefits of treating hyperlipidemia in asymptomatic at-risk men with the results of other risk factor intervention studies. Two such studies have been compared elsewhere (13): 4S, a trial of cholesterol lowering in patients with established CHD (14), and the MRC trial (15), a primary prevention study of the effects of treating hypertension on the risk of stroke. Analyzed in this way (Table 5) approximately three times as many WOSCOPS patients had to be treated to save one CHD death or nonfatal MI, compared to patients with established CHD who were enrolled in 4S. Table 5 also shows that two to four times more hypertensives were required to be treated to save one stroke in the MRC trial than patients with high cholesterol to save an MI in WOSCOPS.

At first glance these results suggest that the absolute benefits of treatment in WOSCOPS are less than have been shown for patients with established CHD, but that they compare favorably with the treatment of mild to moderate hypertension. Further analysis of event rates experienced by the WOSCOPS placebo group, as noted earlier, however, has shown that within a cohort of moderately hypercholesterolemic middle-aged men there exists a continuum of risk from low (<5% coronary event rate over 5 years) to high (20% event rate over 5 years). Low-risk WOSCOPS subjects have isolated hypercholesterolemia and a coronary event rate similar to that recorded for stroke in the MRC trial. High-risk WOSCOPS subjects are older men with overt or silent atherosclerotic disease whose coronary event rates approach those recorded in 4S. The number of low-risk and high-risk WOSCOPS subjects needed to be treated to save one major coronary event may then be estimated by calculating absolute risk reduction for different levels of baseline risk, assuming a comparable relative risk reduction of 30%.

How useful is our comparison of WOSCOPS with the results of the MRC trial? While matching of these two studies permits direct comparison of primary prevention interventions in men of the same age, based on data available in the main results paper of each trial, the following reservations should be noted. The two studies were conducted about 10 years apart. The study populations, key risk factors, and end points were different. Perhaps most importantly, the entry criteria for the MRC trial defined mild hypertension as diastolic BP 90 to 109 mm

Table 5 Benefits of Treatment in WOSCOPS, 4S, and the MRC Hypertension Trial

Trial (event avoided)	Risk of event at 5 years (%)		Relative Risk Reduction (%)	Absolute Risk Risk Reduction (%)	Number treated to avoid one event
	Placebo	Active			
WOSCOPS men, 45–64 years (definite + suspect CHD death or nonfatal MI)	9.3	6.8	29	2.5	40
4S men and women, 35–70 years (definite + suspect CHD death or nonfatal MI)	25.9	18.0	34	7.9	13
WOSCOPS men, 55–64 years (definite CHD death or nonfatal MI)	9.8	7.3	26	2.5	40
MRC men, 55–64 years (definite fatal or nonfatal stroke)	2.6	1.5	42	1.1	91

Hg after three visits, during a run in period of only 2 weeks (15). Accepted definitions and classifications of hypertension have changed since the MRC trial: some patients randomized to active treatment or placebo would not now be regarded as hypertensive, while others in this BP range would be said to have moderate hypertension, according to current guidelines (16). It is likely that fewer hypertensive patients would need to be treated for 5 years to prevent one stroke if treatment was restricted to those in whom BP elevation was sustained over a more prolonged period of pretreatment observation.

A subsequent analysis of placebo group patients in the MRC trial (17) confirms that the slope of the relation between BP and risk is steeper when based on measurements made after 6 months of follow-up on placebo. The 5-year stroke risk for men and women aged 35 to 64 years whose systolic BP remained 160 mm Hg or greater after 6 months was twice that of patients whose systolic BP was 160 mm Hg upon entry to the trial (in many of whom BP spontaneously fell thereafter). However, the magnitude of any uncertainty created by this difference in treatment threshold is not sufficient to invalidate the comparisons we have drawn between WOSCOPS and the MRC trial—it remains the case that fewer hyperlipidemic patients must be treated to prevent one MI than hypertensives of similar age to prevent one stroke. It is certainly possible to increase the number of strokes prevented by treatment of hypertension, but in order to do so older patients must be targeted. Given that proportionate benefit in the stroke trials is comparable at around 40%, it follows that absolute benefit is greatest among patients whose absolute risk of stroke is greatest (i.e., older patients) at the start of treatment (17).

In conclusion, analysis of subgroup data from WOSCOPS permits identification of men whose 5-year risk of a major coronary event is $\geq 10\%$, sufficiently high to justify lipid lowering drug therapy for primary prevention. The absolute benefits of treating such men are less than for men with established CHD but greater than the benefits of treating mild to moderate hypertension in men of similar age to reduce the risk of stroke. It is of course cheaper to treat a hypertensive with bendrofluazide than to prescribe pravastatin for hyperlipidemia, although more expensive drugs are frequently used in hypertension, and it is likely that competitive pressures will force a reduction in the price of lipid-lowering therapy in future. Thus, in light of the widespread acceptance of the value of treating hypertension, we feel it is difficult to justify not offering lipid-lowering pharmacotherapy to at-risk middle-aged men with hypercholesterolemia. The results of WOSCOPS, in which lipid-lowering therapy produced statistically significant reductions in CHD death and nonfatal MI, all cardiovascular death and, for the first time, all-cause mortality, clearly support the use of a statin in such patients.

Summary: Who Should Be Treated?

On the basis of WOSCOPS and the European Joint Guidelines, primary CHD prevention
should focus on
    <55-year-old hypercholesterolemic men with:
        Smoking habit
        Diabetes
        Minor ECG abnormality
        Preexisting vascular disease
        Multiples of the above
    >55-year-old hypercholesterolemic men with any of the following:
        Hypertension
        Diabetes
        Family history of CHD
        Low HDL cholesterol (<1.1 mM; <40 mg/dL)
        Smoking habit
        Preexisting vascular disease
        Minor ECG abnormality
        Multiples of the above

      The above strategy applies to individuals with modest hypercholesterolemia
(<7.8 mM; <300 mg/dL) and does not cover lipid abnormalities of genetic origin.
      A minor ECG abnormality is defined as Minnesota codes 4-2, 4-3, or 5-2,
5-3. This effectively is equivalent to ST depression ≤ 1 mm with a horizontal or
downward-sloping ST segment, or T-wave inversion < 5 mm. It should be noted,
however, that in WOSCOPS, which was a primary prevention study, individuals with more significant ST-T changes were excluded from the study. Thus, none
of the participants had a left bundle branch block on entry or ST depression >1
mm or T-wave inversion >5 mm. It is to be expected that such findings would
constitute an even more significant risk factor.
      Women were not examined in WOSCOPS.

B. What Is the Treatment Strategy?

Both the Lipid Research Clinics Coronary Primary Prevention Trial (3,4) and the
Helsinki Heart Study (5) showed that lipid-lowering strategies were effective in
reducing the risk of ischemic heart disease. However, the drugs used in these
projects (cholestyramine and gemfibrozil) differ substantially in their effects on
the lipoprotein profile; not surprisingly, there was enthusiastic interest at that time
in identifying the elements in the plasma lipid spectrum that had the greatest
bearing on coronary risk and whose modification produced the best cardioprotective outlook. It transpired that both LDL and HDL levels in the plasma

influenced risk—a reduction in the former or a rise in the latter being protective. The drugs employed in these trials, however, in contrast to the WOSCOPS project, were relatively ineffectual or disagreeable to patients and so the changes which they managed to induce in plasma lipids and lipoproteins were, in general, modest. The WOSCOPS study is therefore the first primary prevention trial with adequate statistical power to examine the relationship between plasma lipid levels and their influence on CHD events.

The WOSCOPS study plasma lipid analysis, which is the subject of a recent publication (18), addressed three fundamental questions: (1) How do baseline lipids impact on the treatment benefits of statin therapy? (2) Is there an association between LDL cholesterol reduction and event reduction? and (3) Does LDL reduction alone account for all of the cardioprotection offered by pravastatin? Each of these issues is addressed in detail below.

1. How Do Baseline Lipids Impact on the Treatment Benefits of Statin Therapy?

In the WOSCOPS study, plasma lipid and lipoprotein concentrations (triglyceride, cholesterol; VLDL, LDL, and HDL cholesterol) were measured according to the Lipid Research Clinics Protocol (19). Measurements were made twice during screening, and individuals were included in the study if they had an LDL ≥ 4.0 mmol/L (155 mg/dL) on both occasions and ≥ 4.5 mmol/L (174 mg/dL) on one. If LDL exceeded 6.0 mmol/L (232 mg/dL) at both screening visits, the subject was excluded. Men were randomized to receive placebo or pravastatin 40 mg/d, and subsequent visits were conducted every 3 months. Fasting lipid profiles were obtained at 6-month intervals during the follow-up period. In the present analysis, baseline plasma lipid levels were taken as the mean of the values observed at the two screening visits. As noted in the publication of baseline characteristics (12), there was no significant difference in any lipid variables between these visits.

The endpoint used in this review is all-cardiac events, defined as the occurrence of fatal MI, other cardiovascular death, nonfatal MI, coronary artery bypass graft (CABG), or percutaneous transluminal coronary angioplasty (PTCA) as a first event. This provided approximately 50% more events than the primary endpoint and thus enhanced power to detect associations and differences. CABG and PTCA as separate endpoints showed similar risk reduction on pravastatin to the primary endpoint.

All 6595 randomized patients were used to calculate quintiles of mean baseline LDL cholesterol, HDL cholesterol, and plasma triglyceride. The Kaplan-Meier 5-year risk of an all-cardiac event was then determined separately for each quintile of the placebo and pravastatin groups. Baseline lipids as continuous variables were related to risk of an all-cardiac event in the two groups separately using

Cox recession (20) both univariately and then multivariately with other baseline covariates (as described in Ref. 12) to test their independence as predictors. The covariates employed in the adjustment were age, smoking, blood pressure, ECG abnormality, self-reported angina, self-reported hypertension, diabetes, family history of premature death from CHD, and nitrate use.

Baseline LDL cholesterol was a weak predictor of risk in both treatment groups largely because of the artificially narrow band of LDL cholesterol dictated by the entry criteria of the trial. Individuals in the top quintile of the placebo group experienced a rate for the all-cardiac endpoint of 12% per 5 years compared with 9% in the bottom quintile (Fig. 2, *top left*). The proportionate reduction in risk of an event was similar across all quintiles in patients taking pravastatin. Baseline HDL cholesterol exhibited a clear negative association with event rate (Fig. 2, *bottom*) and was a major predictor of CHD risk in both treatment arms of the study. Again, the relative risk reduction was similar for all quintiles of this lipid fraction. The plasma triglyceride level at baseline was positively related to the risk of CHD (Fig. 2, *top right*). Patients receiving placebo

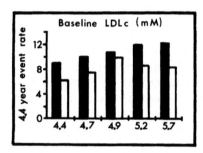
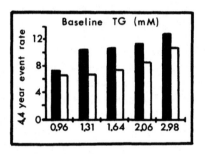
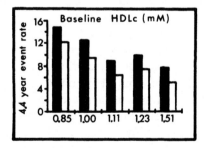

CONCLUSION

Risk reduction with pravastatin
is independent of baseline
lipid phenotype.

Figure 2 Baseline lipids and treatment benefits from pravastatin. Subjects in the placebo and pravastatin-treated groups in WOSCOPS were divided into quintiles according to their mean baseline LDL cholesterol (*top left*), HDL cholesterol (*bottom*), or plasma triglyceride (*top right*). Kaplan-Meier 5-year estimated event rates were determined for each quintile. Solid columns = placebo; open columns = pravastatin. To convert mmol/L cholesterol to mg/dL, multiply by 38.7; to convert mmol/L triglyceride to mg/dL, multiply by 88.5.

who had a baseline triglyceride level of $\geq$ 2.3 mmol/L (204 mg/dL) had almost twice the event rate of patients with an initial triglyceride of <1.2 mmol/L (106 mg/dL) despite having similar baseline LDL levels (mean of 5.0 mmol/L; 194 mg/dL) in all quintiles of plasma triglyceride). On the basis of univariate analysis, the starting triglyceride value was a highly significant predictor of risk in both groups. In line with previous observations (21), inclusion of baseline HDL in multivariate models led to a loss of significance of baseline plasma triglyceride as a predictor.

The above results permit us to conclude that pravastatin therapy confers the same benefit to hypercholesterolemic (plasma cholesterol = 6.5 to 7.8 mmol/ L; 250 to 300 mg/dL) individuals, independently of baseline LDL, HDL, and triglyceride. By the same token, and as reported previously (1), the proportionate gain from the therapy is uninfluenced by age, smoking status, or presence of symptoms of CHD. So, within the constraints of the lipid inclusion criteria in WOSCOPS, all subjects were equally helped by the drug.

2.   Is There an Association Between LDL Cholesterol Reduction and Event Reduction?

There is heated ongoing debate over the value of aggressive total and LDL cholesterol reduction in the prevention of coronary heart disease. The protagonists assert that increasing benefit will accrue from greater lipid lowering, implying that there is a linear association between plasma cholesterol values and coronary risk. But the Multiple Risk Factor Intervention Trial (22), the largest observational study of its kind, and other, similar large-scale prospective projects make it clear that populations experience a curvilinear rise in coronary events with increasing plasma cholesterol. By inference, then, cholesterol-lowering strategies would most effectively reduce events in individuals with more severe hypercholesterolemia and would become attenuated in their capacity to do so as individual plasma cholesterol values fell. In other words, a patient will receive greater cardioprotection from the first, say, 25% cholesterol reduction than from efforts to lower cholesterol by 25% more. This principle of diminishing returns was tested in the present WOSCOPS analysis, using the following approach.

First, the percentage reduction of LDL cholesterol from baseline was calculated for each individual in the pravastatin treated group. To provide the most accurate measure of plasma lipid concentration during follow-up, on-treatment lipid values were calculated at the mean of all lipid measurements made after randomization until the patient had an event or reached the end of the study. If a lipid value was missing at a visit but study medication had been issued at the previous visit (3 months earlier), the most recent measurement that had been preceded by a medication issue was carried forward. If before the visit no such on-treatment measurement existed, then the baseline value was imputed. Baseline value was also imputed if no medication had been issued at the previous visit and the present lipid level was missing.

Second, the Kaplan-Meier 4.4-year risk of a cardiac event (as specified in Section 1 above) was determined for each quintile of percent LDL reduction. Each quintile was then compared with the whole placebo group in a Cox model. In this analysis, the first six months of follow-up in both groups were excluded, since no on-treatment lipid values were available for patients who had an event before this time had elapsed. The difference between treatments was assessed with and without adjustment for potential baseline covariate imbalance and expressed as risk ratios relative to placebo with 95% confidence limits. To determine if quintiles differed from each other with respect to risk of an all-cardiac event, quintiles 1 through 4 were compared in Cox multivariate models with quintile 5 (highest percent change in LDL).

The percentage fall in LDL cholesterol during treatment varied (Fig. 3) even in patients who complied with the treatment regimen (i.e., they attended and had study medication issued on at least 75% of the scheduled visits). When the pravastatin group was divided into quintiles of percentage LDL reduction (based on measured plus imputed values), it was observed that the mean change varied from 0% to –39% (Fig. 3). We expected that a decrease in LDL would be the

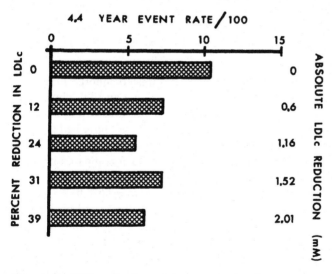

Figure 3 LDL reduction versus event reduction in WOSCOPS. Subjects receiving pravastatin were divided into quintiles of percentage decrease from baseline in LDL as a result of treatment. Kaplan-Meier 4.4-year risk of an event was calculated for each quintile. Similar calculations relating event reduction to absolute LDL cholesterol reduction gave comparable results since baseline LDL cholesterol values for each quintile were essentially the same.

major determinant of risk reduction and that a strong, graded association would be present between these two variables. However, Figure 3 illustrates that in quintiles 2 through 5 there was no obvious correlation between percent LDL reduction and event rate. When adjustment was made for baseline covariates, the relative risk in quintile 4 versus 5 was 1.39 ($P = .14$); in quintile 3 versus 5, 1.09 ($P = .72$); in quintile 2 versus 5, 1.43 ($P = .11$); and in quintile 1 versus 5, 2.24 ($P = .001$). Thus, quintiles 2 through 4 did not differ significantly from quintile 5 in terms of cardiovascular risk reduction achieved, and individuals in quintile 1, who achieved no LDL reduction, had the same CHD risk as the entire placebo group over the 5 years of the study. So, if LDL levels were not reduced by treatment, there was no cardiovascular gain, but a fall of 24% was sufficient to produce maximum benefit. Whatever additional gain (if any) came from driving LDL lower did not translate into further measurable cardioprotection.

3. Does LDL Reduction Alone Account for All of the Cardioprotection Offered by Pravastatin?

The inexplicably early appearance of benefit in WOSCOPS which resulted within 6 months in visible divergence of the cumulative event curves in the placebo and pravastatin-treated cohorts raised the possibility that change in lesion size and, by inference, LDL cholesterol reduction alone, might not fully explain the actions of the drug in reducing coronary events. Two independent approaches were made to examine this issue: *overlap analysis*, and *comparison with Framingham*.

Overlap Analysis. There was considerable overlap in the distribution of mean LDL cholesterol in the placebo and pravastatin-treated cohorts, 1120 individuals on placebo and 1071 on pravastatin therapy having an LDL cholesterol between 3.62 and 4.65 mmol/L (140 to 180 mg/dL). To be included in the overlap analysis, patients had to be >75% compliant and should not have had an event in the first 6 months of follow-up. Risk of a cardiac event (fatal MI or other cardiovascular death, nonfatal MI, CABG, or PTCA) was then compared between the groups. Event rates for the two subgroups differed markedly. Pravastatin treatment was associated with a 36% (95% confidence interval = 56% to 9%) lower risk ($P = .014$; Fig. 4), a finding that did not appear to be due to an imbalance in baseline risk factors or to differences in on-treatment LDL (or on-treatment plasma triglyceride, VLDL cholesterol, or HDL cholesterol). Similar differences were obtained on examination of a narrower 3.88 to 4.39 mmol/L (150 to 170 mg/dL) overlap (Fig. 4), a region where the on-treatment LDL values were virtually equal in the two groups (4.23 versus 4.15 mmol/L (164 versus 161 mg/dL) for placebo and pravastatin-treated subjects respectively).

Comparison with Framingham. The equation published by the Framingham investigators (23–25) permits calculation of the risk of a CHD event based

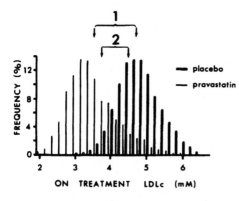

OVERLAP		NUMBER OF		P
		Subjects	Events	value
1	Placebo	1120	108	0.002
	Pravastatin	1071	67	
2	Placebo	517	48	0.014
	Pravastatin	546	31	

Figure 4 Overlap analysis in WOSCOPS. The frequency distribution for LDL cholesterol in the placebo and pravastatin-treated subjects overlap substantially. Subjects in each cohort with LDL cholesterol between 3.62 and 4.65 mmol/L (overlap 1) and 3.88 and 4.39 mmol/L (overlap 2) were compared. Despite having the same LDL cholesterol values as their placebo counterparts, pravastatin-treated individuals exhibited significantly fewer cardiovascular events (definite or suspect CHD death or nonfatal MI, other cardiovascular death, or revascularisation) over the 4.4 years of observation.

on gender, age, plasma cholesterol, HDL cholesterol, smoking habit, systolic blood pressure, and presence of diabetes. This model was employed as a further approach to test the hypothesis that the event reduction seen in patients taking pravastatin could not be explained entirely by changes in the measured plasma lipid levels. To maintain compatibility with the Framingham coronary event definition, the endpoint in this instance was taken as definite nonfatal MI or CHD death plus revascularization (PTCA and CABG). Again, patients were omitted from the analysis if they had experienced a coronary event, had cancer, or had undergone angiography within 6 months of randomization, and in order to maintain compatibility with the Framingham cohort, men with baseline angina, as determined by nitrate use or positive Rose questionnaire, were excluded. Inclusion necessitated that patients fall into the plasma cholesterol (4.13 to 7.23 mmol/ L; 160 to 280 mg/dL) and blood pressure (diastolic, 70 to 105 mm Hg; systolic 110 to 170 mm Hg) ranges that characterized the Framingham population from which the risk equation was derived. They also had to comply with the treatment regimen (as described above) and to have at least one on-treatment lipid measurement available. Prediction of risk was again made from the point at which on-treatment lipid levels were available (6 months after randomization) over the remaining period of the trial (4.4 years, since the mean total length of follow-up was 4.9 years).

Predicted event rates were derived for each subject using his mean (with imputation when necessary) on-treatment level for plasma cholesterol and HDL cholesterol. After individuals were grouped into quintiles of predicted risk, a Kaplan-Meier 4.4-year risk of an event was determined from the observed rates for each quintile. The number of predicted and observed events across the quintiles were compared for placebo and pravastatin groups using a chi-square test with 5 degrees of freedom. A total of 1251 patients in the placebo group and 1803 patients in the pravastatin group met the inclusion criteria for this analysis. There was remarkable agreement between the observed CHD event rate in the placebo group and the value predicted from the Framingham model (Fig. 5a). The predicted overall 4.4-year rate of 7.4 per 100 subjects was close to that observed in WOSCOPS (7.0 per 100). Treatment with pravastatin reduced total plasma cholesterol values and, as expected, diminished the risk of coronary events over the duration of the study (Fig. 5b). However, in contrast to the placebo cohort, those receiving pravastatin exhibited an observed reduction in events which, overall, was significantly ($P = .026$) greater than that predicted from the Framingham risk equation. According to the Framingham model, the cholesterol reduction achieved should have lowered the risk of a coronary event by 24%. In fact, the observed reduction was 36%.

There are a number of possible explanations for the findings of the overlap analysis and the Framingham comparison. First, patients who experience a sudden reduction in LDL cholesterol may benefit from, at least for a time, a lower risk than those who naturally have an LDL at that concentration. Second, in addition to lowering LDL, pravastatin has been shown to promote the removal of triglyceride-rich remnant particles from the bloodstream (26). These lipoprotein species have been linked to the progression of atherosclerotic lesions, and their clearance, which is known to occur through receptor-mediated pathways (27), may lead to stabilization of plaques whose rupture might otherwise have given rise to clinical events. Third, pravastatin may, through pathways not involving lipid lowering (28), beneficially affect atherosclerosis (e.g., by decreasing a tendency for thrombosis or by inhibiting the inflammatory response which characterizes atheromatous plaques). The latter two possibilities could account for the relatively early benefit seen during pravastatin therapy in WOSCOPS. It is noteworthy that other lipid-lowering therapies (3,4,29) that work by stimulating receptor-mediated catabolism of LDL (i.e., bile acid sequestrant resins and surgical biliary diversion) did not show an early treatment effect, although they did provide long-term risk reduction. Possibly this difference in response arises because these therapeutic approaches enhance VLDL production in the liver and do not have the same impact on the plasma concentration of remnants of triglyceride-rich lipoproteins as does pravastatin.

The results described here are derived from post hoc analysis and therefore must be viewed cautiously. Nevertheless, they indicate that the benefit seen with pravastatin treatment, although obviously linked to a decrease in LDL, cannot be explained by this alone.

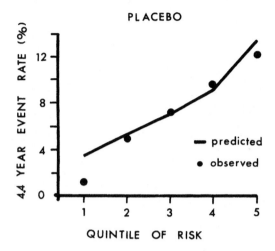

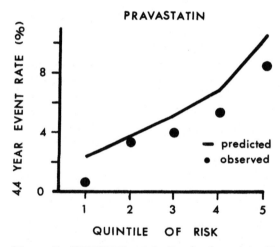

Figure 5   WOSCOPS and the Framingham study. The predicted cardiovascular risk over 4.4 years for each WOSCOPS participant was derived from the Framingham risk equation (23–25). Subjects were then ranked separately in quintiles of predicted risk (continuous line) in the placebo (a) and pravastatin (b) cohorts and the data plotted against the 4.4 year Kaplan-Meier estimate of an event derived from the study results. Only patients who fell into the range of plasma lipid levels and blood pressure readings seen in Framingham were included in the analysis. In the placebo group, cardiovascular risk, predicted from the Framingham equation (7.4 events/100 over 4.4 years), matched the observed values (7.0 events/100). Conversely, the pravastatin-treated cohort expressed significantly fewer events (4.5 events/100; $P < .026$) than predicted from Framingham (5.6 events/100). A chi-square test was used to compare predicted and observed rates across the quintiles.

Summary: What Is the Treatment Strategy?
_____

Detailed analysis of the lipid data in WOSCOPS shows that

Baseline levels of triglyceride and LDL and HDL cholesterol were predictive of CHD risk

All subjects gain the same relative benefit from pravastatin, independently of their baseline lipid profile and within the framework of the WOSCOPS lipid inclusion criteria

There is no cardiovascular gain from pravastatin therapy unless LDL cholesterol levels are reduced, but a fall of about 25% is sufficient to produce maximum benefit in modestly hypercholesterolemic men

LDL reduction alone does not appear to account entirely for the benefit of pravastatin therapy

C.   How Much Will It Cost?

So far we have considered whom to treat and how aggressively to approach their problem. The third issue we must explore is the question of how much the prevention strategy will cost. With this objective in mind, an economic model of cardiovascular disease prevention was created (30) based on the premise that the first cardiovascular event (be it the appearance of angina, a nonfatal myocardial infarction, or the need for coronary revascularization) is seen by the individual and by society as a whole as an irreversible transition from health to sickness, and therefore worthy of prevention.

To estimate the effects of pravastatin on this process, the WOSCOPS database was interrogated to identify the various events that constitute the transition. Pravastatin treatment was compared to the effect of the placebo, each intervention being added to basal dietary advice. The process allowed for a progression of events occurring in the same individual and was conducted at monthly intervals for 60 months (the mean duration of the trial). The net consequence of not using pravastatin was the cumulative difference in transitions (Table 6). The number needed to treat to prevent one transition differed in its definition from the usual approach in that it identified those required to *start* treatment (rather than those who continued on treatment) to avoid a transition.

Calculation of the number of life years gained by those on treatment made use of the unique Scottish Record Linkage System (31), a database containing details on more than 460,000 cardiovascular events that were managed in Scotland between 1981 and 1995. This database allowed forward projection of what would have happened to individuals who would have succumbed to an event had it not been for pravastatin therapy. The cost of any transition was based on the average direct 1996 price of managing each type of event, averaged from information supplied by 200 hospitals in Scotland. All costs were discounted at 6% to satisfy Treasury recommendations. The cost of the pravastatin treatment it-

Table 6 Vascular Events Avoided Through Use of Pravastatin Therapy in WOSCOPS

Event	Events avoided/10,000 subjects over 5 years
Sudden deaths	33
Nonfatal MI	138
	(35 ultimate deaths)
Angina hospitalization	68
Revascularizations	33
Strokes/TIAs	47
TOTAL	319

self (£1.66 per 40 mg tablet) was calculated from the number of people receiving one tablet daily; incidental costs of monitoring the subject's progress were estimated on the basis of current practice guidelines (GP visit, liver function tests, lipoprotein analysis, etc.).

Finally, in order to assess the effect of varying risk in different patient subgroups on the price of intervention, individuals were collected into groupings based on the risk factors (12) of age, diagnosis of hypertension, diastolic blood pressure, baseline HDL cholesterol, smoking habit, ECG abnormality, family history of CHD, widowhood, positive Rose questionnaire, previous vascular disease, diabetes, and use of glyceryl trinitrate.

Pravastatin treatment of the hypercholesterolemic asymptomatic WOSCOPS men avoided 319 transitions in 10,000 men treated over 5 years (Table 6); that is, to prevent one transition, 31.4 men would require to receive treatment for this time. There are several ways of measuring the implications of avoiding these transitions. For example, eliminating the events would have saved 2017 hospital bed days or extended life in the 10,000 individuals by 2460 years. The cost of the drug for the 10,000 men would have been £23,340,984, offset by £529,214 in savings from treatment of the prevented diseases. So, overall, discounting at 6% would have led to a cost of £20,375 per life year gained. This assumes that all shades of risk in the WOSCOPS trial were accommodated. If one limits treatment, in line with the European Joint Guidelines (11), to those individuals with a 20% risk of an event over the next 10 years (approximately 2% per annum), then the numbers requiring treatment to avoid an event would fall to 22.5 and the cost of extending life by 1 year would fall (discounted at 6% per annum) to £13,995. Finally, if a more rigorous approach were taken, and only individuals with a 3% risk of an event were selected, the discounted cost of each life year gained would be £9,680. Selection of this target would mean treating 8% of the WOSCOPS population (Table 7). So, although widespread use of pravastatin for primary prevention may seem (31) like an unjustified extravagance

Table 7 Treatment Costs per Life Year Gained Based on Coronary Risk
Stratification in the WOSCOPS Cohort

Average annual risk of a CHD event (%) population	Cost (£) per life year gained		% of study requiring treatment
	Undiscounted	Discounted at 6%	
1.6	8121	20,375	100
2.0	5601	13,995	23
3.0	3900	9,682	8

imposed on an already strained health care system, the WOSCOPS results provide good evidence that this intervention, if judiciously applied, is economically
sound.

Summary: How Much Will It Cost?

From the WOSCOPS results
   Starting 10,000 men on pravastatin would help 319 of them avoid a transition from
      health to cardiovascular disease
   If all WOSCOPS men were treated, the cost of adding 1 year to life would be £20,375
      (discounted at 6%)
   If, according to the Joint European Guidelines, those with an annual risk of 2% were
      selected for treatment, the cost per life year gained would be £13,995
   If only those with an annual risk of 3% were selected for treatment, the cost per life
      year gained would be £9680. Eight percent of the WOSCOPS population would
      require to be treated

REFERENCES

1. Shepherd J, Cobbe SM, Ford I, Isles CG, Lorimer AR, Macfarlane PW, McKillop
    JH, Packard CJ. Prevention of coronary heart disease with pravastatin in men with
    hypercholesterolemia. N Engl J Med 1995; 333:1301–1307.
2. A co-operative trial in the primary prevention of ischaemic heart disease using clofi-
    brate: report from the Committee of Principal Investigators. Br Heart J 1978;
    40:1069–1118.
3. Lipid Research Clinics Coronary Primary Prevention Trial results. I. Reduction in
    incidence of coronary heart disease. JAMA 1984; 251:351–364.
4. Lipid Research Clinics Coronary Primary Prevention Trial results. II. The relation-
    ship of reduction in incidence of coronary heart disease to cholesterol lowering.
    JAMA 1984; 251:365–374.
5. Frick MH, Elo O, Haapa K, et al. Helsinki Heart Study: primary prevention trial

with gemfibrozil in middle-aged men with dyslipidemia: safety of treatment, changes in risk factors, and incidence of coronary heart disease. N Engl J Med 1987; 317:1237–1245.

6. Oliver MF. Might treatment of hypercholesterolaemia increase non-cardiac mortality? Lancet 1991; 337:1529–1531.

7. Hulley SB, Walsh JMB, Newman TB. Health policy on blood cholesterol: time to change directions. Circulation 1992; 86:1026–1029.

8. Davey Smith G, Pekkanen J. Should there be a moratorium on the use of cholesterol lowering drugs? BMJ 1992; 304:431–434.

9. Rose G, McCartney P, Reid DD. Self administration of a questionnaire on chest pain: intermittent claudication. Br J Prev Social Med 1977; 31:42–48.

10. West of Scotland Coronary Prevention Study Group. Compliance and adverse event withdrawal: their impact on the West of Scotland Coronary Prevention Study. Eur Heart J 1997; 18:1718–1724.

11. Pyorala K, DeBacker G, Poole-Wilson P, Wood D. Prevention of coronary heart disease in clinical practice. Recommendations of the Task Force of the European Society of Cardiology, European Atherosclerosis Society and European Society of Hypertension. Atherosclerosis 1994; 110:121–161.

12. West of Scotland Coronary Prevention Study Group. Baseline risk factors and their association with outcome in the West of Scotland Coronary Prevention Study. Am J Cardiol 1977; 79:756–762.

13. West of Scotland Coronary Prevention Study Group. West of Scotland Coronary Prevention Study: identification with high risk groups and comparison with other cardiovascular intervention trials. Lancet 1996; 348:1339–1342.

14. Scandinavian Simvastatin Survival Study Group. Randomised trial of cholesterol lowering in 4444 patients with coronary heart disease: the Scandinavian Simvastatin Survival Study (4S). Lancet 1994; 344:1383–1389.

15. Medical Research Council Working Party. MRC trial of treatment of mild hypertension: principal results. BMJ 1985; 291:97–104.

16. Sever P, Beevers G, Bulpitt C, Lever A, Ramsay L, Reid J, Swales J. Management guidelines in essential hypertension: report of the second working party of the British Hypertension Society. BMJ 1993; 306:9830987.

17. Lever AF, Ramsay LE. Treatment of hypertension in the elderly. J Hypertens 1995; 13:571–579.

18. West of Scotland Coronary Prevention Study Group. Influence of pravastatin and plasma lipids on clinical events in the West of Scotland Coronary Prevention Study. Circulation. 1998; 97:1440–1445.

19. Lipid Research Clinics Manual of Laboratory Operations. DHEW publication No. (NIH) 85-268. Washington: Government Printing Office, 1975.

20. Collett D. Modelling Survival Data in Medical Research. London: Chapman Hall, 1994.

21. Austin MA. Plasma triglyceride and coronary heart disease. Arterioscler Thromb 1990; 11:1–14.

22. Martin MJ, Hulley SB, Browner WS, Keeler LH, Wentworth D. Serum cholesterol, blood pressure and mortality: implications from a cohort of 361,662 men. Lancet 1986; 2:933–936.

23.   Anderson KM, Odell PM, Wilson PWF, Kannel WB. Cardiovascular disease risk profiles. Am Heart J 1990; 121:293–298.
24.   Anderson KM, Wilson PWF, Odell PM, Kannel WB. An updated coronary risk profile. Circulation 1991; 83(1):357–363.
25.   Abbott RD, McGee D. *The Framingham Study, Section 37. The Probability of Developing Certain Cardiovascular Diseases in Eight Years at Specified Values of Some Characteristics.* NIH Publication No. 87-2284. Washington: Government Printing Office, 1987.
26.   Knopp RH, Illingworth DR, Stern EA, Ginsberg HN, Broyles EE, Behounek BD. Effect of pravastatin in the treatment of patients with type III hyperlipoproteinemia. Am J Ther 1996; 3:755–762.
27.   Gaw A, Packard CJ, Murray EF, Lindsay GM, Griffin BA, Caslake MJ, Vallance BD, Lorimer AR, Shepherd J. Effects of simvastatin on apoB metabolism and LDL subfraction distribution. Arterioscler Thromb Vascul Biol 1993; 13:170–189.
28.   Shepherd J. Pleiotrophism among the statins. Br J Cardiol 1997; 4(suppl 1):S28–S31.
29.   Buchwald H, Varco RL, Matts JP, Long JM, Fitch LL, Campbell GS, Pearce MB, Yellin AE, Edmiston WA, Smink RD Jr et al., and the POSCH Group: Effect of partial ileal bypass surgery on mortality and morbidity from coronary heart disease in patient with hypercholesterolemia: report of the Program on the Surgical Control of the Hyperlipidemias (POSCH). N Engl J Med 1990;323:946–955.
30.   Caro J, Klittich W, McGuire A, Ford I, Norrie J, Pettitt D, McMurray J, Shepherd J, for the West of Scotland Coronary Prevention Study: Economic benefit analysis of primary prevention with pravastatin. Br Med J 1997; 315:1577–1582.
31.   Kendrick S, Clarke J. The Scottish record linkage system. Health Bull (Edinb) 1993; 51:72–79.
32.   Pharoah PDP, Hollingworth W. Cost effectiveness of lowering cholesterol concentration with statins in patients with and without pre-existing coronary heart disease: life table method applied to health authority population. Br Med J 1996; 312:1443–1448.

5

Program on the Surgical Control of the Hyperlipidemias (POSCH) Trial

A Pivotal 25-Year Study

Henry Buchwald
The University of Minnesota and The POSCH Group,* Minneapolis, Minnesota

I. INTRODUCTION

The Program on the Surgical Control of the Hyperlipidemias (POSCH), launched in the United States a quarter-century ago, was a multiclinic randomized, prospective secondary intervention trial. It was designed to ascertain whether the effective reduction of plasma total cholesterol and, in particular, low-density lipoprotein (LDL) cholesterol levels induced by the partial ileal bypass operation had a favorable impact on overall mortality and on mortality and morbidity due to atherosclerotic coronary heart disease (ACHD). Using both clinical and arteriographic endpoints, POSCH was the first atherosclerosis intervention trial to correlate changes observed on sequential coronary arteriograms with clinical atherosclerosis events.

POSCH was funded by the U.S. National Heart, Lung, and Blood Institute (NHLBI) in 1973. Between 1975 and 1983, a total of 838 survivors of a single myocardial infarction (documented by electrocardiograms and changes in enzyme values) were entered into the study. The formal trial ended on July 19, 1990, with a mean patient follow-up of 9.7 years (range, 7.0 to 14.8 years). The POSCH long-term follow-up study ended on July 19, 1997, with a minimum patient follow-up of 14 years; many patients were followed for more than 20 years. No patient in POSCH, whether in the control group or in the intervention group, was lost to follow-up. The POSCH databank and over 3000 patient arteriograms are avail-

---

*See Appendix for the POSCH Group.

able to interested investigators. Currently, 48 POSCH papers have been published, as well as 12 book chapters on different aspects of the POSCH trial. Several additional publications by the POSCH group are expected before 2000.

The lipid/atherosclerosis theory has been substantially confirmed. Observational studies started in the 19th century (1,2), experimental studies of the early 20th century (3–5), and post–World War II epidemiologic studies (6–12) gave rise to the concept that the higher the plasma cholesterol level of a given population and its constituent individuals, the greater the incidence and severity of ACHD. Since this concept was supported by most of the available data, proof of its corollary became the goal of innumerable scientists: namely, demonstration that lowering the plasma total cholesterol, or more precisely lowering the LDL cholesterol level and modifying the ratio of LDL to high-density lipoprotein (HDL) cholesterol (13), would decrease the incidence of ACHD and favorably alter the outcome of established ACHD. Altered outcome was to be manifest by the prevention or delay of an ACHD death, a nonfatal myocardial infarction (MI), or another ACHD event, all resulting in a prolongation of life expectancy.

Yet, the first 13 lipid/atherosclerosis intervention trials, in the 1960s and 1970s, actually provided little or no evidence to reject the null hypothesis that individuals do not derive clinical benefits from lipid intervention (14). Then followed the suggestive reports of the 1980s from the Cholesterol Lowering Atherosclerosis Study (CLAS) (15), the Helsinki Heart Study (16), the NHLBI Type II Coronary Intervention Study (17), and the Lipid Research Clinics—Coronary Primary Prevention Trial (LRC-CPPT) (18). In 1990, POSCH reported its in-trial findings, ushering in 5 years of multiple independent reports from controlled clinical trials—all of which strongly confirmed the validity of the lipid/atherosclerosis theory (19).

Also in 1990, there followed affirmative reports from the Familial Atherosclerosis Treatment Study (FATS) (20) and the Arteriosclerosis Specialized Center of Research (SCOR) study (21), both of which tracked sequential coronary arteriography changes as surrogate endpoints for clinical events. In 1992, the positive clinical findings of the St. Thomas' Atherosclerosis Regression Study (STARS) (22) were published, as well as the favorable results of the changes in life style study by Gould et al. (23). In 1993, the Mevinolin Atherosclerosis Regression Study (MARS) (24) provided further arteriographic evidence for a favorable change of prognosis secondary to marked lipid intervention, as did the 1994 Stanford Coronary Risk Intervention Project (SCRIP) (25).

The year 1994 marked the start of the reporting of the statin drug trials, beginning with the landmark Scandinavian Simvastatin Survival Study (4S) (26)—the first trial ever to demonstrate a statistically significant reduction in overall mortality secondary to lipid intervention. In rapid succession came complementary reports from the Multicentre Antiatheroma Study (MAAS) (27), the Canadian Coronary Atherosclerosis Intervention Trial (CCAIT) (28), the

Asymptomatic Carotid Artery Progression Study (ACAPS) (29), the Regression Growth Evaluation Statin Study (REGRESS) (30), and the West of Scotland Coronary Prevention Study (WOSCOPS) (31). The West of Scotland study was the first long-term primary prevention trial with statistically significant evidence of clinical benefits from LDL cholesterol lowering.

In the chronology of lipid/atherosclerosis intervention trials, POSCH was a pivotal study. Performed in the United States entirely with NHLBI funding, it clearly demonstrated that effective plasma total cholesterol and LDL cholesterol lowering will inhibit atherosclerosis progression and enhance patient prognosis.

II. POSCH GOALS

1. To conduct a clinical secondary ACHD intervention trial of the impact of effective lipid modification.
2. To conduct a sequential coronary arteriographic ACHD intervention trial of the impact of effective lipid modification.
3. To demonstrate that changes in sequential coronary arteriograms can be used as surrogate endpoints for clinical ACHD events in lipid/atherosclerosis trials.
4. To study concurrently clinical and arteriographic peripheral vascular disease changes within a secondary ACHD intervention trial of the impact of effective lipid modification.
5. To include women in a lipid/atherosclerosis intervention trial and to analyze the impact of effective lipid modification in women in POSCH and in other lipid/atherosclerosis intervention trials.

III. STUDY DESIGN

A. Patient Distribution

The four POSCH clinics were selected to provide a wide geographic sampling of the U.S. population. A total of 838 patients were entered into the POSCH trial protocol: 378 at the University of Minnesota (Minneapolis), 135 at the University of Arkansas Medical Center (Little Rock), 141 at the University of Southern California (Los Angeles), and 184 at the Lankenau Hospital and Research Center (Philadelphia). The first patient was enrolled in September 1975 and the last in July 1983. Based on calculations of the study's power made in 1982, closure of the formal trial was scheduled for July 1990. Long-term follow-up was continued through July 1997.

B. Recruitment

The design and methods of the POSCH trial (32) and the enrollment of patients (33) have been detailed elsewhere. Recruitment employed three approaches: (1) contacting local and area hospitals for the names of potentially suitable patients and their physicians, explaining the trial to the responsible physicians, and only with the physicians' approval communicating with the patients; (2) regional media campaigns, in the press and on the radio, with space and time donated as a public service; and (3) hand-outs sponsored by chain outlets (notably by Dairy Queen, Inc.) in the recruitment neighborhood. After a brief telephone screening interview, potentially suitable patients and their spouses were invited to visit one of the POSCH clinics to learn about the study in detail; to initiate intensive screening (medical history, physical examination, laboratory profile); and, if interested and eligible, to give fully informed consent and enter the study protocol.

C. Eligibility and Exclusion Criteria

Eligibility and exclusion criteria are summarized in Table 1. A more detailed description of these criteria is available from the author on request. Patients with the risk factor of cigarette smoking and other potential atherosclerosis risk factors were distributed by the randomization process.

Table 1   Eligibility and Exclusion Criteria

Eligibility criteria
  Age 30–64 years
  Male or female
  Survival of a single MI, documented by electrocardiographic and enzymatic changes, 6–60 months before the randomization date.
  Plasma total cholesterol level of at least 5.69 mmol/L (220 mg/dL), or if between 5.17 and 5.66 mmol/L (200 and 219 mg/dL) an LDL cholesterol of at least 3.62 mmol/L (140 mg/dL)—after following the AHA phase II diet for a minimum of 6 weeks.

Exclusion criteria
  Hypertension (systolic blood pressure ≥180 mm Hg or diastolic blood pressure ≥105 mm Hg).
  Body weight 40% above the ideal weight.
  Diabetes mellitus.
  Impairments that could influence study outcome (e.g., cancer occurring within 5 years of randomization).
  Prior cardiac surgery or implantation of a permanent cardiac pacemaker.
  Stenosis of the left main coronary artery ≥75% or no measurable coronary stenosis on the prerandomization arteriogram.

D. Screening and Randomization

During the screening visit, eligibility was determined and most of the baseline values for the study variables were obtained. Before randomization, all patients received instruction in the American Heart Association (AHA) phase II diet, which prescribes consuming <25% of total daily calories as fat (one-third as saturated, one-third as monounsaturated, and one-third as polyunsaturated fat) and no more than 250 mg of cholesterol daily. All patients taking hypocholesterolemic drugs were asked to discontinue them at least 6 weeks before baseline plasma lipid measurement. All patients were encouraged not to resume or start hypocholesterolemic drugs during their participation in the formal trial. Immediately after the baseline coronary arteriogram was obtained and while the patient remained in the hospital, randomization was performed by the POSCH Coordinating Center. Control group patients were discharged from the hospital. Intervention group patients remained in the hospital and underwent partial ileal bypass surgery.

E. Intervention Modality

The POSCH trial was initiated before the statin drugs were introduced. Lipid-lowering therapy was limited to diet and to drugs with only minimum or moderate efficacy, often associated with intolerable side effects. When chosen as the POSCH intervention modality, partial ileal bypass was a well-proven, effective, lasting, relatively safe, and obligatory means markedly to lower plasma total cholesterol and LDL cholesterol and to raise HDL cholesterol. This operation has been described in detail (32,34–39). It consists of bypass of either the distal 200 cm or the distal one-third of the small intestine, whichever is greater, with restoration of bowel continuity by an end-to-side ileocectomy (Fig. 1). One surgeon at each clinic performed all the study's partial ileal bypass procedures.

F. Endpoints

The primary endpoint of the trial was overall mortality. Secondary endpoints included cause-specific mortality, in particular ACHD mortality (determined by blinded review of all available records by the POSCH Mortality Review Committee). A key secondary endpoint was the combination of ACHD mortality and confirmed nonfatal MI. Other secondary endpoints were overall and ACHD mortality as a function of the left ventricular ejection fraction (LVEF) (<50% vs. $\geq$ 50%), determined by the method of Sandler and Dodge (40); recurrent myocardial infarctions, whether confirmed or suspected; performance of coronary artery bypass grafting (CABG) or percutaneous transluminal coronary angioplasty

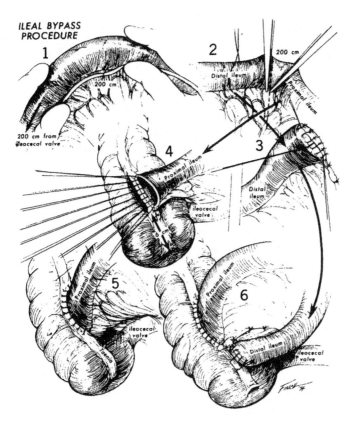

Figure 1 Schematic of the partial ileal bypass operation. Operative technique: (1) measurement of the bowel along the mesenteric border; (2) division of the bowel; (3) closure of the distal end leaving the middle and corner sutures uncut; (4) end-to-side anastomosis of the proximal bowel into the anterior taenia of the cecum; (5) completed anastomosis; (6) suturing of the closed end of the bypassed bowel to the anterior taenia of the cecum by using the uncut sutures and closure of the mesenteric defects. (From Ref. 37.)

(PTCA); development of peripheral vascular disease; and documentation of other atherosclerosis events. Diagnosis of a nonfatal MI required a combination of electrocardiographic and cardiac enzyme changes, as well as ischemic cardiac pain. MIs were assessed by a cardiologist blinded to the patient's assigned treatment. Assessments of other health-related events were recorded, as were all adverse effects of partial ileal bypass.

G. Sequential Assessments

1. Lipid Measurements

Sequential lipid profiles consisted of measurement of the plasma total choles-terol, LDL cholesterol, very-low-density lipoprotein (VLDL) cholesterol, HDL cholesterol, plasma total triglycerides, and lipoprotein phenotyping (41–43). From 1985 on, the levels of the HDL_2 and HDL_3 subfractions and apolipoprotein A-1 and B-100 were determined as well (44,45). All lipid samples were obtained af-ter a fast of at least 14 hours. The lipid analyses were performed in the POSCH Central Lipid Laboratory, whose procedures had been standardized and certified by the Lipid Standardization Laboratory of the Centers for Disease Control (At-lanta). The POSCH Central Lipid Laboratory was closed on November 1, 1990. Subsequently, lipid determinations were requested from patients' personal phy-sicians.

2. Arteriography

Sequential coronary arteriograms were assessed by two-member teams from the POSCH Arteriography Review Panel; no team member was from the clinic at which the arteriograms were obtained. The films were interpreted in pairs, with the readers blinded to the patient's assigned treatment and the sequence of the films. The evaluation protocol was identical to that employed in the Cholesterol Lowering Atherosclerosis Study (46). Global evaluation of the severity of the coronary artery disease was derived by consensus (32). An 8-point scale was used to grade the change between two films (-3, -2, -1, -0, +0, +1, +2, and +3; with -3 = much worse, and +3 = much better).

Peripheral vascular disease was defined arteriographically as a reduction of 20% or more in the luminal diameter of any of seven arterial segments visu-alized (the distal 4 cm of the abdominal aorta, the right and left common and external iliac arteries, and the right and left common femoral arteries). The baseline and follow-up pelvic arteriograms were interpreted, unpaired and inde-pendently, in blinded fashion by a single reader. No radiopaque markers were used during partial ileal bypass, to prevent unblinding of treatment assignment when the pelvic arteriograms were evaluated.

3. Electrocardiography

Resting and exercise electrocardiograms were obtained at each clinic visit. Ex-ercise electrocardiography was performed according to a modified Bruce proto-col, with the addition of a 3-min stage zero at 1.7 mph and a grade of 0% (47). The electrocardiograms were evaluated in blinded fashion by the POSCH Cen-tral Electrocardiography Laboratory.

4. Doppler Ultrasonography

From April 1, 1981, on, the peripheral pulse pressures were determined by Doppler ultrasonography at each clinic visit, and the ankle-brachial index (ABI) was calculated—i.e., the ratio of the systolic blood pressure in the ankle (posterior tibial or dorsalis pedis) to that in the arm (brachial). An index ratio of 0.95 or more indicated the absence of peripheral vascular disease.

H. Follow-up Protocol

All patients were followed by means of clinic visits and telephone calls, according to a uniform protocol (32). Lipid analyses were performed at baseline, 3 months after randomization, and at every clinic visit (annual visits during the first 5 years, then one visit at 7 or 10 years). Follow-up coronary and peripheral arteriograms were obtained at 3 and 5 years, then at either 7 years (for patients enrolled on or after June 1, 1980) or 10 years (for patients enrolled before June 1, 1980).

After the formal POSCH trial ended on July 19, 1990, all surviving patients underwent scheduled close-out interviews at their home clinic. During those interviews, the trial results and each patient's individual data were reviewed with the patient and subsequently sent to the patient's physician. All patient records were consolidated at the Minnesota Clinic, and the other three clinics were closed. Each surviving patient was contacted by the Minnesota Clinic and signed an informed-consent form indicating willingness to participate in the long-term follow-up protocol.

Subsequently, all surviving POSCH patients have been contacted annually, preferably within a 60-day window of the anniversary of their randomization, for an extended telephone interview. During this structured interview, a medical history update—designed to identify all health changes within the past year—was completed. Arrangements were made for the documentation of reported events from hospital records and outpatient office files. The patient's medication record, smoking history, bowel habits, and quality of life were recorded. The cholesterol level and other lipid parameters, weight, and blood pressure were requested from the patient's personal physician. All deaths were documented by death certificates, hospital records, autopsy reports, and physician and family member statements.

I. Statistical Analysis

All analyses were based on randomization assignment (intention-to-treat), whether or not the treatment was actually carried out. Relative risks and their confidence intervals were calculated (48), adjusting for baseline covariates by

Cox regression (49). The POSCH baseline covariates have previously been defined (50). In the Kaplan-Meier life-table analyses (51), the randomization date was used as the starting point. To account for differences in the length of time to events, both the Mantel-Haenszel statistic (52) and the Gehan statistic (53) were calculated. The Mantel-Haenszel statistic, having greater power to detect differences in the Kaplan-Meier curves at later time intervals, is the one used throughout this report in determining the cited P value.

To evaluate data from the sequential coronary arteriography, patients were divided into three groups: those whose findings were worse (-3, -2, or -1), unchanged (-0 or +0), and better (+1, +2, or +3). Analyses were then performed with a 3-by-2 chi-square test and a chi-square test for linear trend (54). A two-sided Fisher's Exact Test was used to assess the sequential peripheral arteriograms (55). Other data analyses were performed with standard chi-square tests, paired and unpaired Student's t-tests, and Spearman correlation coefficients (for measures of agreement).

A two-sided P value $<.05$ was used to define statistical significance.

IV. PRINCIPAL FINDINGS

A. Clinical Trial Results

1. Patient Characteristics and Compliance

The mean age of the patients at randomization was 51 years. Of the 838 patients, 90.7% were men and 97.9% were white. The mean interval between the qualifying myocardial infarction and entry into the trial was 2.2 years. Cigarette smokers were equitably distributed by the randomization process: 83.8% had smoked cigarettes and 35.0% still did at baseline. The patient characteristics at baseline have been presented in detail elsewhere (56).

At 1, 5, 7, and 10 years of follow-up, 5.3%, 16.4%, 25.1 %, and 31.5% of the control group patients were taking at least one cholesterol-lowering medication. In the intervention group, the corresponding figures were 0.5%, 3.0%, 6.2%, and 3.7%. In 1995, 5 years after formal trial closure, 145 (34.8%) control group patients and 51 (12.1%) intervention group patients were taking at least one cholesterol-lowering medication.

Of the 421 patients assigned to surgical intervention, 22 refused to undergo the operation and 23 eventually underwent reversal of the partial ileal bypass procedure through July 1990, primarily for intolerable diarrhea or recurrent nephrolithiasis. Since July 1990, another four patients have undergone reversal. Through July 1990, no control group patient underwent partial ileal bypass; one subsequently did so. Endpoint analyses excluding patients who refused or reversed their randomization assignment do not reveal statistically significant differences from the intention-to-treat analyses. Therefore, all analyses of clinical results, except the LVEF analyses, are based on the total POSCH randomization numbers of 417 control group and 421 intervention group patients.

2. Lipid Profiles

The baseline mean plasma total cholesterol level was 6.49 mmol/L (251 mg/dL), with a mean LDL cholesterol level of 4.62 mmol/L (179 mg/dL) and a mean HDL cholesterol level of 1.04 mmol/L (40 mg/dL); the baseline mean triglyceride level was 2.30 mmol/L (419 mg/dL). At 5 years after randomization, the intervention group, as compared with the control group, had a 23.3 ± 1.0% (mean ± SE) lower plasma total cholesterol level ($P < .001$), a 37.7 ± 1.2% lower LDL cholesterol level ($P < .001$), a 4.3 ± 1.8% higher HDL cholesterol level ($P = .02$), an 18.3 ± 7.5% higher VLDL cholesterol level ($P = .02$), a 19.8 ± 6.5% higher triglyceride level ($P = .003$), a 37.8 ± 2.8% higher ratio of HDL cholesterol to total plasma cholesterol ($P < .001$), and a 71.8 ± 4.3% higher ratio of HDL cholesterol to LDL cholesterol ($P < .001$). Detailed reports have been published of the in-trial changes in the lipid and lipoprotein concentrations, as well as an analysis of the predictors of the 5-year in-trial changes in plasma total cholesterol and LDL cholesterol levels (57–60). Complete analyses have also been published of the plasma lipid values in the control and intervention groups for up to 10 years of follow-up, including the following lipid parameters: total cholesterol, LDL cholesterol, HDL cholesterol, VLDL cholesterol, triglycerides, HDL/total cholesterol, HDL/LDL cholesterol, HDL-2 cholesterol, HDL-3 cholesterol, apolipoprotein A-1, and apolipoprotein B-100 (19). Plasma total cholesterol levels in the control and the intervention groups are graphed in Figure 2.

The 5-year posttrial lipid data, as provided by the patients' physicians, include plasma total cholesterol levels for 203 (65.3%) of the surviving control group and 188 (55.8%) of the surviving intervention group patients. In this cohort, the mean plasma total cholesterol level was 5.66 mmol/L (219 mg/dL) in the control group and 4.9 mmol/L (189 mg/dL) in the intervention group, for a 14.0% difference ($P < .003$).

3. Endpoints

Overall Mortality. At formal trial closure in 1990, overall mortality was reduced, but not significantly so: 62 control group vs. 49 intervention group patients had died ($P = .16$). At the 5-year posttrial follow-up, 106 control group (25.4%) vs. 84 intervention group (20.0%) patients had died. The relative risk in the intervention group was 0.75 (95% CI, 0.056 to 1.00). Using the Mantel-Haenszel statistic, the difference between the control and intervention groups now

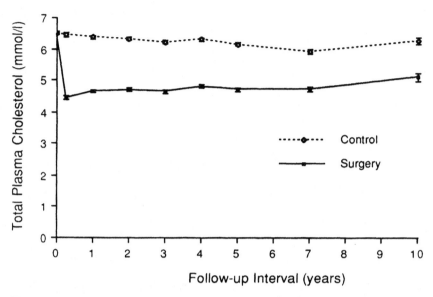

Figure 2 Plasma total cholesterol levels in the control and intervention groups. Values are means with 95% confidence intervals. The difference between the groups at each follow-up interval was significant ($P < .001$). (From Ref. 19.)

was statistically significant ($P = .049$) (Table 2, which compares 5-year posttrial to in-trial results; Fig. 3a).

In the subgroup with LVEF $\geq$ 50%, in 1990, intervention group patients had a 36.0% lower overall mortality than control group patients (39 vs. 24; $P = .05$). At the 5-year posttrial follow-up, 67 (23.0%) of 292 control group patients and 42 (14.9%) of 281 intervention group patients had died. The relative risk in the intervention group was 0.61 (95% Cl, 0.41 to 0.89), with $P = .01$ (Table 2; Fig. 3b). In 1995, overall mortality in the subgroup with an LVEF <50% was 33 (30.8%) of 107 in the control group and 41 (32.3%) of 127 in the intervention group ($P = .75$).

ACHD Mortality. In 1990, the difference in ACHD mortality, like overall mortality, did not reach statistical significance: 44 control group vs. 32 intervention group patients had died of ACHD ($P = .13$). But again, at 5-year follow-up, the difference was statistically significant ($P = .03$); 70 (16.8%) vs. 49 (11.6%). The relative risk in the intervention group was 0.66 (95% Cl, 0.46 to 0.96) (Table 2; Fig. 3c).

At the 5-year posttrial follow-up in the subgroup with an LVEF $\geq$50%, 37 (12.7%) of 292 control group patients and 23 (8.2%) of 281 intervention group

Table 2 Clinical Results

	Control group	Intervention group	Relative risk	95% Confidence Interval	Mantel-Haenszel P value
Overall mortality					
In-trial	62	49	0.77	0.53–1.12	.16
5 years posttrial	106	84	0.75	0.57–1.00	.049
Overall mortality EF ≥50					
In-trial	39	24	0.61	0.37–1.01	.05
5 years posttrial	67	42	0.61	0.41–0.89	.01
ACHD mortality					
In-trial	44	32	0.71	0.45–1.12	.13
5 years posttrial	70	49	0.66	0.46–0.96	.03
ACHD mortality EF≥ 50					
In-trial	24	15	0.62	0.32–1.18	.13
5 years posttrial	37	23	0.60	0.36–1.02	.05
ACHD mortality and confirmed nonfatal MI					
In-trial	125	82	0.60	0.45–0.79	<.001
5 years posttrial	157	105	0.60	0.47–0.77	<.001

Confirmed nonfatal MI				
In-trial	93	56	0.57	0.41–0.79	<.001
5 years posttrial	109	68	0.57	0.42–0.77	<.001
ACHD mortality, confirmed and suspected MI, unstable angina				
In-trial	222	160	0.62	0.50–0.76	<.001
5 years posttrial	279	229	0.68	0.57–0.81	<.001
CABG or PTCA				
In-trial	170	67	0.35	0.26–0.46	<.001
5 years posttrial	201	106	0.41	0.32–0.53	<.001
Onset peripheral vascular disease				
In-trial	71	52	0.70	0.49–1.00	.04
5 years posttrial	93	68	0.68	0.50–0.93	.02

EF=baseline left ventricular ejection fraction; ACHD=atherosclerotic coronary heart disease; MI=myocardial infarction; CABG=coronary artery bypass grafting; PTCA=percutaneous transluminal coronary angioplasty.

In-trial refers to formal trial closure (1990); posttrial refers to 5-year follow-up assessment (1995).

(From Ref. 70.)

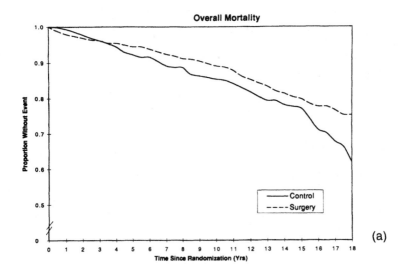

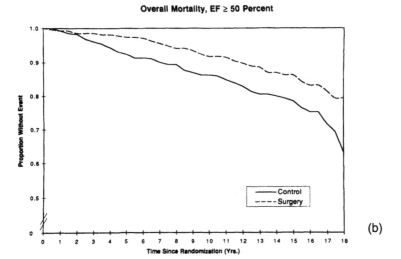

Figure 3 Life and event tables for (a) overall mortality; (b) overall mortality, left ventricular ejection fraction (EF) ≥50%; (c) atherosclerotic coronary heart disease (ACHD) mortality; (d) ACHD mortality, EF ≥50%; (e) ACHD mortality and confirmed nonfatal myocardial infarction (MI); (f) confirmed nonfatal MI; (g) ACHD mortality, confirmed and suspected MI, and unstable angina; and (h) coronary artery bypass grafting (CABG) or percutaneous transluminal coronary angioplasty (PTCA). The difference between the groups was significant for all eight endpoint analyses ($P < .05$ to $P < .001$). (From Ref. 70.)

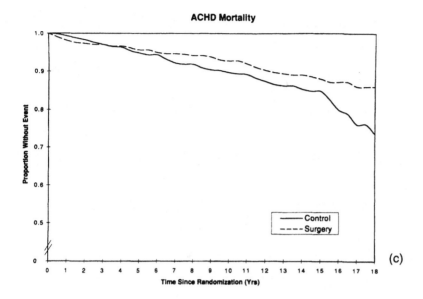

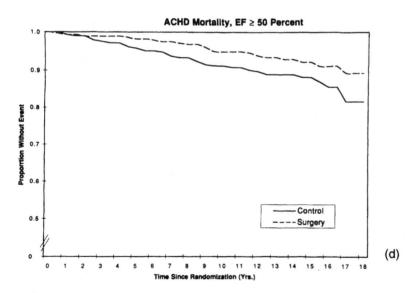

Figure 3 Continued

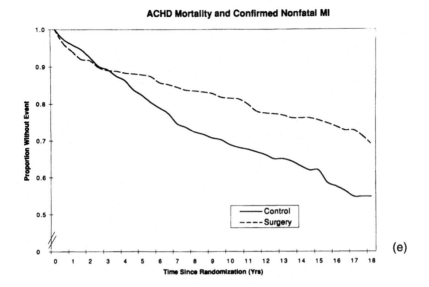

(e)

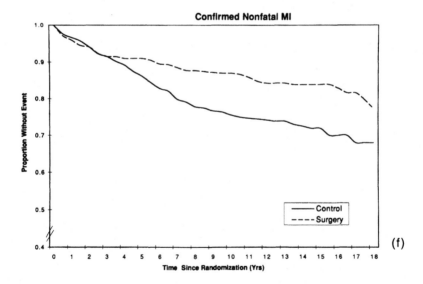

(f)

Figure 3   Continued

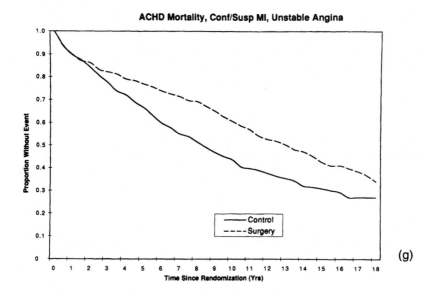

(g)

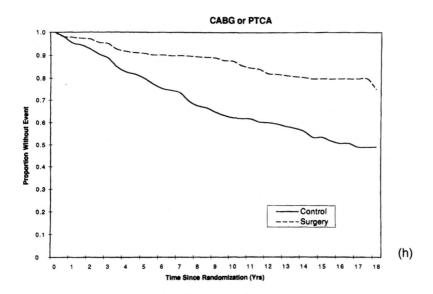

(h)

Figure 3   Continued

patients had died of ACHD. The relative risk in the intervention group was 0.60 (95% Cl, 0.36 to 1.02), with $P = .05$ (Table 2; Fig. 3d). In the subgroup with an LVEF <50%, 28 (26.2%) of 107 in the control group and 26 (20.5%) of 127 in the intervention group patients had died of ACHD ($P = .43$).

ACHD Mortality and Confirmed Nonfatal Myocardial Infarction. This universal endpoint in 1990 was reduced by 35.0% in the intervention group in comparison with the control group; this difference was highly statistically significant ($P < .001$). At the 5-year posttrial follow-up, there were 157 occurrences of this combined endpoint in the control group and 105 in the intervention group patients. The relative risk in the intervention group was 0.60, (95% Cl, 0.47 to 0.77; $P < .001$) (Table 2; Fig. 3e).

Other ACHD Events. Comparable to the combined endpoint of ACHD mortality and confirmed nonfatal MI, all the POSCH endpoints differences were statistically significant at formal trial closure and at the 5-year posttrial follow-up. Table 2 presents the in-trial and the 5-year posttrial relative risks, with 95% Cl, and the Mantel-Haenszel P values for confirmed nonfatal myocardial infarction (Fig. 3f), ACHD mortality, confirmed and suspected myocardial infarction and unstable angina (Fig. 3g), and the incidence of CABG or PTCA (Fig. 3h). The reduced occurrence of CABG or PTCA in intervention group patients indicates not only a clinical benefit but also a considerable economic advantage.

Cerebrovascular Events. At the 5-year posttrial follow-up, a cerebrovascular accident (CVA) was confirmed in 28 control group patients and 31 intervention group patients ($P = .88$). When all cerebrovascular events and transient ischemic attacks were combined, there were 68 events in the control group (16.3%) and 85 in the intervention group (20.2%) ($P = .24$). The absence of a statistically significant difference in cerebrovascular events between the two groups has been consistent in POSCH at every interval of assessment.

Non-ACHD Mortality. At the 5-year posttrial follow-up, 36 control group (8.6%) and 35 intervention group (8.3%) patients had died of non-ACHD causes ($P = .73$). The absence of a statistically significant difference in non-ACHD mortality between the two groups has been consistent in POSCH at every interval of assessment. The significant difference in overall mortality in POSCH was therefore the result of the significant decrease in ACHD mortality, uninfluenced by non-ACHD mortality.

Cancer Incidence. At the 5-year posttrial follow-up, 58 cancers were detected in control group (13.9%) and 61 in intervention group (14.5%) patients ($P = .99$). There were 8 (1.9%) colorectal cancers in the control group and 10 (2.4%) in the intervention group ($P = .69$). This lack of a significant difference in colorectal cancers between the two groups is an important finding, because the intervention group with a partial ileal bypass was subjected to an increased concentration of bile acids in the colon and rectum. Thus, in POSCH, there was no compensatory increase in cancer incidence or in cancer mortality to offset the reduction in ACHD mortality and events.

B. Arteriographic Results

The changes observed on coronary arteriography from baseline to follow-up at 3, 5, 7, and 10 years are summarized in Table 3. The percentage of patients with disease progression increased in both the control and intervention groups, but was consistently higher in the control group. At each follow-up interval, the difference between the two groups in the percentage of patients with definite disease progression (a score of -1, -2, or -3) was statistically significant ($P < .001$): control vs. progress intervention, 41.4% vs. 28.1 % at 3 years, 65.4% vs. 37.5% at 5 years, 77.0% vs. 48.1 % at 7 years, and 85.0% vs. 54.7% at 10 years. The precision of the interpretation of the films was assessed by 109 blinded repeat readings (Spearman $r = .84$).

In addition, sequential coronary artery arteriography revealed regression of atherosclerosis lesions in both groups, but statistically significant greater disease regression ($P < .01$) in the intervention group at 5 and 7 years: control vs. intervention, 4.7% vs. 12.3% at 5 years, 6.3% vs. 14.4% at 7 years (61).

C. Correlation of Arteriographic Changes and Clinical Endpoints

Changes in the overall disease assessment (ODA) score between the baseline and the 3-year coronary arteriograms were statistically significantly associated with subsequent overall and ACHD mortality ($P < .01$). The overall mortality rate was 12% for patients with an ODA score indicating progression, and only 7% for patients with an ODA score indicating regression or no change. The ACHD mortality rate was 10% for patients with an ODA score indicating progression, and only 4% for patients with an ODA score indicating regression or no change.

For the combined endpoint of ACHD mortality and confirmed nonfatal myocardial infarction, a significant relationship between the ODA score and subsequent clinical events was independently found both in the control group ($P < .0001$) and in the intervention group ($P = .04$). Interestingly, for this combined endpoint, statistically significant differences between the control and the intervention groups were found within the same strata of coronary atherosclerotic ODA ($P < .001$). In the control group, this combined endpoint occurred in 33% of those patients with an ODA score indicating progression, and in 15% of the patients with an ODA score indicating regression or no change; within the intervention group, the corresponding event rates were 16% and 8%.

Table 3 Changes on Coronary Arteriography in the Study Groups During In-Trial Follow-Up (percent of patients with score)

Follow-up interval	Score for change between films[a]								P value[b]
	-3	-2	-1	-0	+0	+1	+2	+3	
From baseline to 3 yr									
Control (n = 333)	3.3	14.4	23.7	32.1	19.2	6.3	0.9	0.0	
Intervention (n = 363)	1.9	9.1	17.1	33.3	29.5	6.9	2.2	0.0	.0008
From baseline to 5 yr									
Control (n = 301)	8.6	25.2	31.6	22.3	7.6	3.0	1.7	0.0	
Intervention (n = 333)	2.7	11.4	23.4	26.1	23.7	11.7	0.6	0.3	<.0001
From baseline to 7 yr									
Control (n = 174)	12.6	31.0	33.3	11.5	5.2	4.6	1.7	0.0	
Intervention (n = 187)	1.6	21.4	25.1	22.0	15.5	9.6	4.8	0.0	<.0002
From baseline to 10 yr									
Control (n = 80)	26.2	35.0	23.8	7.5	3.8	2.5	0.0	1.3	
Intervention (n = 95)	4.2	21.1	29.5	26.3	12.6	5.3	1.1	0.0	.0002

[a]For explanation of these scores, see Methods.
[b]By chi-square test for a linear trend.
Source: Ref. 19.

D. Peripheral Vascular Results

At formal trial closure, claudication or limb-threatening ischemia was exhibited in 71 of 417 control group patients and 52 of 421 intervention group patients (relative risk, 0.70; 95% CI, 0.49 to 1.00; $P = .049$) (Table 2). At 5-year posttrial follow-up, clinical peripheral arterial disease was evident in 93 control group patients and in 68 intervention group patients (relative risk, 0.68; 95% CI, 0.50 to 0.93; $P = .02$). At the 5-year posttrial follow-up evaluation, an ABI < 0.95 was present in 41 of 120 control group patients and in 24 of 126 intervention group patients, all of whom had an ABI ≥ 0.95 at baseline (relative risk, 0.58; 95% CI, 0.36 to 0.86; $P < .01$). No appreciable differences between the two groups were noted in the progression or regression of arteriographic peripheral arterial disease. These results have been detailed elsewhere (62).

E. Results In Women

For the 78 women in POSCH, there was no evidence of clinical benefit in the intervention group (63) even though there may have been a difference in arteriographic progression between the control group and the intervention group (64).

We reviewed the seven primary or secondary lipid/atherosclerosis intervention trials—including POSCH—that have published results in the English-language literature that included women and that analyzed results in women separately from those in men or in the entire trial population (63). A meta-analysis of these seven studies for overall mortality demonstrated a statistically significant reduction in this clinical endpoint in men but not in women. This finding has at least two possible interpretations: either the mechanism of coronary obstruction is different in men from that in women, or the mortality rate in the women in these studies was too low for a statistically significant result. Whatever the basis for this finding, however, the available clinical trial data in women do not provide convincing evidence of clinical benefits derived from effective lipid intervention.

F. Major Collateral Findings

1. Disease-Free Intervals Analysis

Analyses of the relative disease-free intervals in control and intervention groups of clinical endpoints in atherosclerosis intervention trials, in particular in lipid/atherosclerosis intervention trials, are absent from published reports. In contrast, intervention trials in oncology routinely report findings as survival rates and disease-free intervals. For the clinician and the patient, estimation of disease-free intervals may be more relevant than assessment of differences in incidence rates and risk ratios.

The POSCH results were subjected to disease-free interval analysis and calculation of the derived freedom from atherosclerosis events (65). The overall mortality rate was 10% at 6.7 years in the control group and 9.4 years in the intervention group, for a gain in disease-free interval of 2.7 years in the intervention group (P = .032). The ACHD mortality rate was 8% at 7.2 years in the control group and 11 years in the intervention group, for a gain of 3.8 years (P = .046). The combined endpoint of ACHD mortality and confirmed nonfatal myocardial infarction occurred in 20% of patients at 5.9 years in the control group and at 11.4 years in the intervention group, for a gain of 5.5 years (P < .001). CABG, PTCA, or heart transplantation occurred in 20% of patients at 5.4 years in the control group and at 12.4 years in the intervention group, for a gain of 7.0 years (P < .001).

2. Aspirin Use in Cigarette Smokers

Since the clinical impact of aspirin intake in cigarette smokers and former cigarette smokers has not been well studied, cohorts of cigarette smokers in the POSCH control group, free of marked lipid modification, were evaluated (66). In current cigarette smokers at baseline (n = 90), with a mean follow-up of 8.3 years, the overall mortality rate was 45.2% in patients with no aspirin use and 10.4% in patients who reported even infrequent aspirin use (relative risk, 4.3; 95% CI, 2.4 to 10.6; P < .001, by Fisher's Exact Test). For ACHD mortality, the comparable relative risk was 17.1 (35.7% vs. 2.1 %; 95% CI, 1.4 to 125.0; P < .001), for the combined endpoint of ACHD mortality and confirmed nonfatal myocardial infarction, the comparable relative risk was 2.4 (40.5% vs. 16.7%; 95% CI, 1.2 to 5.1; P = .018). In former cigarette smokers with no aspirin use at baseline (n = 92), with a mean follow-up of 8.8 years, the relative risk of overall mortality was 3.1 (20.0% vs. 6.5%; 95% CI, 1.0 to 10.2; P = .07); of ACHD mortality, 3.4 (16.7% vs. 4.8%; 95% CI, 0.9 to 13.5; P = .11), and of ACHD mortality and confirmed nonfatal myocardial infarction 1.1 (23.3% vs. 21.0%; 95% CI, 0.5 to 2.5; P = .79). After adjustment for covariates by Cox regression, none of these relative risks changed appreciably. Further, in current cigarette smokers, a statistically significant (P = .004) relationship was found between the percentage of follow-up time during which aspirin was used and the overall mortality rate. For each 10% increase in the time of aspirin use, the odds of overall mortality decreased by 30.2%. In the former cigarette smokers, a similar 10% increase in the time of aspirin use reduced the odds of overall mortality by 15.9%; however, this result was not statistically significant (P = .18)

3. Subgroup Analyses

We analyzed 105 subgroups in 35 variables in POSCH, chosen predominantly for their potential relationship to the risk of ACHD (50). We defined potential differential effects as those with (1) an absolute z value ≥ 2.0 for the subgroup, if the absolute z value for the overall effect was <2.0; and (2) an absolute z value ≥ 3.0 for the subgroup and a relative risk ≤ 0.5, if the absolute z value for the overall effect was ≥ 2.0. For each of the three major POSCH endpoints—overall mortality, ACHD mortality, and ACHD mortality and confirmed nonfatal MI—we found seven subgroups with a differential risk reduction in the intervention group compared with the control group. Allowing for identical subgroups for more than one endpoint, there were 13 individual subgroups with differential effects. Of these, seven demonstrated internal consistency across endpoints, and five of these seven displayed external consistency with known ACHD risk factors and for biological plausibility.

An additional outcome of these subgroup analyses was the demonstration of a greater risk reduction, compared with the overall treatment effect, when a single risk factor, hypercholesterolemia, was reduced in patients with at least two major ACHD risk factors. An hypothesis-generating outcome of this analysis is that the reduction of a single risk factor in patients with multiple risk factors may follow a reverse exponential curve.

4. Quality of Life

The patients' perception of their quality of life in POSCH was evaluated immediately before and shortly after disclosure to them of the trial results (67). The instrument used was the McMaster's Health Index Questionnaire (MHIQ); in addition four study-specific questions were asked. The MHIQ findings before disclosure showed a difference ($P = .07$, Wilcoxon rank test) favoring the control group in the social function index. After disclosure, this difference was larger ($P < .05$). For the four study-specific questions, all differences favored the control group ($P < .01$, standard chi-square test) before and after disclosure, except for satisfaction with randomization allocation in the intervention group ($P = .08$). The intragroup MHIQ indices before and after disclosure showed no significant differences, except in the intervention group, whose emotional function index improved after disclosure ($P = .03$). The intragroup responses to the study-specific questions before and after disclosure again showed no significant differences, except in the intervention group, whose satisfaction with randomization allocation improved after disclosure ($P = .04$). For the intervention group, the side effects of partial ileal bypass (see below) undoubtedly negatively influenced their perception of quality of life and the knowledge of salutary results heightened it. The knowledge of a favorable outcome for the intervention group, however, did not seem to alter the control group's perception of their quality of life.

5. Percent Arteriographic Stenosis and Myocardial Infarction

It is unresolved whether the susceptible coronary artery culprit lesion responsible for an acute MI is relatively large ($\geq 50\%$ arteriographic stenosis) and hemodynamically significant ($\geq$ 70% stenosis), or small (<50% stenosis) and asymptomatic. Certain necropsy and arteriography studies support the large-progenitor lesion concept, and other arteriography studies support the small-lesion hypothesis. The coronary arteriogram immediately preceding a Q-wave (transmural) myocardial infarction was analyzed in POSCH for the degree of stenosis of the suspected culprit lesion, which was selected by visual inspection of the coronary circulation supplying the ECG-defined area of myocardial infarction (68). The time interval from the preceding coronary arteriogram closest to the index myocardial infarction ranged from 0 days to 10 years; however, 64.6% of the argeriograms were performed 2 years or less prior to the myocardial infarction. Only 5.1 % of the patients in the control and intervention groups combined had a culprit vessel stenosis $\leq$ 50%, while 88.6% of the patients in both groups combined had a culprit vessel stenosis $\geq$ 70%. The currently popular composite null hypothesis that $\geq$ 50% of patients with an acute MI have as their arteriographic culprit lesion a vessel stenosis <50% was rejected by the POSCH analysis ($P <$ .0001). The POSCH data strongly favor the large-lesion hypothesis of causation for myocardial infarction.

G. Analyses in Progress

Multiple POSCH data analyses are in progress for future publication:

1.   Analysis of specific lipid changes with clinical events and changes in arteriograms.

  a.   Relationship to baseline and, in particular, to in-trial levels of plasma total cholesterol, LDL cholesterol, HDL cholesterol, various cholesterol ratios, and triglycerides.
  b.   Relationship to in-trial degree of change (flux) in levels of plasma total cholesterol, LDL cholesterol, HDL cholesterol, various cholesterol ratios, and triglycerides.

2.   Analysis of the relationship of triglycerides to risk and prognosis, when the LDL cholesterol influence is excluded, i.e., analysis of the association of the triglyceride concentration after successful reduction of the LDL cholesterol level.

3.   Analysis of the effect on patient prognosis of the use of hypocholesterolemic drugs in the control group and the intervention group.

4.   Analysis of the POSCH in-trial ACHD mortality and confirmed nonfatal myocardial infarction cohort in the control group and in the intervention group for:

  a.   Risk factors.
  b.   Impact on prognosis for myocardial infarction survivors.

5.   Analysis of the CABG and PTCA cohort in the control group and the intervention group, compared with the non-CABG and non-PTCA patients, for:

a. Risk factors.
b. Lipids.
c. Prognosis.

6.   Analysis of the relationship of resting/exercise electrocardiograms to the sequential arteriograms, to the difference in prognosis of silent myocardial infarctions vs. clinical myocardial infarctions, and to overall prognosis.

7.   Analysis of the nutrient profile—e.g., levels of b-carotene, zinc—on patient prognosis.

8.   Analysis of the effect of antioxidant use, in particular vitamin E, on patient prognosis.

9.   Analysis of the effect of the documented use of alcohol on prognosis.

10.   Further analysis of cigarette smoking and its effect on prognosis.

11.   Further analysis of aspirin use and its effect on prognosis.

12.   Analysis of the POSCH trial methods and mechanics responsible for a successful long-term follow-up trial with no patient lost to follow-up.

V.   SIDE EFFECTS OF PARTIAL ILEAL BYPASS

No immediate, in-hospital deaths occurred after partial ileal bypass (19). The 30-day mortality after surgery was limited to two patients, at day 23 and day 29. One death was due to an event related to ACHD and was so attributed. The other death was due to sepsis secondary to complications of a bowel obstruction shortly after reversal of the partial ileal bypass.

The principal side effect of partial ileal bypass was diarrhea. At each follow-up visit, the intervention group patients reported having an average of >3.0 bowel movements per day, whereas the control group patients had <1.5 movements per day ($P < .0001$) (19). In addition to more frequent bowel movements, the intervention group patients had looser stools. During the first 5 years of follow-up, 6% to 8% of the intervention group patients had watery or frothy stools, compared with 0% to 1% of the control group patients ($P < .0001$).

During the course of the trial, a higher incidence of kidney stones and gallstones was observed among the patients who had undergone partial ileal bypass (19). The incidence rate of kidney stones was 4.0% per year in the intervention group, compared with 0.8% in the control group ($P < .0001$). Of the 286 control group patients and 320 intervention group patients whose gallbladder was found to be free of gallstones at baseline or immediately after partial ileal bypass, four control group patients and 14 intervention group patients underwent cholecystectomy during the first 5 years of follow-up (19). An additional 10 control group patients and 40 intervention group patients had gallstones that were detected by oral cholecystography or ultrasonography. The difference between the two groups in the 5-year rate of gallstone formation was statistically significant ($P < .0001$).

The incidence of symptoms of bowel obstruction (at least one episode) was higher in the intervention group: 57 patients (13.5%) had obstructive symptoms, 15 (3.6%) of whom required operative intervention (19). Most patients who had symptoms of bowel obstruction (n = 39) had them within the first year after operation. The intervention group had a mean weight loss of 5.3 kg ($P < .0001$) (19).

VI. UNIQUE CONTRIBUTIONS OF POSCH

A. To Lipid/Atherosclerosis Trials

As stated, POSCH was the pivotal study in the chronology of lipid/atherosclerosis intervention trials. The first truly affirmative study for most of the ACHD endpoints, it demonstrated in the intervention group a highly statistically significant reduction in the event rate of ACHD mortality combined with confirmed nonfatal MI, a concurrent and statistically correlated decrease in arteriographic progression and actual arteriographic regression of coronary atherosclerotic lesions, a statistically and economically significant decrease in the incidence of CABG and PTCA, and a simultaneous reduction in peripheral arterial disease. POSCH foreshadowed the current decade of trials, primarily using the statin drugs for intervention, that confirmed the POSCH results. In demonstrating that over time (up to 20 years) overall mortality and ACHD mortality also were statistically significantly lowered, the POSCH long-term follow-up study completed a single study proof of the lipid/atherosclerosis theory.

B. To Current Strategy for Cholesterol Management

Although there were no in-hospital deaths after partial ileal bypass in the POSCH trial, this lipid intervention modality does involve an operative procedure and was associated with inconvenient, lifestyle-compromising, and well-documented side effects. At the same time, partial ileal bypass is highly effective in markedly reducing plasma total and LDL cholesterol levels, increasing the HDL cholesterol levels and favorably modifying the lipid ratios and subfractions. The operative effect has been shown to be durable. In a study independent of POSCH, lipid modification lasted >20 years, with up to a 30% reduction in the plasma total cholesterol level (69). Partial ileal bypass therapy is obligatory, as long as the operation is not reversed. In POSCH, 94% of patients undergoing the procedure had operative integrity for an average of 9.7 years. No other trial-tested therapy, diet or drug, comes close to that degree of patient compliance.

Is there any clinical role for partial ileal bypass today? We believe this therapeutic option should be considered for young patients facing the alternative of many years of expensive, easy-to-neglect drug therapy. After 7 years, partial ileal bypass becomes cost-effective, compared with the current cost of statin drug therapy. Partial ileal bypass should also be considered for patients for whom various drug regimens have failed or for whom statins, or combination drug protocols, have proven to be toxic. Moreover, in a limited number of patients, the combination of a partial ileal bypass with dose titration of a statin drug has been an eminently successful means to achieve a specified low LDL cholesterol level.

ACKNOWLEDGMENTS

As the author of this chapter, I express my deep appreciation to the many individuals in the POSCH group who gave this study life, precision, perseverance, results, and success, including my fellow principal investigators, clinic managers, nurses, Coordinating Center staff, and members of the Central Lipid Laboratory, the Central ECG Laboratory, the Mortality Review Committee, the ECG Review Panel, and the Arteriography Review Panel. Special gratitude goes to the members of the POSCH Data Monitoring Committee for their untiring assistance and dedication, including travel to several annual meetings and their personal support at NHLBI site visits: J. Ward Kennedy, M.D. (Chairman), Jacob E. Bearman, Ph.D., Gerald R. Cooper, M.D., Ph.D., Samuel W. Greenhouse, Ph.D., Paul Meier, Ph.D., Curtis L. Meinert, Ph.D., Jeremiah Stamler, M.D., and D. Eugene Strandness, M.D. In particular, we owe an immeasurable debt to the late Thomas C. Chalmers, M.D., who chaired the POSCH Data Monitoring Committee from its inception until his death on December 27, 1995. His drive and spirit, his integrity, and his belief in the value of the POSCH trial for patients and scientists worldwide are integral to, and responsible for, the POSCH data.

Thanks also to the NHLBI-appointed Policy and Data Monitoring Board for their careful oversight and support: Antonio M. Gotto Jr., M.D., Ph.D. (Chairman), C. Morton Hawkins, Sc.D., James J. Leonard, M.D., Floyd D. Loop, M.D., Elliot Rapaport, M.D., David L. Sylwester, Ph.D. and Doris Tulcin, B.A. POSCH appreciates the work of the several NHLBI project officers.

We in POSCH have always considered our patients our coinvestigators. Their contributions were the ultimate basis for the longevity and success of this study. They gave of themselves, they volunteered for randomization, and half of them underwent an operative intervention. They submitted to up to four coronary and peripheral arteriograms, and they remained faithful to the clinic follow-up and to the study protocol. It was the POSCH patients who were responsible

for the singular POSCH record for over 20 years of having lost no patient to follow-up.

REFERENCES

1. Rayer PFO. Traité Théorique et Pratique des Maladies de la Peau. Paris: J.B. Ballillière, 1835.
2. Addison T, Gull W. On a certain affection of the skin, vitiligoidea—a. plana, b. tuberosa. Guys Hosp Rep 1851; VII(2):267–277.
3. Ignatovski AI. Zur Frage über den Einfluss der animalischen Nahrung auf den Kaninchenorganismus. Izviest Imp Voyenno-med Akad S Petersb 1908;16:154–176.
4. Anitschkow N, Chalatow S. Ueber experimentelle Cholesterinsteatose und ihre Bedeutung für die Entstehung einiger pathologischer Prozesse. Centtralb Allg Pathol Pathol Anat 1913; 24:1–9.
5. Wacker L, Hueck W. Ueber experimentelle Athnerosklerose und Cholesterinämie. Münchener Med Wochenschr 1913;60:2097–2100.
6. Keys A, Aravanis C, Blackburn H, et al. Coronary heart disease in seven countries. Circulation 1970;41(suppl 1):1–211.
7. Kannel WB, Castelli WP, Gordon T, McNamara PM. Serum cholesterol, lipoproteins, and the risk of coronary heart disease: the Framingham Study. Ann Intern Med 1971; 74:1–12.
8. Pooling Project Research Group. Relationship of blood pressure, serum cholesterol, smoking habit, relative weight and ECG abnormalities to the incidence of major coronary events: final report of the Pooling Project. J Chron Dis 1978;31:201–306.
9. Stamler J, Wentworth D, Neaton JD. Is the relationship between serum cholesterol and risk of premature death from coronary heart disease continuous and graded? Findings in 356,222 primary screenees of the Multiple Risk Factor Intervention Trial (MRFIT). JAMA 1986;256:2823–2828.
10. Castelli WP, Garrison RJ, Wilson PWF, Abbott RD, Kalousdian S, Kannel WB. Incidence of coronary heart disease and lipoprotein cholesterol levels: the Framingham study. JAMA 1986;256:2835–2838.
11. Ragland DR, Brand RJ. Coronary heart disease mortality in the Western Collaborative Group Study: follow-up experience of 22 years. Am J Epidemiol 1988; 127:462–475.
12. Katz LN, Stamler J, Pick R. Nutrition and Atheroscslerosis. Philadelphia: Lea & Febiger, 1958.
13. Expert Panel on Detection, Evaluation, and Treatment of High Blood Cholesterol in Adults. Summary of the second report of the National Cholesterol Education Program (NCEP) Expert Panel on detection, evaluation, and treatment of high blood cholesterol in adults (Adult Treatment Panel II). JAMA 1993;269:3015–3023.
14. Buchwald H, Fitch L, Moore RB. Overview of randomized clinical trials of lipid intervention for atherosclerotic cardiovascular disease. Controlled Clin Trials 1982;3:271–283.
15. Blankenhorn DH, Nessim SA, Johnson RL, Sanmarco ME, Azen SP, Cashin-

Hemphill L. Beneficial effects of combined colestipol-niacin therapy on coronary atherosclerosis and coronary venous bypass grafts. JAMA 1987;257:3233–3240.

16. Frick MH, Elo O, Haapa K, et al. Helsinki Heart Study: primary-prevention trial with gemfibrozil in middle-aged men with dyslipidemia: safety of treatment, changes in risk factors, and incidence of coronary heart disease. N Engl J Med 1987;317:1237–1245.

17. Brensike JF, Levy RI, Kelsey SF, et al. Effects of therapy with chylostyramine on progression of coronary arteriosclerosis: results of the NHLBI Type II Coronary Intervention Study. Circulation 1984;69:313–324.

18. Lipid Research Clinics Program. The Lipid Research Clinics Coronary Primary Prevention Trial results. I. Reduction in the incidence of coronary heart disease. JAMA 1984;251:351–364.

19. Buchwald H, Varco RL, Matts JP, et al. Effect of partial ileal bypass surgery on mortality and morbidity from coronary heart disease in patients with hypercholesterolemia. Report of the Program on the Surgical Control of the Hyperlipidemias (POSCH). N Engl J Med 1990;323:946–955.

20. Brown G, Alpers JJ, Fisher LD, et al. Regression of coronary artery disease as a result of intensive lipid lowering therapy in men with high levels of apolipoprotein B. N Engl J Med 1990;323:1289–1298.

21. Kane JP, Malloy MJ, Ports TA, et al. Regression of coronary atherosclerosis during treatment of familial hypercholesterolemia with combined drug regimens. JAMA 1990;264:3007–3012.

22. Watts GF, Lewis B, Brunt JNH, et al. Effects on coronary artery disease of lipid-lowering diet, or diet plus cholestyramine, in the St Thomas' Atherosclerosis Regression Study (STARS). Lancet 1992;339:563–569.

23. Gould KL, Ornish D, Kirkeeide R, et al. Improved stenosis geometry by quantitative coronary arteriography after vigorous risk factor modification. Am J Cardiol 1992;69:845–853.

24. Blankenhorn DH, Azen SP, Kramsch DM, et al. Coronary angiographic changes with lovastatin therapy: the Mevinolin atherosclerosis regression study (MARS). Ann Intern Med 1993; 119:969–976.

25. Haskell WL, Alderman EL, Fair JM, et al. Effects of intensive multiple risk factor reduction in coronary atherosclerosis and clinical cardiac events in men and women with coronary artery disease: the Stanford Coronary Risk Intervention Project (SCRIP). Circulation 1994;89:975–990.

26. Scandinavian Simvastatin Survival Study Group. Randomized trial of cholesterol lowering in 4,444 patients with coronary heart disease: the Scandinavian Simvastatin Survival Study (4S). Lancet 1994;344:1383–1389.

27. MAAS Investigators. Effect of simvastatin on coronary atheroma: a multicentre antiatheroma study. Lancet 1994;344:633–638.

28. Waters D, Higginson L, Gladstone P, et al. Effects of monotherapy with an HMG-CoA reductase inhibitor on the progression of coronary atherosclerosis as assessed by serial quantitative arteriography: the Canadian Coronary Atherosclerosis Intervention Trial. Circulation 1994;89:959–968.

29. Furberg CD, Adams HP Jr, Applegate WB, et al. Effect of lovastatin on early carotid atherosclerosis and cardiovascular events. Circulation 1994;90:1679–1687.

30. Jukema JW, Bruschke AVG, van Boven AJ, et al. Effects of lipid lowering by pravastatin on progression and regression of coronary artery disease in symptomatic men with normal to moderately elevated serum cholesterol levels: the Regression Growth Evaluation Statin Study (REGRESS). Circulation 1995;91:2528–2540.
31. Sheperd J, Cobbe SM, Ford I, et al. Prevention of coronary heart disease with pravastatin in men with hypercholesterolemia. N Engl J Med 1995;333:1301–1307.
32. Buchwald H, Matts JP, Fitch LL, et al. Program on the Surgical Control of the Hyperlipidemias (POSCH): design and methodology. J Clin Epidemiol 1989;42:1111–1127.
33. Buchwald H, Matts JP, Hansen BJ, Long JM, Fitch LL, POSCH Group. Program on Surgical Control of the Hyperlipidemias (POSCH): recruitment experience. Controlled Clin Trials 1987;8(suppl 4):94S–104S.
34. Buchwald H. Lowering of cholesterol absorption and blood levels by ileal exclusion: experimental basis and preliminary clinical report. Circulation 1964;29:713–720.
35. Buchwald H, Moore RB, Varco RL. Surgical treatment of hyperlipidemia. Circulation 1974;49(suppl 1):1–37.
36. Buchwald H. Intestinal bypass for hypercholesterolemia. In: Nyhus LM, Baker RJ, eds. Mastery of Surgery. Vol. 3. Boston: Little, Brown, 1997:1358–1365.
37. Buchwald H, Campos CT. Partial ileal bypass for control of hyperlipidemia and atherosclerosis. In: Sabiston DC Jr, Spencer FC, eds. Surgery of the Chest. 5th ed. Philadelphia: W.B. Saunders, 1990:1799–1819.
38. Koivisto P, Mieettinen TA. Long-term effects of ileal bypass on lipoproteins in patients with familial hypercholesterolemia. Circulation 1984;70:290–296.
39. Schouten JA, Beynen AC. Partial ileal bypass in the treatment of heterozygous familial hypercohlesterolemia: a review. Artery 1986;13:240-263.
40. Sandler H, Dodge HT. The use of single plan angiocardiograms for the calculation of the left ventricular volume in man. Am Heart J 1968;75:325–334.
41. National Heart and Lung Institute. Lipid Research Clinics Program. Manual of Laboratory Operations. Vol. 1. (DHEW publication No. (NIH) 75-628). Bethesda, MD: National Institutes of Health, 1974.
42. Hatch FT, Lees RS. Practical methods for plasma lipoprotein analysis. Adv Lipid Res 1968;6:1–68.
43. Classification of hyperlipidaemias and hyperlipoproteinaemias. Bull WHO 1970;43:389–915.
44. Gidez LI, Miller GJ, Burstein M, Slagle S, Eder HA. Separation and quantitation of subclasses of human plasma high density lipoproteins by a simple precipitation procedure. J Lipid Res 1982;23:1206–1223.
45. Mancini G, Carbonara AO, Heremans JF. Immunochemical quantitation of antigens by single radial immunodiffusion. Immunochemistry 1965;2:235–254.
46. Blankenhorn DH, Johnson RL, Nessim SA, Azen SP, Sanmarco ME, Selzer RH. The Cholesterol Lowering Atherosclerosis Study (CLAS): design, methods, and baseline results. Controlled Clin Trials 1987;8:356–387.
47. Bruce RA. Exercise testing of patients with coronary heart disease: principles and normal standards for evaluation. Ann Clin Res 1971;3:323–332.

48. Lee ET. Statistical Methods for Survival Data Analysis. Belmont, CA: Lifetime Learning Publ., 1980.
49. Cox DR. Regression models and life-tables. J R Stat Soc (B) 1972;34:1887–1220.
50. Matts JP, Buchwald H, Fitch LL, et al. Subgroup analyses of the major clinical endpoints in the Program on the Surgical Control of the Hyperlipidemias (POSCH): overall mortality, atherosclerotic coronary heart disease (ACHD) mortality, and ACHD mortality or myocardial infarction. J Clin Epidemiol 1995;48:389–405.
51. Kaplan EL, Meier P. Nonparametric estimation from incomplete observations. J Am Stat Assoc 1958;53:457–481.
52. Mantel N. Evaluation of survival data and two new rank order statistics arising in its consideration. Cancer Chemother Rep 1966;50:163–170.
53. Gehan EA. A generalized Wilcoxon test for comparing arbitrarily singly-censored-samples. Biometrika 1965;52:203–223.
54. Armitage P. Tests for linear trends in proportions and frequencies. Biometrics 1955;11:375–386.
55. Conover WJ. Practical Nonparametric Statistics. New York: John Wiley, 1971:163.
56. Matts JP, Buchwaid H, Fitch LL, et al. Program on the Surgical Control of the Hyperlipidemias (POSCH): patient entry characteristics. Controlled Clin Trials 1991;12:314–339.
57. Campos CT, Matts JP, Fitch LL, et al. Lipoprotein modification achieved by partial ileal bypass: five-year results of the POSCH trial. Surgery 1987;102:424–432.
58. Campos CT, Matts JP, Fitch LL, et al. Normalization of lipoproteins following partial ileal bypass in individual WHO lipoprotein phenotypes. Curr Surg 1988;45:380–382.
59. Campos CT, Matts JP, Fitch LL, et al. Comparisons of lipoprotein results in men and women after partial ileal bypass for hypercholesterolemia. Surg Forum 1988;39:193–195.
60. Campos CT, Matts JP, Santilli SM, et al. Predictors of total and LDL cholesterol change following partial ileal bypass. Am J Surg 1988; 155:138–146.
61. Campos CT, Buchwald H, POSCH Group. The Program on the Surgical Control of the Hyperlipidemias (POSCH): demonstration of the beneficial effects of treatment of hypercholesterolemia. Cardiovasc Risk Factors 1992;2(4):261–275.
62. Buchwald H, Bourdages HR, Campos CT, et al. Impact of cholesterol reduction on peripheral arterial disease in the Program on the Surgical Control of the Hyperlipidemias (POSCH). Surgery 1996;120:672–679.
63. Buchwald H, Campos CT, Boen JR, et al. Gender-based mortality follow-up from the Program on the Surgical Control of the Hyperlpidemias (POSCH) and meta-analysis of lipid intervention trials. Ann Surg 1996;224:486–500.
64. Buchwald H, Campos CT, Matts JP, et al. Women in the POSCH trial. Effects of aggressive cholesterol modification in women with coronary heart disease. Ann Surg 1992; 216:389–398.
65. Buchwald H, Campos CT, Boen JR, et al. Disease-free intervals after partial ileal bypass in patients with coronary heart disease and hypercholesterolemia: report from the Program on the Surgical Control of the Hyperlipidemias (POSCH). J Am Coll *Cardiol* 1995;26:351–357.

66. Fitch LL, Buchwald H, Matts JP, et al. Effect of aspirin use on death and recurrent myocardial infarction in current and former cigarette smokers. Am Heart J 1996;129:656–662.
67. Buchwald H, Fitch LL, Matts JP, et al. Perception of quality of life before and after disclosure of a trial results: a report of the Program on the Surgical Control of the Hyperlipidemias (POSCH). Controlled Clin Trials 1993;14(6):500–510.
68. Buchwald H, Hunter DW, Tuna N, et al. Myocardial infarction and percent arteriographic stenosis of culprit vessel: report from the Program on the Surgical Control of the Hyperlipidemias (POSCH). Atherosclerosis 1998; 138:391–401.
69. Buchwald H, Stoller DK, Campos Ct, Matts JP, Varco RL. Partial ileal bypass for hypercholesterolemia: 20- to 26-year follow-up of the first 57 consecutive cases. Ann Surg 1990;232(3):318–331.
70. Buchwald H, Varco RL, Boen JR, et al. Effective lipid modification by partial ileal bypass reduced long-term coronary heart disease mortality and morbidity: five-year posttrial follow-up repart from the POSCH. Arch Intern Med 1998; 158:1253–1261.

APPENDIX: THE POSCH GROUP

Principal Investigators
Henry Buchwald, MD, PhD
Richard L. Varco, MD, PhD
University of Minnesota Clinic
Christian T. Campos, MD
Arthur S. Leon, MD
Jean Rindal, RN, MA
Rebecka A. Hagen, RN, MS
University of Arkansas Clinic
Gilbert S. Campbell, MD, PhD
Malcolm B. Pearce, MD
Joseph K. Bissett, MD
Meredith R. Stuenkel, RN
University of So. California Clinic
Albert E. Yellin, MD
W. Allan Edmiston, MD
Dorothy C. Fujii
Julie A. Hatch, RN
Lankenau Hospital
Robert D. Smink Jr., MD
Henry S. Sawin Jr., MD
Frederic J. Weber, MD, PhD
Helene B. Brooks, BS
Rebecca F. Carins, RN, MS
Margaret E. Trobovic, RN
Central ECG Laboratory
Naip Tuna, MD, PhD

James N. Karnegis, MD, PhD
James E. Stevenson, MD
Regina Brykovsky
Mark A. Linssen
Central Lipid Laboratory
Jane C. Speech, MS
Central Arteriography/Radiology Laboratory
Kurt Amplatz, MD
Miguel E. Sanmarco, MD
Wilfredo R. Castaneda-Zuniga, MD
David W. Hunter, MD
Nancy P. Wehage, BS
ECG Review Panel
Naip Tuna, MD, PhD (Chairman)
Joseph K. Bissett, MD
W. Allan Edmiston, MD
James N. Karnegis, MD, PhD
Arthur S. Leon, MD
Malcolm B. Pearce, MD
Henry S. Sawin, Jr., MD
James E. Stevenson, MD
Arteriography Review Panel
Miguel E. Sanmarco, MD (Chairman)
Kurt Amplatz, MD
Joseph K. Bissett, MD
Wilfredo R. Castaneda-Zuniga, M
W. Allan Edmiston, MD

David W. Hunter, MD
Malcolm B. Pearce, MD
Henry S. Sawin Jr., MD
Frederic J. Weber, MD, PhD
Coordinating Center
John M. Long, EdD
John P. Matts, PhD
Laurie L. Fitch, MPH
James W. Johnson, MS
James R. Boen, PhD
Stanley E. Williams, MHA
Phuong Nguyen, BA
James M. Vagasky
Administration
Betty J. Hansen, RN
Data Monitoring Committee
Thomas C. Chalmers, MD
  (Chairman-deceased)
J. Ward Kennedy, MD (Chairman)
Jacob E. Bearman, PhD
Gerald R. Cooper, MD, PhD
Samuel W. Greenhouse, PhD
Paul Meier, PhD
Curtis L. Meinert, PhD
Jeremiah Stamler, MD
D. Eugene Strandness, MD

Mortality Review Committee
Jesse E. Edwards, MD (original Chairman)
Jack L. Titus, MD, PhD (Chairman)
Lawrence S. C. Griffith, MD
Arthur J. Moss, MD
Policy and Data Monitoring Board
Antonio M. Gotto Jr., MD, DPhil
  (Chairman)
C. Morton Hawkins, ScD
James J. Leonard, MD
Floyd D. Loop, MD
Elliot Rapaport, MD
David L. Sylwester, PhD
Doris Tulcin, BA
Consultants
David H. Blankenhorn, MD (deceased)
Linda H. Cashin-Hemphill, MD
Jerome Cornfield, MA (deceased)
William L. Holmes, PhD (deceased)
Manford D. Morris, PhD
*National Heart, Lung, and Blood
  Institute Project Officers*
Thomas P. Blaszkowski, PhD
Curt D. Furberg, MD
Lawrence M. Friedman, MD
Jeffrey L. Probstfield, MD

6

The Air Force/Texas Coronary Atherosclerosis Prevention Study (AFCAPS/TexCAPS)

Primary Prevention of Acute Major Coronary Events in Men and Women with Average Cholesterol

Stephen Weis and Michael Clearfield
Wilford Hall Medical Center, Lackland Air Force Base, Lackland, Texas

Lt. Col. John R. Downs
U.S. Department of Defense Pharmacoeconomic Center, Fort Sam Houston, Texas

Antonio M. Gotto, Jr.
Weill Medical College of Cornell University, New York, New York

for the AFCAPS/TexCAPS investigators*

INTRODUCTION

Although the relation between coronary heart disease (CHD) mortality and serum cholesterol concentration is well established, epidemiologic evidence suggests that the majority of CHD events occur in persons who do not have severe elevations in blood cholesterol. In the Framingham Heart Study, 40% of participants who developed myocardial infarction had total cholesterol values in the

---

*See Appendix for committee members.

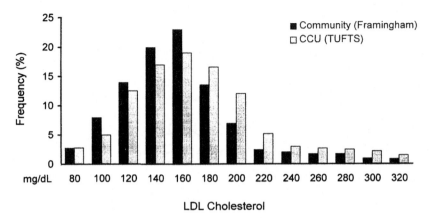

Figure 1   Distribution of LDL-C values in populations with (CCU) and without (Framingham) CHD. (From Ref. 1.)

range of 200 to 250 mg/dL (1,2). Furthermore, the distribution of low-density lipoprotein cholesterol (LDL-C) values in CHD patients shows a surprising overlap with values in patients without coronary disease (Fig. 1). In the Multiple Risk Factor Intervention Trial follow-up of screened men, 69% of CHD deaths in the first 16 years of follow-up were in those with total cholesterol (TC) between 182 and 264 mg/dL (3). Although reduction of elevated serum cholesterol is an important strategy in the primary prevention of CHD (4), before now, there has been no clinical trial evidence that individuals without coronary disease, and with total cholesterol in the range of 180 to 240 mg/dL, benefit from lipid-modifying therapy.

Additional evidence indicates that CHD risk is jointly affected by TC and high-density lipoprotein cholesterol (HDL-C) concentrations (5,6). Low HDL-C is predictive of increased CHD risk independent of total cholesterol (7). In addition, CHD risk is increased when the TC/HDL-C ratio is ≥ 4.5 (6). According to the guidelines of the second Adult Treatment Panel of the National Cholesterol Education Program (NCEP ATP II), an HDL-C of <35 mg/dL is considered a risk factor for coronary disease, whereas an HDL-C ≥60 mg/dL is considered cardioprotective (4). Thus, the prognostic significance of low HDL-C appears to be well accepted (6,7), and there is considerable interest in identifying appropriate risk reduction strategies in patients with lower concentrations of HDL-C. In the ATP II algorithm, LDL-C reduction is the primary goal of therapy in the face of dyslipidemia, thus elucidating the potential benefit of lowering LDL-C in relatively healthy patients with average TC and LDL-C concentrations and below-average HDL-C is an important objective.

The application of such an investigation in the clinical setting cannot be overstated. For most 60-year-old Americans, for whom the average additional life expectancy may be another 20 years (8), preventing the onset of the first CHD event is a critical aspect of promoting healthy longevity. In a cohort of men and women, aged 45 to 73 and 55 to 73 respectively, AFCAPS/TexCAPS explored the effect of LDL-C reduction on the entire spectrum of acute major coronary events, including the most common first nonfatal event—unstable angina (9).

II. STUDY DESIGN

The AFCAPS/TexCAPS trial design and rationale have been previously described in detail (10). Briefly, AFCAPS/TexCAPS was a double-blind, placebo-controlled, primary prevention trial conducted at two sites in Texas: the University of North Texas Health Science Center at Fort Worth, and Wilford Hall USAF Medical Center, Lackland Air Force Base in San Antonio. The primary objective of AFCAPS/TexCAPS was to investigate whether lipid lowering with lovastatin in a population without clinical evidence of cardiovascular disease and with average TC and LDL-C and below-average HDL-C would decrease the rate of first acute major coronary events compared with placebo. The study was designed to continue until either 320 participants had a primary endpoint event or until 5 years after the last participant was randomized, whichever came last. The trial design provided a 90% to 97% power to detect a estimated 30% to 35% risk reduction from lovastatin treatment.

A total of three analyses were planned over the course of the trial. In addition to the final analysis, two interim analyses were planned: one after 120 participants and the other after 240 participants experienced a primary endpoint event. The study was designed with an early-stopping rule which preserved the type I error probability of .05.

A. Screening and Randomization

In summary, 65,216 men and women responded to recruitment efforts and underwent an initial fasting lipid assessment. From those screened, 36,282 came to a risk factor modification counseling session which included dietitian-taught group instruction in the AHA Step I diet. Of these, 16,698 gave informed consent to participate in the study. Men (45 to 73 years old) and postmenopausal women (55 to 73 years) who met the lipid entry criteria and had no prior history or signs or symptoms of definite myocardial infarction, angina, claudication, cerebrovascular accident, or transient ischemic attack were eligible for participation. Lipid entry criteria (TC 180–264 mg/dL, LDL-C 130–190 mg/dL, HDL-C $\leq$45 mg/dL for men or $\leq$47 mg/dL for women, and triglycerides $\leq$400 mg/dL) were to be met at both 4 and 2 weeks prior to randomization, with <15% difference in LDL-C values. In addition, those with LDL-C between 125 and 129 mg/dL were included when the ratio T-C/HDL-C was >6.0. Participants who met the entry criteria were randomized to either lovastatin 20 mg or placebo in the evening. The baseline lipid measurements reported were after a minimum of 12 weeks AHA Step I diet, with a dietitian providing group reinforcement instruction at least 2 and 4 weeks prior to randomization. Participants whose nutrient intake did not meet the guidelines of the AHA Step I diet (>30% total calories as fat or >300 mg dietary cholesterol) during a nutrient analysis at least 4 weeks before randomization were given individualized dietitian instruction.

Participants were excluded if they had uncontrolled hypertension, secondary hyperlipoproteinemia, insulin-dependent diabetes mellitus, or a glycohemoglobin level >20% above the upper limit of normal. Additionally, volunteers who had a body weight of >50% over ideal for height as established by 1983 Metropolitan Life Insurance tables were excluded (10). Use of other lipid-modifying medications was not permitted, but multivitamins, including niacin ≤500 mg/day, and hormone replacement therapy could be continued.

Randomization of 5608 men and 997 women occurred between May 30, 1990, and February 12, 1993: 3737 participants at the Air Force clinic and 2868 at the University of North Texas. Participants in the lovastatin group were titrated from 20 to 40 mg lovastatin daily if their LDL-C at the weeks 6 and 12 postrandomization averaged >110 mg/dL (10). An equivalent number of placebo patients were titrated to keep the study blind (Fig. 2).

B. Endpoints

The study's composite primary endpoint, first acute major coronary events, comprised unstable angina, fatal and nonfatal myocardial infarction (MI), and sud-

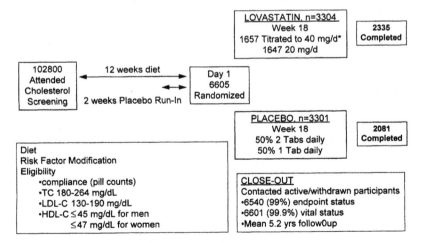

Figure 2 Overview of AFCAPS/TexCAPS trial.

den cardiac death. Including unstable angina in the primary endpoint is a unique design feature of this trial and was intended to reflect the increasing frequency of unstable angina as the initial manifestation of CHD in the United States (9). For the purposes of endpoint adjudication, unstable angina was prospectively defined as new-onset exertional or rest angina with at least one of the following: (1) electrocardiographic findings of ≥1 mm ST-segment changes and reversible defect on stress perfusion study; (2) angiographic findings of ≥90% epicardial vessel stenosis or ≥50% stenosis in the left main coronary artery (without exercise testing); or (3) ≥ 1 mm ST-segment changes with pain on electrocardiographic stress testing and/or rest electrocardiograph and evidence of ≥50% stenosis in a major epicardial vessel (10).

The secondary objective of AFCAPS/TexCAPS was to investigate whether treatment with lovastatin, compared with placebo, would decrease cardiovascular morbidity and mortality across the spectrum of clinical events that can be attributed to progressive atherosclerotic disease, including clinically important morbidity, mortality, or medical intervention. The secondary endpoints were (1) unstable angina, (2) fatal and nonfatal MI, (3) coronary revascularization procedures, (4) cardiovascular events, (5) coronary events, (6) cardiovascular mortality, and (7) CHD mortality. The tertiary objective of AFCAPS/TexCAPS was to investigate the safety of chronic treatment with lovastatin by evaluating the rates of total mortality, noncardiovascular mortality, and discontinuation from the study due to adverse drug effects in the lovastatin and placebo groups. Subset analyses were planned for accidental/violent death, and death from cancer (10).

C. Early Termination for Efficacy and Closeout

Following the second interim analysis, which was based on data from 267 participants who had had a primary endpoint, the Data Safety Monitoring Board recommended early termination of the study. This decision was unanimously endorsed in July 1997 by the Steering Committee, who agreed to end the trial early because of demonstrated efficacy.

Personnel and participants remained blinded to both treatment group assignment and the preliminary results of the study throughout the final closeout visits. At the time of closeout, participants in the study included 2335 of the 3304 participants (71%) in the lovastatin group, and 2081 of the 3301 participants (63%) in the placebo group. As assessed by pill counts, 99% of participants adhered to their study regimen ≥75% of the time that they were on study treatment. Both active and withdrawn participants were contacted during closeout to obtain complete and recent status updates. Endpoint status was determined for 6540 participants (99%) and vital status was determined for 6601 (99.9%) participants within 3 months of the decision to stop. The mean duration of follow-up was 5.2 years.

III. RESULTS

A. Baseline Characteristics of Study Cohort

The two treatment groups were balanced with respect to baseline demographics, risk factors, and medical history. The mean age of all participants at entry into the study was 58 ± 7 years (57 ± 7 years for men and 63 ± 5 years for women). Although the AFCAPS/TexCAPS cohort included both men and women and was multiethnic, the majority of participants were white (5860 [89%]). Table 1 summarizes the baseline characteristics for the entire AFCAPS/TexCAPS cohort compared with values for a reference U.S. population of men age 45–73 and women age 55–73 without clinical evidence of cardiovascular disease, based on NHANES III data (11). Compared with this reference population, women (997 [15%]) and blacks (206 [3%]) were underrepresented in the study cohort. Most

Table 1   Baseline Demographics

	N (%) or mean ± SD	
Baseline characteristic	AFCAPS/TexCAPS	U.S. NHANES III reference population
Men (age 45–73); Women (age 55–73)	5608 (85%); 997 (15%)	25,408,673 (58%); 18,285,565 (42%)
Mean age (yrs): all; men; women	58 ± 7; 57± 7; 63 ± 5	60 ± 8; 57 ± 8; 64 ± 5
N (%) ≥ 65	1416 (21%)	14,342,436 (33%)
Men	1064 (19%)	6,183,762 (24%)
Women	352 (35%)	8,158,674 (45%)
Race		
White	5860 (89%)	37,312,502 (85%)
Black	206 (3%)	3,684,210 (8%)
Hispanic	487 (7%)	2,845,092 (7%)
Current marital status		
Married	5847 (89%)	31,027,149 (71%)
Divorced	379 (6%)	4,686,955 (11%)
Education (highest level attended)		
Some high school	1414 (21%)	27,384,055 (63%)
Some vocational	89 (1%)	NA[a]
Some college	2101 (32%)	7,020,598 (16%)
College graduate or beyond	2999 (45%)	9,221,923 (21%)
Current or former military service	2761 (42%)	15,712,818 (36%)

[a]NA = not applicable/analyzed.

of the study participants were born in the United States (96.8%). The majority were employed full-time (3706 [56%]); 2010 (30%) were retired; 498 (8%) were employed part-time; and 391 (6%) were unemployed at the time of randomization. Additionally, study participants were more likely to be married, to have a history of military service, and to have a higher level of education than the NHANES III reference population.

The AFCAPS/TexCAPS cohort differed from the NHANES III reference population in several respects in terms of baseline risk factors and lifestyle (Table 2). Study participants were more likely than the NHANES III reference population to have a family history of CHD and to have low HDL-C as a positive risk factor. AFCAPS/TexCAPS participants were less likely to have a diagnosis of

Table 2 Baseline Risk Factors

	N (%) or mean ± SD	
Baseline characteristic	AFCAPS/TexCAPS	U.S. NHANES III reference population
Mean weight (lbs); Body mass index		
(kg/m² height)	185 ± 28; 26.9 ± 3.1	170 ± 35; 26.9 ± 4.9
Men	190 ± 25; 27.0 ± 3.0	181 ± 31; 26.4 ± 4.2
Women	156 ± 24; 26.4 ± 3.6	156 ± 34; 27.5 ± 5.7
Mean systolic pressure (mm Hg)	138 ± 17	131 ± 17
Mean diastolic pressure (mm Hg)	78 ± 10	77 ± 10
N (%) who consume alcohol	3098 (47%)	22,245,376 (51%)
NCEP CHD risk factors[a]		
Hypertension	1448 (22%)[b]	15,297,749 (35%)
Diabetes		
Non-insulin-treated diabetes	155 (2%)	1,538,308 (4%)
Non-insulin-treated diabetes or fasting blood glucose ≥ 126 mg/dL	239 (3.6%)	NA[c]
Current smoker	818 (12%)	11,430,887 (26%)
Premature family history of CHD	1035 (16%)	4,107,633 (9%)
Parental	697 (11%)	2,495,726 (6%)
Sibling	443 (7%)	1,935,789 (4%)
HDL-C < 35 mg/dL[d]	2294 (35)	5,463,340 (13%)
HDL-C ≥ 60 mg/dL[d]	1 (<1%)	12,314,797 (28%)

[a]All AFCAPS/TexCAPS participants meet NCEP criteria for CHD age-related risk (age ≥ 45 for men and ≥ 55 for women).
[b]Hypertension includes those reporting history of hypertension and/or those treated with antihypertensive agents for hypertension.
[c]NA = not applicable/analyzed.
[d]Wilford Hall Medical Center Laboratory data.

hypertension or type 2 diabetes, and were less likely to be current smokers than the NHANES III reference population. Regular strenuous exercise was reported by 2314 (35%) participants, with 1996 (30%) reporting that they exercised more than three times a week.

In terms of medication usage at baseline, 21% of AFCAPS/TexCAPS participants reported using one or more of the following antihypertensive medications: ACE inhibitors (501 [8%]), diuretics (406 [6%]), calcium channel blockers (341 [5%]), beta blockers (297 [4%]), and alpha blockers (135 [2%]). Oral hypoglycemics were taken by 84 (1%) participants and thyroid replacement hormone was taken by 239 (4%). Aspirin use during baseline was reported by 1132 (17%) participants, and other NSAIDs were taken by 939 (14%). Detailed data concerning concomitant medication use in the reference U.S. population is not available from NHANES III, although 21% of the NHANES III reference population reported use of any antihypertensive medication and 4% reported use of hypoglycemics. Among AFCAPS/TexCAPS women, 292 (29%) reported taking hormone replacement therapy, and 12 (1%) reported taking estrogens with progestins. For comparison, a 1996 population-based study of 2962 women age 65 or older found that 9% (280) were current users of estrogens and 2% (73) used estrogens with progestin (12).

Baseline lipid values for AFCAPS/TexCAPS participants were also compared with those of the reference NHANES III population (Table 3). The lipid and lipoprotein values at baseline were similar in the lovastatin and placebo groups. In the study cohort, the mean TC was 221 ± 21 mg/dL and median triglycerides 158 ± 76 mg/dL. The baseline mean LDL-C was 150 ± 17 mg/dL. The mean HDL-C for men was 36 ± 5 mg/dL and for women, 40 ± 5 mg/dL. While the AFCAPS cohort had similar baseline TC and LDL-C concentrations to those of the NHANES III reference population, it had much lower baseline concentrations of HDL-C (37 ± 6 mg/dL vs. 50 ± 16 mg/dL).

B. Lipid Response to Treatment

Lipid values were significantly changed by lovastatin treatment ($P < .001$). When baseline levels were compared with posttreatment values at the end of the first year of treatment (year 1), LDL-C was reduced 25%, TC was reduced 18%, and triglycerides were reduced 15%. The HDL-C was increased 6%, and the ratios of TC/HDL-C and LDL-C/HDL-C were decreased 22% and 28%, respectively (Fig. 3) (13). The placebo group had small changes in lipid levels that were not significant. In the lovastatin group, 1657 (50%) were titrated from 20 to 40 mg daily. No participant required subsequent back-titration. At year 1, 42% of the participants in the lovastatin group and 3% in the placebo group reached the study target of LDL-C ≤110 mg/dL. An LDL-C ≤ 130 mg/dL at year 1 was reached by 81% of participants in the lovastatin group and 13% of placebo group.

Table 3 Baseline Lipid Levels (Wilford Hall Data)

Lipid level (mg/dL)	AFCAPS/TexCAPS Average ± SD (mg/dL)	U.S. NHANES III reference population Mean ± SD (mg/dL)
Mean total cholesterol	221 ± 21	225 ± 45
Men	219 ± 21	218 ± 42
Women	230 ± 22	237 ± 47
Mean LDL-C	150 ± 17	142 ± 37
Men	150 ± 17	139 ± 35
Women	154 ± 18	147 ± 40
Mean HDL-C	37 ± 6	50 ± 16
Men	36 ± 5	47 ± 14
Women	40 ± 5	56 ± 17
Median triglycerides	158 ± 76	140 ± 120
Men	155 ± 75	137 ± 129
Women	173 ± 79	144 ± 108
Mean total-C/HDL-C	6.1 ± 1.1	4.9 ± 2.1
Men	6.2 ± 1.1	5.1 ± 1.7
Women	5.9 ± 1.0	4.7 ± 2.6
Mean LDL-C/HDL-C	4.2 ± 4.8	3.1 ± 1.5
Men	4.2 ± 0.8	3.2 ± 1.2
Women	3.9 ± 0.7	2.9 ± 1.9
Apolipoprotein A1	NA	147 ± 27
Men	NA	139 ± 23
Women	NA	158 ± 29
Apolipoprotein B	NA	116 ± 26
Men	NA	115 ± 24
Women	NA	119 ± 27

C. Efficacy of Treatment

Table 4 summarizes efficacy endpoints. There was a 37% reduction in the risk for the first acute major coronary event with lovastatin treatment (Cox model 95% confidence interval, 21% to 50%; $P = .00008$) (13). One or more primary endpoint events occurred in 183 participants who were treated with placebo compared with 116 participants treated with lovastatin. The event rate for first acute major coronary events was 10.9 per 1000 patient-years in the placebo group versus an event rate of 6.8 per 1000 patient-years in the lovastatin group. The life-table plot for the lovastatin and placebo groups illustrates a difference beginning in the first year of treatment and continuing throughout the remainder of the study (Figure 4).

Figure 5 summarizes the effect of lovastatin treatment on the rates of first primary-endpoint events in subgroups defined by sex, age (older defined as above

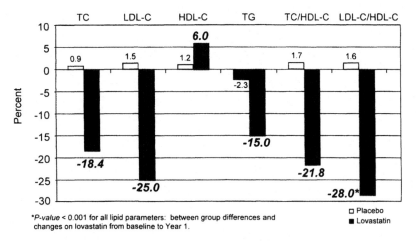

Figure 3 Lipid changes in AFCAPS/TexCAPS trial. (From Ref. 13.)

the median by sex: >57 years for men and >62 years for women), history of hypertension, current cigarette smoking, and family history of CHD. In addition to treatment group, each of these factors demonstrated a significant association with risk (e.g., smoking was positively associated with first acute major coronary

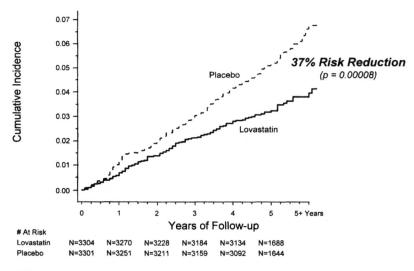

Figure 4 Life-table plot of primary endpoint First Acute Major Coronary Event (composite of unstable angina, fatal and nonfatal MI, and sudden cardiac death). (From Ref. 13.)

Table 4 Efficacy Endpoints

Endpoints	Placebo N = 3301		Lovastatin N = 3304		Relative risk[a] (95% CI), P value[b]
	N	Rates[c]	N	Rates[c]	
Primary endpoint: acute major coronary events defined as fatal and nonfatal MI, unstable angina, or sudden cardiac death	183	10.9	116	6.8	0.63 (0.50, 0.79), P = .00008
Secondary endpoints:					
Revascularizations	157	9.3	106	6.2	0.67 (0.52, 0.85), P = .001
Unstable angina	87	5.1	60	3.5	0.68 (0.48, 0.95), P = .023
Fatal and nonfatal MI	95	5.6	57	3.3	0.60 (0.43, 0.83), P = .002
Fatal and nonfatal cardiovascular events	255	15.3	194	11.5	0.75 (0.62, 0.91), P = .003
Fatal and nonfatal coronary events	215	12.8	163	9.6	0.75 (0.61, 0.92), P = .006
Fatal cardiovascular events	25	1.4	17	1.0	too few for survival analyses
Fatal CHD events	15	0.9	11	0.6	too few for survival analyses

[a]To calculate risk reduction, subtract relative risk from 1.
[b]Relative risk and CI from Cox proportional hazards model; P value from log-rank test. P value adjusted for the interim analysis for the primary endpoint only. P values for secondary endpoints are unadjusted.
[c]Per 1000 patient-years.
Participants are counted only once with a specific endpoint; however the participant may be listed more than once in table because of experiencing different endpoints and/or experiencing an event that is included in more than one endpoint analysis (e.g., the secondary endpoint unstable angina is also a component of the primary endpoint).
(From Ref. 13.)

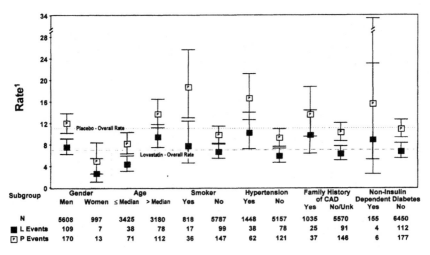

¹Rate of Primary Endpoint Event per 1000 Patient-Years at Risk

Figure 5 Primary endpoint event rate by treatment group in subgroups of the AFCAPS/
TexCAPS cohort. (From Ref. 13.)

events). The baseline triglyceride level ($P = .975$) and diabetes ($P = .335$; 155 diabetics) were not significant predictors of outcomes (13). Within a risk factor, the numerical rate of first acute major coronary events was similar among those treated with lovastatin in the CHD positive-risk subgroup and those treated with placebo who did not have the CHD risk factor (e.g., lovastatin-treated smokers had similar rates to placebo-treated nonsmokers). The effect of lovastatin treatment on the rate of first acute major coronary events was similar in women and in men (46% vs. 37% reduction in relative risk) and there were no statistical differences in treatment effects between sexes. None of the subgroups differed significantly in treatment benefit (e.g, benefit of treatment was not different for current smokers compared with nonsmokers since none of the treatments by subgroup interactions were significant). Relative risk for a primary endpoint across the tertiles of baseline LDL-C and HDL-C are presented in Table 5. Notably, the risk reductions across tertiles appear comparable.

For secondary endpoints, lovastatin treatment resulted in significant, consistent benefits compared with placebo: 33% reduction in revascularizations, 32% reduction in new-onset unstable angina, and 40% reduction in the incidence of fatal and nonfatal MI (Table 4). For coronary and cardiovascular events (total fatal and nonfatal), treatment with lovastatin resulted in highly significant reductions of 25% and 25%, respectively, compared with placebo ($P < .006$). This latter category included all atherosclerotic cardiovascular events, as specified by the endpoint definitions, including stable angina, thrombotic cerebrovascular accidents, transient ischemic attacks, and peripheral arterial vascular disorders. There were too few fatal cardiovascular and fatal CHD events to perform survival analysis based on prespecified criteria (Table 5).

In addition to the protocol-specified rates that considered time to first event for withdrawn and active participants, the total number of events experienced by active and withdrawn participants including multiple events of the same type (e.g., multiple myocardial infarctions experienced by a participant) were analyzed (13). There were 142 and 209 acute major coronary events in the lovastatin and placebo groups, respectively, with rates of 8 and 12 per 1000 patient-years, respectively. There were 137 and 195 coronary revascularizations (8 and 11 per 1000 patient-years) in the lovastatin and placebo groups, respectively. In total there were 279 major coronary events and coronary revascularizations (16 per 1000 patient-years) in the lovastatin group, and 404 (23 per 1000 patient-years) in the placebo group. Based on these findings, treating 1000 men and women with lovastatin for 5 years would prevent approximately 19 acute major coronary events (12 MIs and 7 presentations of unstable angina) and 17 coronary revascularizations (13).

D. Safety and Tolerability

Tertiary endpoints were related to the safety of chronic lovastatin treatment. Overall, lovastatin treatment was well tolerated.

The overall mortality rates were similar in the two groups. There were 80 deaths in the lovastatin group and 77 deaths in the placebo group (4.6 and 4.4 per 1000 patient-years in lovastatin and placebo treatment groups, respectively). The majority of deaths had noncardiovascular causes. There were 17 cardiovascular deaths in the lovastatin group and 25 in the placebo group (1.0 and 1.4 per

Table 5 Relative Risk for a Primary Endpoint Across the LDL-C and HDL-C Tertiles

LDL-C (mg/dL)	≤142	143–156	≥157
% change	−20.1	−26.0	−28.9
RR	0.66	0.64	0.59
(95% CI)	(0.43, 1.00)	(0.42, 0,99)	(0.41, 0.85)
HDL-C (mg/dL)	≤34	35–39	≥40
% change	7.4	6.2	4.3
RR	0.55	0.57	0.85
(95% CI)	(0.38, 0.82)	(0.39, 0.84)	(0.54, 1.32)

1000 patient-years in the lovastatin and placebo groups, respectively), and 63 noncardiovascular deaths in the lovastatin treatment group and 52 in the placebo group (3.6 and 3.0 per 1000 patient-years, respectively). There were a total of four traumatic deaths—three in the placebo group and one the lovastatin group.

Assessing the incidence of fatal and nonfatal cancer did not demonstrate any difference between treatment groups. The overall incidence of fatal and nonfatal cancer, excluding nonmelanoma skin cancers, was 15.1 and 15.6 per 1000 patient-years (252 and 259 cases) in the lovastatin and placebo groups, respectively. The most frequently reported tertiary endpoint cancers are summarized in Table 6. The number of participants reporting nonmelanoma skin cancers, predominantly diagnoses of basal and squamous cell, was 250 (7.6%) in the lovastatin and 243 (7.4%) in the placebo group.

The frequency of discontinuation for other reasons was similar between treatment groups. Placebo-treated participants were more likely to be withdrawn as a result of developing CHD or starting cholesterol-lowering medication (generally at the request of their personal physician). The number of participants with any adverse experience leading to discontinuation was 445 (13.5%) in the lovastatin treatment group and 449 (13.6%) in the placebo group. The two treatment groups had similar number of adverse experiences that were considered serious (i.e., life-threatening, causing death or a permanent disability, resulting in or prolonging hospitalization, any cancer): 1131 (34.2%) and 1126 (34.1%) in lovastatin and placebo groups, respectively. One lovastatin participant was unblinded when he developed study drug-related Stevens-Johnson syndrome after approximately 9 months of 20 to 40 mg daily lovastatin treatment. Follow-

Table 6 Fatal and Nonfatal Cancer[a] by Site

	Placebo (N = 3301)	Lovastatin (N = 3304)	P value
All fatal and nonfatal	259	252	N.S.
Most frequently reported			
Prostate	108	109	N.S.
Melanoma	27	14	.043
Colon	20	25	N.S.
Lung	17	22	N.S.
Lymphomas	11	12	N.S.
Bladder	11	12	N.S.
Breast	9	13	N.S.

[a]Excludes nonmelanoma skin cancer.

ing appropriate treatment and within 2 weeks of discontinuing lovastatin, this participant recovered.[1] No other lovastatin-related life-threatening serious adverse experiences were reported.

Consecutive elevations over three times the upper limit of normal in either aspartate aminotransferase (AST) or alanine aminotransferase (ALT) were rare, and the incidence was similar with lovastatin and placebo treatment: 18 of 3242 (0.6%) and 11 of 3248 (0.3%) for lovastatin and placebo, respectively. Examining these elevations by final dose for those who were titrated also revealed no significant trends: 11 of 1585[2] (0.7%) and 7 of 1657 (0.4%) participants with consecutive elevations over three times the upper limit of normal in either AST or ALT taking lovastatin 20 mg and lovastatin 40 mg, respectively.[3] The numbers of participants with any drug-attributable AST elevation above the upper limit of normal were similar in the two treatment groups (33 [1.0%]) and 34 [1.0%] in the lovastatin and placebo treatment groups, respectively); however, the number with any ALT drug related elevations was significantly ($P = .003$) higher in the lovastatin treatment group (110 [3.3%] and 70 [2.1%] for lovastatin and placebo, respectively). The percentage of participants reporting myalgia leading to discontinuation was 0.3% in both the lovastatin and placebo treatment groups.

Creatine phosphokinase (CPK) elevations >10 times the upper limit of normal were rare, and the incidence was similar with lovastatin and placebo treatment: 11 of 1586 (0.7%), 10 of 1657 (0.6%), and 21 of 3248 (0.6%), receiving lovastatin 20 mg, lovastatin 40 mg, and placebo, respectively. There were no cases of myopathy (defined as muscle symptoms accompanied with CPK elevations >10 times the upper limit of normal). There were three cases of rhabdomyolysis: two occurred in placebo-treated participants. The other case occurred in a lovastatin-treated participant following surgery for prostate cancer. Preadmission CPK, blood urea nitrogen, creatinine, and potassium were normal. The last dose of study medicine was given the day prior to admission. On postoperative day 1, the participant developed rhabdomyolysis associated with transient renal failure that was considered by the investigator to be unrelated to lovastatin treatment. The participant recovered, resumed lovastatin treatment, and completed the trial.

---

[1]Hypersensitivity reactions, including Stevens-Johnson syndrome, are rare and have been reported with agents in this class.

[2]Denominators are patients having postrandomization tests, not all patients.

[3]Unlike the other comparisons of randomized treatment groups, the dose comparisons are of nonrandomized groups.

E. AFCAPS/TexCAPS Resource Utilization

Following the final results analysis, an analysis was performed to evaluate differences in cardiovascular disease (CVD)-related resource utilization in participants on placebo compared with those treated with lovastatin. Resource utilization was measured by determining rates and costs of cardiovascular diagnostic and therapeutic procedures, length of stay, and total hospital duration involving CHD or CVD. The data for the resource utilization analysis were collected prospectively during the study period.

Hospital admission and discharge dates, diagnosis-related group (DRG) assignments, in-hospital medications, and procedure codes were collected for all confirmed endpoints. Costs were estimated from DRG-based costs per case in the United States from MEDSTAT (14). Both data and cost minimization analyses were performed.

According to these analyses, lovastatin therapy resulted in 29% fewer hospitalizations and CVD procedures (including both diagnostic and therapeutic procedures). There were also 32% fewer therapeutic procedures (e.g., heart surgery and angioplasty), 23% fewer diagnostic tests (e.g., electrocardiograms and cardiac catheterizations), and 28% lower total CVD costs (15).

IV. DISCUSSION

AFCAPS/TexCAPS was designed to study the prevention of first acute major coronary event with lovastatin treatment in individuals without clinically evident CHD and for whom optimal treatment of lipids is controversial. The trial demonstrated that men between the ages 45–73 years and women between the ages 55–73 years who are free of CHD, with LDL-C between 130 and 190 mg/dl and HDL-C <50 mg/dl, could achieve substantial benefit with LDL-C reduction with lovastatin therapy, in conjunction with a prudent diet, regular exercise, and risk factor modification. This result is relevant to a substantial portion of the U.S. population who are similar to the AFCAPS/TexCAPS cohort in terms of age and lipid concentrations (approximately 8 million individuals).

A number of characteristics distinguish the AFCAPS/TexCAPS. The AFCAPS/TexCAPS population is unique when compared with prior primary prevention trials (Table 7). AFCAPS/TexCAPS is the first primary prevention trial of statin therapy to demonstrate that lowering LDL-C prevents CHD in persons considered to have average TC and LDL-C and below average HDL-C. Additionally, AFCAPS/TexCAPS is the first primary prevention trial to include women, older individuals, Hispanics, and African-Americans. Both the Lipid Research Clinics Coronary Primary Prevention Trial (LRC-CPPT) and the Helsinki Heart Study included only white, middle-aged men with very high to-

Table 7 Comparison of Primary Prevention Trials

Characteristic	LRC-CCPT	HHS	WOSCOPS	AFCAPS/TexCAPS
No. participants	3806	4081	6595	6605
% women	0	0	0	15
Age: mean (yrs)	48	47	55	59
Age: range (yrs)	35–59	40–55	45–66	45–73 (men)
				55–73 (women)
Race				
% black	NA	0	0	3
% hispanic	NA	0	0	7
% white	96	100	100	89
Lipid profile				
Mean LDL-C/HDL-C	4.8	4.3	4.4	4.2
Mean total-C/HDL-C	NA	NA	NA	6.1
Median triglycerides	159	179	164	158
Risk factors				
% hypertensive	0	15	15	22
% current smokers	38	35	35	12
% diabetics	0	2	1	2
Mean systolic/diastolic blood pressure (mm Hg)	121/80	141/90	136/84	138/78

LRC-CCPT = Lipid Research Clinics - Coronary Primary Prevention Trial (16); HHS = Helsinski Heart Study (17); WOSCOPS = West of Scotland Coronary Prevention Study (18).

NA = not available

tal and LDL cholesterol concentrations (16,17). In the LRC-CPPT, the mean age of participants was 47.8 years with an upper limit of 59 years. The mean total, LDL, and HDL cholesterol concentrations at baseline (prior to diet therapy) were 292 mg/dL, 216 mg/dL, and 45 mg/dL, respectively (16). In the Helsinki Heart Study, the mean age was 47.3 years with an upper limit of 55 years. The mean baseline lipid values were total cholesterol, 270 mg/dL; LDL-C, 188 mg/dL; and HDL-C, 47 mg/dL (17).

The West of Scotland Coronary Prevention Study (WOSCOPS), the first primary prevention trial to use a statin, also included white, middle-aged men with elevated total and LDL cholesterol concentrations. Although none of WOSCOPS participants had a history of myocardial infarction at baseline, 16% had evidence of vascular disease: 5%, evidence of angina pectoris; 3%, evidence of intermittent claudication; and 8%, a minor ECG abnormality. In WOSCOPS, the mean age was 55.2 years with an upper limit of 64 years, and the mean baseline lipid values were total cholesterol, 272 mg/dL; LDL-C, 192 mg/dL; and HDL-C, 44 mg/dL (17). Each of these trials reported statistically significant reductions in the primary endpoint of the combined incidence of nonfatal MI and CHD death: the risk reductions were 19% in the LRC-CPPT, 34% in the Helsinki Heart Study, and 31% in WOSCOPS (16–18). Extrapolation of the results of these three trials of middle-aged, white men with high concentration of cholesterol to the general population, particularly to women, older individuals, and individuals with lower levels of cholesterol, has remained a matter of debate and controversy (19). The positive results with lovastatin treatment in AFCAPS/TexCAPS extends the proven benefit reported in earlier trials in high-risk groups to a much broader segment of the general, healthy population.

That is, the AFCAPS/TexCAPS cohort was generally healthy: 12% smoker, 22% hypertensive. Thirty-four percent of the participants had cholesterol levels <240 without two or more risk factors, and would not have qualified for a fasting lipid measurement by current guidelines of the second Adult Treatment Panel (ATP II) of the National Cholesterol Education Program (NCEP). Only 17% would have been recommended for treatment with combination drug and dietary therapy by NCEP ATP II (Table 8).

The number of persons in the U.S. population meeting the AFCAPS/TexCAPS main eligibility criteria were estimated. The estimate was based on data from the NHANES III cohort of men aged 45–73 and women aged 55–73 with similar lipid values who were free of cardiovascular disease, uncontrolled hypertension, and insulin-treated diabetes. According to this estimate, 8 million Americans met the AFCAPS/TexCAPS inclusion criteria (11). NHANES III data and main study inclusion/exclusion criteria were used to estimate the number of persons in the United States who would be eligible for previous primary prevention trials: 1.1 million Americans met the LRC-CPPT criteria, 4.2 million met

Table 8 Participants Eligible for Specific Treatment Based on NCEP
Adult Treatment Panel II Guidelines

Type of treatment	AFCAPS/TexCAPS N (%)	U.S. Population N (%)
Diet alone	3217 (48)	9,524,188 (22%)
Diet and drug therapy	1140 (17)	8,142,104 (19%)
No treatment	2248 (34)	20,125,594 (46%)
Not classified	—	5,902,354 (14%)[a]

[a]Not classified = LDL-C data not available.

the Helsinki Heart Study criteria, and 6.6 million met the WOSCOPS entry criteria (16–18). Thus AFCAPS/TexCAPS demonstrates the efficacy of lipid lowering in a different population and extends the potential benefit of cholesterol lowering to a much larger portion of the population. If one assumes that only 17% of the 8 million Americans similar to the AFCAPS/TexCAPS cohort would currently qualify for drug treatment, then the positive study results suggest an additional 6.6 million individuals may benefit from pharmacologic lipid modification.

It should be noted that the clinical benefit observed in AFCAPS/TexCAPS in patients with average LDL-C and below average HDL-C reaffirms the ATP II's emphasis on LDL-C reduction as the primary goal of prevention therapy. The reduction in risk associated with lovastatin treatment was independent of baseline LDL-C and was apparent across all LDL-C tertiles. The risk reduction achieved in AFCAPS/TexCAPS was independent of baseline LDL-C and HDL-C. This demonstrates that benefit is conferred even in those subjects in the lowest LDL-C tertile. The benefit was apparent across subgroups, including women, those older than the median age for men and women, diabetics, hypertensives, and smokers. Treatment appeared to neutralize the risk conferred by hypertension, smoking, and low HDL-C (Fig. 5).

The inclusion of unstable angina as a component of the primary endpoint was another novel feature of AFCAPS/TexCAPS and allows the study to be more reflective of the most common initial presentation of CHD than previous studies. The results show a significant benefit across the spectrum of clinical events that are the most frequent initial manifestations of atherosclerotic cardiovascular disease. The early difference between the placebo and lovastatin groups and the proportionally greater reduction of unstable angina than stable angina suggest that lovastatin therapy resulted in stabilization of atherosclerotic plaques in healthy adults with subclinical disease.

Lovastatin treatment was well tolerated, with few reports of drug-related adverse events. The issue of safety and drug tolerance is particularly important in primary prevention, where the risks of chronic drug therapy must be considered in the context of the potential benefit of such therapy. Although longer-term safety data is unavailable, the experience in AFCAPS/TexCAPS suggests that the significant CHD risk reduction and the relatively low occurrence of adverse events may warrant further discussion of drug treatment as a therapeutic option in individuals who are similar to participants in this trial.

V. CONCLUSIONS

To the best of our knowledge, AFCAPS/TexCAPS is the first trial to demonstrate that lipid modification with lovastatin, combined with lifestyle measures, can prevent first acute major coronary events in persons who are at relatively low risk for developing CHD and who do not have severe elevations of blood cholesterol. The risk for such events was reduced by 37% with treatment. Among the secondary endpoints, lipid modification significantly reduced the risk for fatal or nonfatal MI by 40%, unstable angina by 32%, and the need for revascularization by 33%. Total fatal and nonfatal coronary events and fatal and nonfatal cardiovascular events were both reduced by approximately 25%. Treatment was safe and well tolerated over the course of the study. The clinical benefit observed in this cohort with such treatment may prove applicable to a substantial portion of the U.S. population who are generally not considered to be at increased high risk for CHD and for whom the optimal strategy for lipid management is controversial.

ACKNOWLEDGMENTS

The authors wish to acknowledge the editorial assistance of Mr. Jesse Jou in the preparation of this manuscript.

REFERENCES

1. Kannel WB. Range of serum cholesterol values in the population developing coronary artery disease. *Am J Cardiol* 1995;76:69C–77C.
2. Anderson KM, Castelli WP, Levy D. Cholesterol and mortality 30 years of follow-up from the Framingham study. *JAMA* 1987;257:2176–2180.
3. Neaton JD, Blackburn H, Jacobs D, Multiple Risk Factor Intervention Trial research

group. Serum cholesterol level and mortality findings for men screened in the Multiple Risk Factor Intervention Trial. *Arch Intern Med* 1992;152:1490–1500.

4. National Cholesterol Education Program Expert Panel. Summary of the second report of the National Cholesterol Education Program (NCEP) Expert Panel on detection, evaluation and treatment of high blood cholesterol in adults (Adult Treatment Panel II). *JAMA* 1993;269:3015–3023.

5. Castelli WP, Garrison RJ, Wilson PWF, Abbott RD, Kalousdian S, Kannel WB. Incidence of coronary heart disease and lipoprotein cholesterol levels: the Framingham study. *JAMA* 1986;256:2835–2838.

6. Schwartz RS, Jackson WG, Celio PV, Richardson LA, Hickman JR Jr. Accuracy of exercise [201]TI myocardial scintigraphy in asymptomatic young men. *Circulation* 1993;87:165–172.

7. Gordon DJ, Probstfield JL, Garrison RJ, et al. High-density lipoprotein cholesterol and cardiovascular disease: Four prospective American studies. *Circulation* 1989;79:8–15.

8. *Statistical Abstract of the United States* (115th edition). 1996:89.

9. Whitney EJ, Shear CL, Mantell G, et al. The case for unstable angina pectoris as a primary endpoint in primary prevention studies. *Am J Cardiol* 1992;70:738–743.

10. Downs JR, Beere PA, Whitney E, et al. Air Force/Texas Coronary Atherosclerosis Prevention Study (AFCAPS/TexCAPS); design and rationale. *Am J Cardiol* 1997;80:287–293.

11. Department of Health and Human Services (DHHS) National Center for Health Statistics. *Third National Health and Nutrition Examination Survey, 1988-94, NHANES III Examination Data File* (CD-ROM). Public use data file documentation No. 76200. Hyattsville, MD: Centers for Disease Control and Prevention, 1996.

12. Jonas HA, Kronmal RA, Psaty BM, et al. Current estrogen-progestin and estrogen replacement therapy in elderly women: association with carotid atherosclerosis. *Ann Epidemiol* 1996;6:314–323.

13. Downs JR, Clearfield M, Weis S, et al. Extending the benefit: primary prevention of acute major coronary events with lovastatin in men and women with average cholesterol. Results of the Air Force/Texas Coronary Atherosclerosis Prevention Study (AFCAPS/TexCAPS). *JAMA* 1998:279:1615–1622.

14. MEDSTAT Groups Inc. DRG Guide: *Descriptions and Normative Values.* Ann Arbor, MI: 1994.

15. Gotto AM. *AFCAPS/TexCAPS Resource Utilization.* Presented at the American College Cardiology, March 1998.

16. Lipid Research Clinics Program. The Lipid Research Clinics Coronary Primary Prevention Trial Results. I. Reduction in incidence of coronary heart disease. *JAMA* 1984;251:351–364.

17. Frick MH, Elo O, Haapa K, et al. Helsinki Heart Study: primary-prevention trial with gemfibrozil in middle-aged men with dyslipidemia. *N Engl J Med* 1987;317:1237–1245.

18. Shepherd J, Cobbe SM, Ford I, et al. Prevention of coronary heart disease with pravastatin in men with hypercholesterolemia. N Engl J Med 1995;333:1301–1307.

19.   Garber AM, Browner WS, Hulley SB. Clinical guideline, part 2: cholesterol screening in asymptomatic adults, revisited. *Ann Intern Med* 1996;124:518–531.

APPENDIX: COMMITTEE MEMBERS (AFCAPS/TexCAPS)

Steering Committee
Antonio M. Gotto, M.D., D. Phil.,
  Chairman
Peter H. Jones, M.D.
Heremiah Stamler, M.D.
Lt. Col. Daniel J. Donovan, M.D.

Lt. Col. John R. Downs, M.D.,
  Principal Investigator
Michael Clearfield, D.O.
Edwin Whitney, M.D.
Stephen Weis, D.O.
Polly Beere, M.D., Ph.D. (non-voting)

Data and Safety Monitoring Committee
John S. de Cani, Ph.D., chairman
H. Alfred Tyroler, M.D.
Phillip D. Houck, M.D. (non-voting)
Donald B. Hunninghake, M.D.
David Schottenfeld, M.D., M.S.
  (consultant)
Evan A. Stein, M.D.
Deborah R. Shapiro, D.P.H. (non-voting)

Endpoint Classification Committee
William B. Kruyer, M.D., Chairman
John Farmer, M.D.
Steven Minor, M.D.

Publication Committee
Antonio M. Gotto, M.D., D. Phil.,
  Chairman
John de Cani, Ph.D.
Lt. Col. John R. Downs, M.D.
Michael Clearfield, D.O.
Stephen Weis, D.O.
William B. Kruyer, M.D.
Evan Stein, M.D., Ph.D.
Deborah Shapiro, D.P.H.
Polly Beere, M.D., Ph.D.
Alexandra Langendörfer (Secretary)

Central Laboratory
Glen E. Mott, Ph.D.
Lt. Col. Jo Haga
Joyce Gray
Gary L. Myers, Ph.D.
Evan Stein, M.D.
Wei-Chung Joe Shih, Ph.D.

Central ECG Laboratory
Maj. David N. Pederson, M.D.

Clinical Centers
Wilford Hall
Texas College of Osteopathic Medicine

7

Long-Term Intervention with Pravastatin in Ischemic Disease (LIPID) Study

Clinical Implications for Cardiovascular Practice

Andrew M. Tonkin
National Heart Foundation of Australia, West Melbourne, Victoria, and the LIPID Study Group, Australia

David Hunt
Royal Melbourne Hospital, Melbourne, Victoria, and the LIPID Study Group, Australia

I. INTRODUCTION

The Long-term Intervention with Pravastatin in Ischaemic Disease (LIPID) study was designed at a time when cholesterol-lowering therapy was controversial, even for secondary prevention of coronary events in patients with preexisting coronary heart disease (CHD). Most of the early intervention studies had been conducted in patients with modest hypercholesterolemia with interventions tested for short periods. Although these trials collectively demonstrated that an average cholesterol reduction of about 0.6 mmol/L (23 mg/dL) resulted in approximately 10% reduction in CHD deaths, an insignificant increase in non-CHD deaths due to cancer and trauma was also observed (1–3). Considerable doubt was cast on the overall benefits of cholesterol lowering with regard to all-cause mortality in patients with high cholesterol levels (4). There was also a paucity of data concerning the effects of more substantial cholesterol lowering, particularly in patients with relatively normal cholesterol levels.

In the late 1980s, the availability of more potent cholesterol-lowering agents such as the 3-hydroxy-3-methylglutaryl coenzyme A (HMG-CoA) reductase inhibitors led to several large-scale clinical trials, including the LIPID study. In 1989, the LIPID study was initiated to investigate the effects of substantial cholesterol lowering with pravastatin on death from CHD in patients with a history of myocardial infarction (MI) or unstable angina pectoris (UAP) and relatively normal cholesterol levels (5,6). Originally, an enrollment of 7000 patients was planned to detect a statistically significant difference in the CHD mortality rate. After review of the initial patients, the death rate was found to be less than assumed, and the patient cohort was extended to include approximately 9000 patients. In 1997, LIPID was prematurely stopped after an interim analysis demonstrated that pravastatin therapy significantly reduced total mortality (6).

Since the LIPID study began, two other large-scale clinical trials with HMG-CoA reductase inhibitors have been completed and reported: the Scandinavian Simvastatin Survival Study (4S) (7), and the Cholesterol and Recurrent Events (CARE) study (8,9). The 4S study demonstrated a significant reduction in total mortality in patients with CHD with overall higher cholesterol levels than those enrolled in LIPID (7). The CARE study demonstrated a significant reduction in the combined endpoint of CHD death and nonfatal MI with pravastatin therapy (9). The CARE study, however, was not sufficiently powered to detect a significant effect on coronary or total mortality. Hence, the effect of cholesterol lowering with pravastatin on coronary and total mortality in patients with relatively normal cholesterol levels remained unknown. This underlies the value of results from the LIPID study.

II. OBJECTIVES AND ENDPOINTS; STUDY DESIGN; BASELINE CHARACTERISTICS OF PATIENTS

A. Objectives and End Points

The primary objective of LIPID was to assess whether treatment with pravastatin in patients with a history of acute MI or hospitalization for UAP and cholesterol levels of 155 to 271 mg/dL (4.0 to 7.0 mmol/L) would reduce coronary mortality over a period >5 years (5,6). Coronary deaths were further classified as due to fatal MI, sudden cardiac death, death after coronary revascularization, or death in hospital after a possible MI, due to heart failure or other coronary event. It is important to note that LIPID (5,6), unlike 4S (7) and CARE (8,9), is the first clinical trial with an HMG-CoA reductase inhibitor that deliberately included patients with a history of hospitalization for UAP. This condition is a more frequent cause for hospital admission than MI in Western countries (10).

Secondary objectives were to determine the effect of treatment on:

1. Total mortality (further classified as due to CHD, other cardiovascular disease, cancer, trauma, or unspecified cause)
2. Stroke (total and nonhemorrhagic)
3. Cardiovascular mortality

4. Incidence of acute MI (fatal and nonfatal)
5. Coronary revascularization (coronary artery bypass surgery [CABG], percutaneous transluminal coronary angioplasty [PTCA], or both
6. Days in the hospital
7. Serum lipid fractions and the relationship between changes in these fractions to cardiovascular endpoints (5,6).

B. Eligibility Criteria and Study Design

Patients between the ages of 31 and 75 years who had a history of acute MI or hospitalization for UAP within the previous 3 months to 3 years and who gave informed consent were registered in the study (5,6). Patients were excluded if they had had a significant medical or surgical event in the previous 3 months, were unavailable for long-term follow-up, had significant cardiac failure (New York Heart Association class III or IV), significant renal or hepatic disease, or uncontrolled endocrine disease, or if they required treatment with other lipid-lowering agents, cyclosporine, or investigational drugs.

The study design of LIPID is illustrated in Figure 1. Patient recruitment began in April 1990 and continued until December 1992. After registration, patients entered a single-blind, run-in phase for 8 weeks, during which time they received placebo and advice on following a diet low in fat and cholesterol. Following the run-in phase, patients with a fasting serum total cholesterol of 155 to 271 mg/dL (4.0 to 7.0 mmol/L) and serum triglycerides <445 mg/dL (< 5.0 mmol/L) measured at a core laboratory after 4 weeks were eligible to enter the study. If at this point patients wished to continue in the study and demonstrated 80%

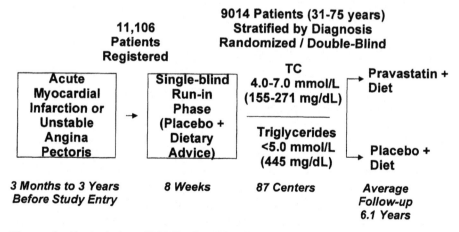

Figure 1 Study design of LIPID. (Modified from Ref. 5.)

compliance with single-blind placebo, and if their usual doctors were uncertain whether lipid-lowering therapy was indicated, they were randomized to treatment with either pravastatin or placebo. Patients were randomized according to a random block design, with stratification based on diagnosis of acute MI or UAP. Patients who were diagnosed with both conditions within the 3-month to 3-year qualifying interval were stratified to the acute MI group. Two-thirds of the cohort were to have acute MI, and one-third was to have UAP.

Of the 11,106 patients who were registered, 9014 patients were randomized (5958 from Australia and 3056 from New Zealand) from a total of 87 centers. This included 473 patients with initial cholesterol levels that were out of range, but who met criteria by continuing on a low-cholesterol diet for an additional 4 weeks.

Patients randomized to drug treatment received either pravastatin 40 mg (two 20-mg tablets) or matching placebo. In addition, all patients were given dietary advice conforming to National Heart Foundation of Australia recommendations, with the goal of restricting fat intake to <30% of total energy intake and dietary cholesterol to <300 mg/day. All lipid assays (for total cholesterol, high-density lipoprotein [HDL] cholesterol, triglycerides, and calculated values of low-density lipoprotein [LDL] cholesterol) were conducted at a central lipid laboratory.

The study design of LIPID was noteworthy in that it was "pragmatic." Patients were not treated simply according to protocol but, rather, received individualized treatment at the discretion of their doctor. This allowed for changes in lipid treatment as new recommendations came to light following the published results of the 4S (7) and CARE (9) studies. Treating doctors were informed of the results of both these studies, and if at any time during the trial they felt it was in the best interest of the patient to start lipid-lowering therapy, they were allowed to do so (11).

1. Treatment Monitoring and Patient Management

A patient review, including measurement of serum alanine transaminase, occurred every 6 months throughout the study. Fasting blood samples for measurement of total cholesterol were collected at 6 months and yearly after randomization. LDL cholesterol, HDL cholesterol, triglycerides, apoprotein (Apo) B, and Apo Al were measured only at years 1, 3, and 5 and at the end of the study. Based on the patient's lipid profile, dose reduction or discontinuation of the study medication was recommended for patients with a persistent low total cholesterol (116 mg/dL [< 3 mmol/L]). For patients that developed a persistently elevated cholesterol (290 mg/L [> 7.5 mmol/dL]), investigators were advised to consider additional dietary measures, other cholesterol-lowering therapy of the clinician's choice, or both. If serum alanine transaminase was elevated significantly (more than three times the upper limit of normal) or the patient complained of muscle pain and had an elevated serum creatine kinase (more than four times the upper limit of normal), treatment with the study medication was to be discontinued immediately (5,6).

2. Classification and Review of Outcomes

All deaths and acute MIs were reviewed by an outcome assessment committee using hospital records, death certificates, autopsy reports, and physicians' notes. The committee had no knowledge of patients' randomized treatment. Definite acute MI was diagnosed by the presence of at least two new pathological Q waves in adjacent electrocardiographic leads or the presence of two of the following three criteria: ischemic chest pain lasting for at least 15 min, an elevated serum creatinine kinase or its MB isoenzyme greater than twice the upper limit of normal, or evolutionary ST-T wave changes over at least 1 day in at least two related electrocardiographic leads. A stroke adjudication committee reviewed records, imaging results, and autopsy reports to confirm stroke and determine subtype of stroke (6).

An independent safety and data monitoring committee reviewed study progress every 6 months. Five formal interim analyses were planned to assess the effects of pravastatin treatment on total mortality and to monitor for serious adverse events. A difference of at least 3 standard deviations ($P < .003$) in either of these two outcomes had to be detected to recommend stopping the trial early (5,6).

3. Statistical Analysis

The trial was initially designed to recruit 7000 patients, assuming that pravastatin would lower cholesterol by 25% compared with placebo. This assumption allowed for the detection of an actual difference of only 18% due to dropouts (discontinuation of study medication) or drop-ins (the initiation of other open-label lipid-lowering treatment in placebo-randomized patients). The original intent was to detect an 18.3% reduction in 5-year coronary mortality with 80% power using a two-sided 5% level test. The 18.3% reduction in coronary mortality translated to a total of 700 deaths due to CHD, at which point the trial was planned to end, unless stopped early by interim analysis. The sample size was increased to 9000 patients after a 1-year review found that the CHD death rate was less than anticipated. Statistical analyses were performed on an intention-to-treat basis, using two-sided P values (5,6).

The effects of pravastatin on the primary endpoint, coronary mortality, and secondary endpoints were analyzed by the log-rank test stratified by qualifying event. A Cox proportional hazards model was used as the basis for estimates of relative risk reduction and 95% confidence intervals. For subgroup analyses, the prespecifed endpoint was the combined endpoint of coronary mortality and non-fatal MI. The intent of subgroup analyses was to examine whether there was variation in treatment effect across the various subgroups, treating age and baseline lipid values as continuous variables (the latter calculated as the average of two lipid values taken during the run-in phase) (5,6).

C. Baseline Patient Characteristics

Baseline demographics, including entry diagnosis, for both the placebo and pravastatin groups are summarized in Table 1. The majority of patients were male, and the median age for both treatment groups was 62 years. A large proportion of the study population was elderly, with more than 3500 patients (39%) >65 years old and more than 1300 (15%) >70 years of age. Women made up 17% (1516) of the LIPID study population, representing the largest cohort of women in any HMG-CoA reductase inhibitor trial to date. The qualifying event was acute MI in 5754 (64%) of patients and hospitalization for UAP in 3260 (36%) patients.

Baseline risk factors and disease states of patients enrolled in LIPID, such as smoking, hypertension, diabetes, and obesity, are summarized in Table 2. Approximately 9.6% of patients were current smokers; 63.6% were ex-smokers. Other manifestations of ischemic heart disease such as stroke, transient ischemic attack, and claudication were present in similar numbers of patients in both treatment groups.

Forty-one percent of patients had had a coronary revascularization procedure (CABG, PTCA, or both). This large percentage of patients reflects current

Table 1 Baseline Demographics/Entry Diagnosis in LIPID

Baseline demographics	Placebo (n = 4502)	Pravastatin (n = 4512)
Age (years)		
Median (25%, 75%)	62 (55, 68)	62 (55,67)
65–69 years	24%	24%
>70 years	15%	15%
Gender		
Male	83%	83%
Female	17%	17%
Qualifying event		
AMI	64%	64%
UAP	36%	36%

AMI = acute myocardial infarction; UAP = unstable angina pectoris.
Source: Ref. 6.

Table 2 Baseline Risk Factors and Disease States in LIPID

Baseline risk factors and disease states	Placebo (n = 4502)	Pravastatin (n = 4512)
Smoker	10%	9%
Ex-smoker	63%	65%
Hypertension	42%	41%
Diabetes	9%	9%
Obesity (BMI>30)	18%	18%
Stroke	4%	4%
TIA	4%	3%
Claudication	10%	10%

BMI = body mass index; TIA = transient ischemic attack.
Source: Ref. 6.

medical practice in both Australia and New Zealand and is similar to that observed in CARE (54% of patients) (9).

Many patients in LIPID were also taking concurrent cardiovascular medications for preexisting CHD. Aspirin, beta blockers, nitrates, and ACE inhibitors were being given to 82%, 47%, 37%, and 16%, respectively, of patients at baseline.

Baseline lipid values (median values and median values for the top and bottom quartiles) for both the placebo and pravastatin treatment groups are summarized in Table 3. Baseline lipid values were calculated as the average of two prerandomized values obtained during the run-in phase of the protocol. The median values in Table 3 represent the cholesterol levels of the majority of patients in the general population having CHD. A major difference between LIPID (5,6) and the 4S trial (7) is that 42% of patients randomized in LIPID had a total cholesterol level <213 mg/dL (< 5.5 mmol/L), the lower entry level for 4S.

III. DESIGN OF LIPID AND OTHER MAJOR HMG-COA REDUCTASE INHIBITOR TRIALS

LIPID is the largest clinical trial conducted with an HMG-CoA reductase inhibitor. Whereas LIPID was designed primarily to assess the effects of pravastatin on CHD death, CARE (9) examined the effects of pravastatin in post-MI patients with average cholesterol levels (serum total cholesterol <240 mg/dL [6.2 mmol/L]) on the combined endpoint of fatal CHD and nonfatal MI. The other recently completed major secondary prevention study, 4S, evaluated the effects of simvastatin on total mortality in patients with elevated cholesterol and previous MI or angina pectoris (7).

Table 3 Qualifying Serum Lipid Levels of 9014 Patients Randomized to Pravastatin or Placebo in LIPID

| Baseline Lipid Values | Median (25%, 75%), mg/dl [mmol/L] | |
	Placebo (n = 4502)	Pravastatin (n = 4512)
Total cholesterol	218 (196, 240) [5.64 (5.07, 6.21)]	218 (196, 241) [5.64 (5.07, 6.23)]
LDL-C	150 (131, 170) [3.88 (3.39, 4.40)]	150 (130, 170) [3.88 (3.36, 4.40)]
HDL-C	36 (31, 42) [0.93 (0.80, 1.09)]	36 (31, 41) [0.93 (0.80, 1.06)]
Triglycerides[a]	138 (105, 188) [1.56 (1.19, 2.12)]	142 (104, 196) [1.60 (1.17, 2.21)]
TC/HDL-C ratio	6.07 (5.12, 7.14)	6.11 (5.13, 7.16)

[a]Except for baseline triglyceride levels ($P = .03$), there were no significant differences between the groups.
LDL-C = low-density lipoprotein cholesterol; HDL-C = high-density lipoprotein cholesterol; TC = total cholesterol.
Source: Ref. 5.

Two landmark primary prevention studies using HMG-CoA reductase inhibitors are the West of Scotland Coronary Prevention Study (WOSCOPS) (12) and the Air Force/Texas Coronary Atherosclerosis Prevention Study (AFCAPS/ TexCAPS) (13,14). WOSCOPS examined the effects of pravastatin in men with no history of MI with hypercholesterolemia on the combined incidence of nonfatal MI and CHD death (12). The AFCAPS/TexCAPS study evaluated the effects of lovastatin on the prevention of a first acute major coronary event, defined as sudden cardiac death, fatal or nonfatal MI, and UAP, in men and women with average total cholesterol levels and LDL cholesterol levels, below-average HDL cholesterol levels, and no clinical evidence of atherosclerosis at study entry (13,14).

LIPID (5,6), like CARE (8,9), was designed to assess the effects of pravastatin on patients having relatively normal cholesterol levels. This was intentional, since most patients with CHD have cholesterol levels that are not significantly elevated, enabling the results of LIPID to be extrapolated to the general population seen in clinical practice (15).

Unlike 4S (7) and CARE (9), LIPID was the only trial including patients with a history of UAP (Fig. 2). This is meaningful because UAP is becoming

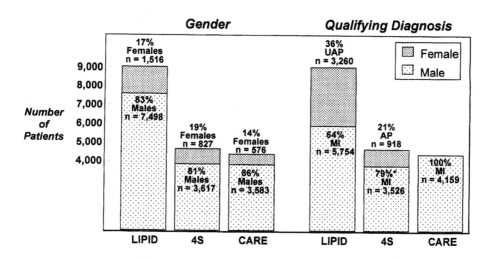

*62% AMI only, 17% both AMI and angina

Figure 2 Breakdown of patients in LIPID (6), 4S (7), and CARE (9) by qualifying diagnosis.

increasingly important as a cardiovascular health issue (6). LIPID also had the largest number of women in its study population (Fig. 2) and included a large number of diabetics and older patients. Including these patient subgroups provided a more accurate representation of patients with CHD seen in clinical practice.

IV. RESULTS OF LIPID

A. Effects on Lipid Parameters

Treatment with pravastatin over a 5-year period produced beneficial changes in the lipid profile (6). With analysis on an intention-to-treat basis and averaging changes in lipid fractions over 5 years, treatment with pravastatin resulted in a decrease in plasma total cholesterol from 218 mg/dL (5.6 mmol/L) of 39 mg/dL (1.0 mmol/L) (an 18% reduction when compared with placebo); a decrease in LDL cholesterol from 150 mg/dL (3.9 mmol/L) of 25%, a decrease in triglycerides from 142 mg/dL (1.6 mmol/L) of 11%, and an increase in HDL cholesterol from 36 mg/dL (0.9 mmol/L) of 5%.

B. Effects on Clinical Events

On an intention-to-treat basis, pravastatin treatment resulted in a 24% relative risk reduction in CHD mortality, the primary endpoint ($P = .0004$). Among the secondary endpoints, pravastatin reduced the relative risk of total mortality by 22% ($P = .00002$) (Fig. 3) and total stroke by 19% ($P = .048$) (Fig. 4). For the combined endpoint of fatal CHD and nonfatal MI, pravastatin resulted in a 24% risk reduction in CHD events, which was highly statistically significant ($P = .000001$) (Fig. 5). In terms of absolute risk reduction and clinical benefit, 30 deaths, 28 nonfatal MIs, and 9 nonfatal strokes were avoided (allowing for multiple events per patient) in 48 patients for every 1000 patients treated with pravastatin over an average of 6.1 years.

For other secondary endpoints, pravastatin resulted in a lower rate of coronary revascularization procedures (22% reduction in CABG [$P = .0003$], 19% reduction in PTCA [$P = .024$], and 20% reduction in the risk of either procedure [$P < .001$]). Treatment with pravastatin also reduced the need for hospitalization for unstable angina by 12% ($P = .005$).

These results provide strong evidence that treating patients with a history of MI or UAP and relatively normal cholesterol levels with pravastatin reduces the risk of death from CHD and from cardiovascular disease as well as from all causes combined. Pravastatin also significantly reduces the risk of MI and stroke.

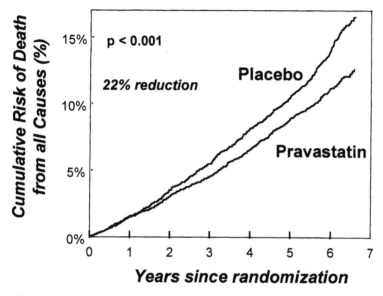

Figure 3   Kaplan-Meier estimate of total mortality from all causes. (From Ref. 6.)

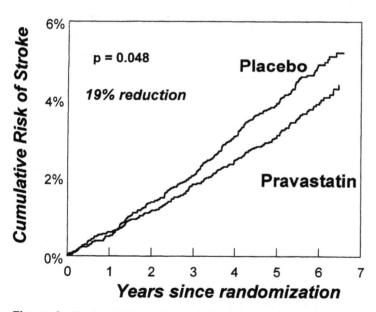

Figure 4   Kaplan-Meier estimate of risk reduction for total stroke. (From Ref. 6.)

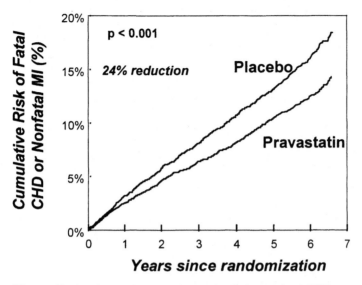

Figure 5   Kaplan-Meier estimate of death due to fatal CHD or nonfatal MI. (From Ref. 6.)

C.   Effects in Patient Subgroups

Subgroup analyses of the combined endpoint of CHD mortality and nonfatal MI were conducted for prespecified subgroups defined by sex, qualifying event (AMI or UAP), age, entry lipid levels, and other risk factors for CHD (6). No evidence of heterogeneity was detected for any of these subgroups ($P > .08$ for all analyses of heterogeneity of treatment effect). The reduction in risk with pravastatin in each subgroup was consistent with the risk reduction for the entire cohort. Pravastatin treatment was associated with a significant reduction in CHD events whether the qualifying diagnosis was acute MI or UAP. There was a significant reduction in CHD events in patients with baseline plasma total cholesterol <213 mg/dL (5.5 mmol/L) (6).

D.   Safety and Tolerability

Over the 6.1 years of follow-up, pravastatin treatment appeared to be safe and well tolerated with no evidence of an increased risk of death due to non-cardiovascular causes, such as cancer, trauma, accidents, or violence (fatal or hospitalized episodes) (6). As for cancer, there were 403 newly diagnosed primary cancers in the pravastatin group (in 379 patients) and 417 newly diagnosed cancers in the placebo group (in 399 patients), a difference that was not statistically significant (6). The incidence of specific types of cancer, including breast cancer, also was not significantly different.

Overall there was no significant difference in the incidence of adverse events (3.2% vs. 2.7% attributed to pravastatin and placebo, respectively) or in the incidence of any serious adverse events (6).

Specifically, with respect to laboratory parameters measured to ascertain safety, an increase in serum alanine aminotransferase greater than three times the upper limit of normal was reported in 96 (2.1%) of patients in the pravastatin group, compared with 80 (1.9%) in the placebo group (6). This difference was not statistically significant ($P = .41$). Similarly, there were no significant differences in the numbers of patients with elevation of serum creatine kinase or skeletal myopathy or with an adverse event from hepatic disease (6).

E. Factors Influencing the Applicability of the Results of LIPID

Possible biases that may have reduced or inflated the treatment effect observed with pravastatin must be considered when evaluating the relative and absolute estimates of treatment effect. In LIPID, the large number of patients randomized to pravastatin who discontinued treatment (dropouts) and those patients randomized to placebo who commenced treatment with open-label cholesterol-lowering medication (drop-ins) would have reduced the average difference in cholesterol levels between the two groups (6). This reduction could have resulted in a reduced difference in the incidence of clinical events between two groups, such that the observed benefits with pravastatin treatment were actually diluted. At the midpoint of the trial, the crossover rate was 20% (a total of 11% dropouts and 9% drop-ins). The relative risk reduction in CHD events of 24% may have actually been closer to 30% if patients had not been allowed to cross over from either the placebo or treatment arm. However, one caveat may be greater compliance than in usual practice, which would increase the estimate of benefit in usual practice.

A second possible bias is that patients in LIPID with MI or UAP could represent a group of patients with a lower risk than the general population of patients with MI or UAP. The rate of death from CHD in LIPID patients in the placebo group was 1.4% per year (1.5% in patients with a qualifying diagnosis of MI and 1.1% in patients with UAP) (6). If the rate in nonvolunteer patient groups was higher (e.g., 2% per year), as was initially expected for LIPID, a greater absolute benefit would be observed, assuming that the relative treatment effect would be similar for this population. The estimate of the absolute treatment effects may therefore significantly underestimate the actual effects of pravastatin treatment in broad clinical practice.

However, since the study was discontinued early because of a clear benefit with pravastatin, the observed relative risk reduction may have been inflated by chance. There is only a small probability that this occurred, however, since the final analysis was based on 660 CHD deaths. This is only 40 fewer than the total initially planned. It is unlikely that study continuation would have significantly changed the estimates of relative risk (6).

At least 3 months elapsed after the qualifying diagnosis of acute MI or UAP before patients were randomized into the study. A similar time frame was followed in CARE (8,9) and 4S (7). LIPID does not provide data on the effects of pravastatin <3 months after an acute coronary event. However, because the early increase in hazard after acute coronary syndromes subsides after 3 months, it is reasonable to extrapolate the results of LIPID to patients with stable CHD seen in clinical practice (6).

V. CHD IN LIPID

CHD is the leading cause of premature death in the United States and throughout the Western world (16,17). In 1998, the number of patients with CHD in the United States is expected to approach 13.9 million (16). The economic impact is significant, and the direct and indirect costs of CHD in the United States are predicted to be $95.6 billion in 1998 (16). LIPID is the largest endpoint trial with an HMG-CoA reductase inhibitor conducted in patients with a history of CHD, providing invaluable information about the benefits of pravastatin treatment in this patient population. The findings of LIPID support and extend those of CARE, which only included patients who had had an MI (6).

The marked reduction in coronary mortality in LIPID is particularly impressive, since many of the patients enrolled in LIPID were receiving aspirin and other cardiovascular medications (e.g., beta-blockers, ACE inhibitors), reflecting current treatment for patients with CHD. Also, a large percentage had undergone PTCA, CABG, or both at baseline. Thus, the benefits of pravastatin observed in the study were in addition to those provided by contemporary medical and interventional therapies (6).

A. Unstable Angina in LIPID

Approximately 7.2 million persons in the United States have angina pectoris (16), and, based on data from the Framingham Heart Study, it is estimated that 350,000 new cases will occur each year (16). Unstable angina is associated with a poorer prognosis, and approximately 4% to 5% of patients hospitalized with UAP will die or suffer acute MI within the first month after hospitalization. In addition to being an increasingly important discharge diagnosis, UAP poses a significant economic burden to the United States, with costs per patient estimated at $43,065 in 1995 (18).

LIPID is the only endpoint trial with an HMG-CoA reductase inhibitor that deliberately enrolled patients with a history of hospitalization for UAP. The *P* value for heterogeneity in the analysis of CHD events in subgroups indicates that there was no significant interaction between treatment effect and qualifying diagnosis. Patients with a history of UAP derived a similar benefit from pravastatin as patients with a history of MI (6).

Treatment with pravastatin reduced the incidence of UAP in LIPID patients by 12% (24.6% in the pravastatin group vs. 22.3% in the placebo group, *P* = .005) (6). It is likely that the risk reductions in coronary revascularization procedures (PCTA and CABG) observed with pravastatin treatment in LIPID related at least partly to the reduction in incidence of UAP, since PTCA and CABG are commonly used in the management of patients with UAP.

B. Stroke and LIPID

In the United States, stroke is the third leading cause of death, following death due to CHD and cancer (16). In 1993, stroke accounted for 149,740 deaths in the United States, or 1 of every 15 deaths (16). Based on data from the Framingham Heart Study, 500,000 people are expected to experience a new or recurrent stroke every year. In addition to being highly prevalent, stroke is associated with a high mortality rate (26.6% in 1992) and is the leading cause of long-term disability in the United States (16).

Stroke also poses a severe economic burden. In 1998, the cost of stroke in the United States is predicted to approach $43.3 billion (16). The largest part of this expense ($23.3 billion) comes from hospitalizations or nursing home services (16). Stroke accounts for approximately half of all patients hospitalized for acute neurological disease, and in 1995, 424,000 males and 501,000 females were hospitalized for acute stroke (16). Their average length of hospital stay was 6.7 days (16).

In LIPID, the significant 19% reduction in total stroke (4.5% in the placebo group vs. 3.7% in the pravastatin treatment group) (Fig. 4) (6) was in the context of concommitant use of aspirin and antihypertensive therapy.

These results in LIPID confirm the findings from 4S (7) and from CARE (9), demonstrating the benefits of certain HMG-CoA reductase inhibitors in reducing stroke.

VI. CLINICAL IMPLICATIONS OF LIPID

Cardiovascular disease is already the major public health problem in industrialized nations and, within the next 15 to 20 years, is expected to replace nutritional deficiencies and transmissible diseases as the major global public health problem. For these reasons, preventive cardiology has become a major focus of health care.

Some of the questions to consider when translating the results of the recent HMG-CoA reductase inhibitor trials into clinical practice include:

1. How representative are the trial patients?
2. Are background therapies appropriate?
3. What are absolute rather than relative risk reductions?
4. Is the treatment safe? (19)

A. How Representative Are the Trial Patients?

Data from the Framingham study and other epidemiological studies have provided strong evidence of a continuous, curvilinear association between serum cholesterol levels and risk of CHD (20). Despite this evidence, most patients with CHD have cholesterol levels that are only modestly elevated or are within the normal range (15). Whereas the 4S study (7) had previously demonstrated the benefits of cholesterol lowering in CHD patients with elevated cholesterol levels, the inclusion of CHD patients with a broad "average" range of cholesterol levels in the CARE (8,9) and LIPID (5,6) studies extends the benefits of cholesterol lowering to this group of patients with lower levels, representing the majority seen in clinical practice.

B. Are Background Therapies Appropriate?

Background therapies in patients reflected contemporary medical practice in patients with CHD. In LIPID, for example, 82% of the patients were taking aspirin; 47% beta-blockers, and 16% ACE inhibitors; 41% had undergone a revascularization procedure (PTCA, CABG, or both) before being randomized into the study (5,6). Therefore the observed benefits of pravastatin in LIPID were in addition to those achieved with these other therapies.

C. What Are the Absolute Rather Than Relative Risk Reductions?

The absolute risk in the population under study determines the impact of medical therapies on public health. For cardiovascular disease, event rates are higher in patients with preexisting CHD than in those without established CHD (primary prevention). In approximately one in four patients, however, the outcome of a first coronary event (MI or stroke) is fatal. For this reason, cholesterol lowering must be considered for high-risk patients in both primary and secondary prevention (19).

Because atherosclerosis and CHD relate to multiple variables, the "global" assessment of risk in a patient should encompass all known risk factors for CHD in the future. These include age, sex, smoking, diabetes, being overweight, and other risk factors, as well as whether or not a previous CHD event has occurred.

D. Is the Treatment Safe?

Concerns that cholesterol-lowering therapy may increase the risk of non-CHD death have been allayed by the findings of the recent large-scale HMG-CoA reductase inhibitor trials. Of these trials, LIPID provides the largest database on the safety of one of these agents. More than 85% of the LIPID cohort is being followed, over an extended period of time, on open-label pravastatin to further exclude any associated significant adverse events.

VII. CONCLUSION

Cholesterol-lowering drug therapy has been significantly underutilized in patients with CHD. The results of the recent large-scale clinical trials with HMG-CoA reductase inhibitors suggest that the vast majority of patients with previous acute coronary syndromes should at least be considered for treatment with cholesterol-lowering therapy. LIPID provides the most substantive data to this time showing the benefits of cholesterol-lowering therapy in a cohort representing the majority of patients in clinical practice with CHD (19).

REFERENCES

1. Roussouw JE, Lewis B, Rifkind BM. The value of lowering cholesterol after myocardial infarction. *N Engl J Med* 1990;323:1112–1119.
2. Muldoon MF, Manuck SB, Mathews KA. Lowering cholesterol concentrations and mortality: a quantitative review of primary prevention trials. *Br Med J* 1990; 301:309–314.
3. Peto R, Yusuf S, Collins R. Cholesterol lowering trial results in their epidemiological context. *Circulation* 1985;72(suppl III):III–451.
4. Keech AC. Does cholesterol lowering reduce total mortality? *Postgrad Med J* 1992; 68:870–871.
5. LIPID Study Group. Design features and baseline characteristics of the LIPID (Long-Term Intervention with Pravastatin in Ischaemic Disease) study: a random-

ized trial in patients with previous acute myocardial infarction and/or unstable angina pectoris. *Am J Cardiol* 1995;76:474–479.

6. LIPID Study Group. Prevention of cardiovascular events and death with pravastatin in patients with coronary heart disease and a broad range of initial cholesterol levels. *N Engl J Med* 1998;339:1349–1357.

7. Scandinavian Simvastatin Survival Study Group. Randomized trial of cholesterol lowering in 4444 patients with coronary heart disease: the Scandinavian Simvastatin Survival Study (4S). *Lancet* 1994;344:1383–1389.

8. Sacks FM, Pfeffer MA, Moyé L, et al. Rationale and design of a secondary prevention trial of lowering normal plasma cholesterol levels after acute myocardial infarction: the Cholesterol and Recurrent Events Trial (CARE). *Am J Cardiol* 1991; 68:1436–1446.

9. Sacks FM, Pfeffer MA, Moyé et al. The effect of pravastatin on coronary events after myocardial infarction in patients with average cholesterol levels. *N Engl J Med* 1996;335:1001–1009.

10. Whitney EJ, Shear CL, Mantell G, Lydick E, Wolf R. The case for unstable angina pectoris as a primary end point in primary prevention studies. *Am J Cardiol* 1992;70:738–743.

11. Tonkin AM on behalf of LIPID Study Group. Management of the Long-term Intervention with Pravastatin in Ischemic Disease (LIPID) Study after the Scandinavian Simvastatin Survival Study (4S). *Am J Cardiol* 1995;76:107C–112C.

12. Shepherd J, Cobbe SM, Ford I, et al. Prevention of coronary heart disease with pravastatin in men with hypercholesterolemia. *N Engl J Med* 1995;333:1302–1307.

13. Downs JR, Beere PA, Whitney W, et al. Design and rationale of the Air Force/Texas Coronary Atherosclerosis Prevention Study (AFCAPS/TexCAPS). *Am J Cardiol* 1997;80:287–293.

14. Downs JR, Clearfield M, Weis S, et al. Primary prevention of acute coronary events with lovastatin in men and women with average cholesterol levels: results of AFCAPS/TexCAPS. *JAMA* 1998;279:1615–1622.

15. Kannel WB. Range of serum cholesterol values in the population developing coronary artery disease. *Am J Cardiol* 1995;76:69C–77C.

16. American Heart Association. *1998 Heart and Stroke Statistical Update.* Dallas: American Heart Association, 1997.

17. Reckless JPD. Cost implications of lipid lowering treatments. *Pharmacoeconomics* 1994;6:310–323.

18. Conti CR. Unstable angina: cost of conservative and invasive strategies using TIMI 3 as a model. *Clin Cardiol* 1995;18:187–188.

19. Tonkin AM. Putting the evidence into practice. Presentation at the 20th Congress of the European Society of Cardiology, Vienna, Austria, Aug. 24, 1998.

20. Stamler J, Wentworth D, Neaton JD. Is the relationship between serum cholesterol and the risk of premature death from coronary heart disease continuous and graded? Findings in 356,222 primary screenees of the multiple risk factor intervention trial (MRFIT). *JAMA* 1986;256:2823–2828.

8

Angiographic Results of Lipid-Lowering Trials

A Systematic Review and Meta-Analysis

Cheryl L. Holmes, Michael Schulzer, and G. B. John Mancini
Vancouver Hospital and Health Sciences Centre, University of British Columbia, Vancouver, British Columbia, Canada

Coronary atherosclerosis leads to myocardial infarction, heart failure, and sudden coronary death. Interventions to reduce morbidity and mortality from heart disease can be targeted to reduce the rate and degree of progression of atherosclerosis, stabilize plaque, and induce regression. Interventions to date have included lipid-lowering medications, surgery, lifestyle changes, calcium channel blockers, and combinations of these. Future trials include interventions to raise HDL cholesterol, estrogen replacement in postmenopausal women, antioxidant therapy, and angiotensin-converting enzyme (ACE) inhibitors. The most promising interventions to date are targeted at lipid lowering.

This chapter is a critical appraisal of the available angiographic trials and includes a summary of trial design, a meta-analysis of the treatment effects, and estimates of the relationship between coronary angiographic change and subsequent clinical events.

I. INTRODUCTION

Quantitative coronary angiography (QCA) is an observer-analyzed or computer-generated means of quantifying the inner dimensions of the coronary artery tree. Computer-assisted quantitative methods are more reliable than other approaches. For example, the standard deviation of measuring percent diameter stenosis

(%DS) can be cut in half or less compared to visual approaches (1). The disadvantage of coronary arteriography is that it is invasive and it yields essentially a "lumenogram." The coronary vessel wall adapts to the development of atherosclerosis by displacing the bulk of the atheroma in an outward direction, preserving the arterial lumen. More recently, vessel wall "shrinkage" has also been noted (2). These mechanisms lead to potential ambiguity in interpretation of angiographic trials purporting to show progression and regression of coronary atherosclerosis (3). Nevertheless, this angiographic approach to data analysis is amenable to both categorical and patient-based classification of patients as progressors or nonprogressors, or to analysis of continuous variables such as the rate of progression. More important, evidence is available linking surrogate, angiographic endpoints with clinical events and prognosis.

Early angiographic trials used a global change score to assess change of paired angiograms: score of 0 meant no change; 1, discernible change; 2, intermediate change; and 3, extreme change. The angiographic endpoints that are now more customarily measured include minimal lumen diameter, mean lumen diameter, and percent diameter stenosis. Both minimum lumen diameter and mean lumen measurements are simple and reproducible. Minimum diameters are perhaps most sensitive but have the drawback in that regression will tend to be minimized if the location of the minimum diameter within a segment is not constant between the baseline and follow-up films. Mean lumen diameter begins to take into account the planar area of atheroma change and can be extrapolated to reflect the volume of atheroma change in the entire coronary tree. But again, it may underestimate regression of atherosclerosis if some areas progress and others regress. Percent diameter stenosis change is more appreciated clinically, but ambiguity of the reference diameter (i.e., where and how to select it and whether it is diseased or normal) makes this less reproducible and also more difficult to interpret than absolute diameter measures. Thus, no single measurement is ideal. The complex vascular remodeling that takes place in the face of atherosclerosis suggests that trials should measure and report as many quantitative parameters as possible. The purpose of this chapter is to try to synthesize the results of multiple angiographic "regression trials" and to show evidence that links angiographic changes with clinical outcome. Other chapters will develop the specific mechanisims by which various interventions may impart changes in clinical outcomes.

II. METHODOLOGY

A Medline search from 1976 to October 1997 was conducted using search parameters "atherosclerosis/ regression/ quantitative angiography/ lipid-lowering drug/ lifestyle/ and randomized controlled trial." In addition, specific journals were hand-searched to find all relevant studies. Seventy-three items were retrieved.

The trials were assessed for the following criteria: blinding; equality of baseline characteristics of study groups; equal treatment of groups other than with respect to the intervention being studied; all subjects accounted for at the end of the study; assessment of dropout rate; intention-to-treat analysis; and clear reporting of statistical methodology. The results of this search are presented in Table 1.

Descriptive and analytic data were extracted and tabulated for each trial: change from baseline for control and treatment groups in lipid profiles, angiographic endpoints and clinical event rates are shown in Tables 1 through 6.

III. DESCRIPTIVE SUMMARY OF THE TRIALS

Our search yielded 23 trials that specifically looked at lipid lowering as an intervention and two trials using calcium channel blockers, which will be included in parts of this analysis. Table 1 outlines the breakdown of the trials by intervention. Note that many of the trials were plagued by the problem of dropout in both the control and the intervention groups. In planning an angiographic endpoint trial it seems conservative to assume that 20% of the randomized subjects will not complete a second angiogram.

The specific interventions for each trial are summarized in Table 2. All of the trials are secondary intervention trials; all of the subjects had clinically or angiographically determined coronary artery disease (CAD). The length of trial ranges from 1 to 9 years. Most of the trials were conducted on men; only one randomized more women than men—the SCOR trial (4).

The definition of angiographic progression and regression are presented in Table 3. Most of the trials measured minimum lumen diameter; some reported percent diameter stenosis, and the earlier trials used a global change score to assess progression and regression.

In the Lifestyle Heart Trial (5) (Lifestyle) published by Ornish et al. in 1990, the experimental group was prescribed a 10% fat, vegetarian diet and moderate aerobic exercise, stress management training, smoking cessation, and group support. In spite of the small numbers in this trial, the authors were able to show that regression of even severe coronary atherosclerosis was possible after only 1 year, without the use of lipid-lowering drugs. A criticism of this trial is that self-selection may have occurred, spoiling the randomization because of the high dropout rate in both the intervention and the control groups (Table 1).

The St. Thomas' Atherosclerosis Regression Study (6) (STARS), published in 1992, had three arms. Men with coronary heart disease were randomized to

Table 1 The Trials

Trial	Year	Validity criteria met	Completed/ randomized	Dropout controls	Dropout treatment
Lifestyle intervention					
Heidelberg (7)	1992	single blind	92/113	9%	28%
STARS (6)	1992	not blinded	50/60	20%	13%
Lifestyle (5)	1990	no: self-selected groups	41/94	56%	58%
Combination drug therapy					
HARP (17)	1994	single blind	79/91	17%	9%
SCRIP (14)	1994	yes	246/300	18%	18%
SCOR (4)	1990	not blinded	72/97	35%	17%
FATS (11)	1990	yes	120/146	13%	27%
CLAS (9)	1987	single blind, diet differed	162/188	13%	15%
Resin trials					
STARS (6)	1992	not blinded, diet differed	48/60	20%	20%
NHLBI (8)	1984	no: baseline differences	116/143	21%	17%
Fibrate trials					
DAIS (27)	1998	yes (results not available)	418		
LOCAT (26)	1997	yes	372/395	5%	6%
BECAIT (25)	1996	yes	81/92	13%	11%

Statin trials					
LCAS (24)	1997	yes	340/429	21%	20%
CIS (28)	1997	yes	205/254	11%	12%
CARS (18)	1997	single blind	80/90	9%	13%
Post CABG (21)	1997	yes	1093/1351	11%	15%
REGRESS (19)	1995	yes	653/885	24%	28%
PLAC I (20)	1995	yes	320/408	22%	21%
CCAIT (15)	1994	yes	299/331	8%	11.5%
MAAS (16)	1994	yes	345/381	11%	8%
MARS (13)	1993	yes	247/270	9%	8%
Surgical therapy					
POSCH (10)	1990	blinding not possible	634/838	28%	21%
Calcium channel blocker monotherapy					
Montreal (30)	1990	yes	335/383	8%	12.5%
INTACT (29)	1990	no: see text	348/425	17%	19%

Table 2 Trial Features

Trial	Treatment	Intervention Control	Y	Inclusion criteria	%M
Lifestyle intervention					
Heidelberg	AHA phase 3 diet, intensive exercise	usual care	1	CAD, LDL <210	100
STARS	diet	usual care	3.3	CAD, TC > 230	100
Lifestyle	vegetarian 10% fat diet, exercise, smoking cessation	usual care	1	CAD	88
Combination drug therapy					
HARP	stepwise drug therapy	diet	25	CAD, TC 180–250	89
SCRIP	low fat diet, exercise, weight loss, step-wise drug therapy	usual care	4	CAD	86
SCOR	colestipol to 30 g, niacin to 7.5 g ± lovastatin to 60 mg	diet ± colestipol	2	CAD, FH, LDL > 200	43
FATS	colestipol 30 g + lovastatin 20 mg; colestipol 30 g + niacin 4g	diet ± colestipol	2.5	Familial CAD, apo B > 125	100
CLAS	intense diet + colestipol 30 g + niacin to 12 g	diet	2	CABG, TC 185–350	100
Resin trials					
STARS	diet + cholestyramine 16 g	usual care	3.3	CAD, TC > 230	100
NHLBI	cholestyramine 24 g + diet	diet	5	CAD, LDL-C > 90th % ile	81

	Treatment	Control	Y	Entry criteria	% M
Fibrate trials					
DAIS	fenofibrate 200 mg + diet	diet	3	NIDDM, LDL 3.5–4.5 mmol	73
BECAIT	bezafibrate 600 mg + diet	diet	5	MI under age 45, dyslipidemia	100
LOCAT	slow release gemfibrozil 1200 mg + diet	diet	2.8	postbypass, HLD < 1.1, LDL < 4.5	100
Statin trials					
LCAS	diet, fluvastatin 40 mg ± cholestyramine	diet ± cholestyramine	2.5	CAD, LDL 115–190	81
CIS	diet, simvastatin, 20–40 g ± resin	diet ± resin	2.3	CAD, TC, 207–350, TG < 330	100
CARS	pravastatin 10 mg	usual	2	CAD, TC 160–220	81
Post CABG	lovastatin to 80 mg ± cholestyramine	lovastatin to 5 mg ± cholestyramine	4.3	CABG, LDL < 130–175	92
REGRESS	diet, pravastatin 40 mg ± cholestyramine	diet ± cholestyramine	2	CAD, TC 155–310	100
PLAC I	pravastatin 40 mg + diet	diet	3	CAD, LDL 130–190	78
CCAIT	lovastatin to 80 mg + diet	diet	2	CAD, TC 220–300	82
MAAS	simvastatin 40 mg + diet	diet	4	CAD, TC 215–310	88
MARS	lovastatin 80 mg + diet	diet	2	CAD, TC 195–295	91
Surgical therapy					
POSCH	ileal bypass	diet	9	Previous MI, TC > 220, LDL > 140	91
Calcium channel blocker monotherapy					
Montreal	nicardipine 90 mg	placebo	2	CAD	82
INTACT	nifedipine 80 mg	placebo	3	CAD	?

LEGEND: Y = Years, AHA = American Heart Association, CAD = coronary artery disease, CABG = coronary artery bypass surgery, M = males, FH = familial hypercholesterolemia, MI= myocardial infarction, TC = total cholesterol (mg/dL), NIDDM = non-insulin-dependent diabetes mellitus, LDL = low density lipoprotein (mg/dL, except where stated).

Table 3 Endpoint Definitions

Trial	Method	Progression/regression
Lifestyle intervention		
Heidelberg	QCA	10 point change in percent diameter stenosis (2 SD)
STARS	QCA	0.17 mm change in mean lumen diameter (2 SD)
Lifestyle	QCA	any change in percent diameter stenosis
Combination drug and diet interevention		
HARP	QCA	7.8 point change in percent diameter stenosis
SCRIP	QCA	0.2 mm change in minimum lumen diameter (3 SD)
SCOR	QCA	10 point change in percent diameter stenosis (3 SD
FATS	QCA	10 point change in percent diameter stenosis (3 SD)
CLAS	panel	Global change score of -1 (discernible regression), or +2, +3 (intermediate or extreme progression)
Resin trials		
STARS	QCA	0.17 mm change in mean lumen diameter
NHLBI	panel	definite change in minimum lumen diameter
Fibrate trials		
LOCAT	QCA	patients not defined as regressors or progressors
BECAIT	QCA	0.4 mm change in minimum lumen diameter

Statin trials		
LCAS	QCA	0.4 mm change in minimum lumen diameter
CIS	QCA + panel	Global change score
CARS	QCA	15 point change in percent diameter stenosis
Post CABG	QCA	0.6 ,, change in minimum lumen diameter (3 SD)
REGRESS	QCA	0.4mm change in minimum lumen diameter (2 SD)
PLAC I	QCA	0.4 mm change in minimum lumen diameter
CCAIT	QCA	0.4 mm change in minimum lumen diameter
MAAS	QCA	15 point change in percent diameter stenosis
MARS	panel	12 point change in percent diameter stenosis (2 SD)
Surgical therapy		
POSCH	panel	change in global change score of 1 or more
Calcium channel blocker monotherapy		
Montreal	QCA	0.4 mm change in minimum lumen diameter or 15 point change in percent diameter stenosis
INTACT	QCA	0.4 mm change in minimum lumen diameter 20 point change in percent diameter stenosis

(1) control, (2) a 27% fat, weight reduction diet supervised by dietitians, or (3) diet plus cholestyramine. Again, the numbers were small but dietary change alone was shown to retard overall progression and even to induce regression of coronary artery disease (Table 4).

Also published in 1992 was the diet and exercise trial from Heidelberg by Schuler et al. (7). The important intervention was rigorous exercise and a 20% fat diet. Regression of coronary lesions was not seen, but a significant slowing of progression was demonstrated in the intervention group. Unfortunately, intensive physical exercise was shown to be associated with increased risk of cardiac arrest, particularly in well-motivated but high-risk young men.

The National Heart, Lung and Blood Institute Type II Coronary Intervention Study (8) (NHLBI) was a trial of cholestyramine in patients with coronary artery disease and type II hyperlipoproteinemia. Published in 1984, it was the first trial to demonstrate retardation in the rate of progression of atherosclerosis after 5 years of treatment. The inability to demonstrate regression of coronary artery lesions was attributed to the weak efficacy of cholestyramine and the small sample size.

The Cholesterol-Lowering Atherosclerosis Study (9) (CLAS), published in 1987, was the first angiographic study to show that lowering LDL and increasing HDL with combination drug and diet therapy in high-risk men with previous coronary artery bypass surgery resulted in a clear treatment benefit on atherosclerotic lesions. Evidence of atherosclerotic regression occurred in 16.2% of the intervention group compared with 2.4% of the control group. New lesions were less common in native arteries (14% in the intervention group vs. 40% in the control group, $P < .001$) and bypass grafts (15% vs. 38%, $P < .006$).

The Program on the Surgical Control of the Hyperlipidemias (10) (POSCH) used the unique intervention of partial ileal bypass surgery as a means to reduce LDL cholesterol in 838 survivors of myocardial infarction (MI). Coronary arteriograms were obtained at baseline and at 3, 5, 7, and 10 years. Sustained improvements in blood lipid patterns and cardiovascular morbidity and mortality were achieved as well as a demonstration of less progression of coronary artery disease arteriographically. A specific goal of POSCH was to examine the validity of the use of changes observed on sequential coronary arteriograms as a surrogate endpoint for clinical coronary events. For the combined endpoints of atherosclerotic coronary heart disease mortality or confirmed nonfatal MI, a significant relationship between the overall disease assessment and subsequent clinical events was found in the control group ($P < .0001$) and in the surgery group ($P = .04$) (11), offering credible evidence for this concept.

The Familial Atherosclerosis Treatment Study (12) (FATS) by Brown et al. demonstrated atherosclerotic regression in a specific, high-risk population: men with elevated apolipoprotein B (apo B), the surface component of LDL;

documented coronary artery disease; and a family history of premature coronary artery disease. Two intensive strategies for the modification of lipid levels were compared with a more conventional approach. The conventional approach resulted in frequent progression of coronary disease, infrequent regression, and a substantial number of cardiovascular events. By comparison, more intensive therapy to modify lipid levels halved the frequency of progression, tripled that of regression, and reduced by 73% the frequency of cardiovascular events (Tables 4 and 5).

The San Francisco Arteriosclerosis Specialized Center of Research (4) (SCOR), noted for its high recruitment of women (Table 2), used quantitative coronary angiography to demonstrate regression of coronary artery lesions in young, asymptomatic men and women with heterozygous familial hypercholesterolemia randomized to stepwise drug treatment vs. usual care. The difference in the dimensions of lesions between controls and treated patients over the 2-year period, slightly over 2% of the cross-sectional luminal area, was comparable to that observed by Brown et al. This study was the first to demonstrate a trend toward regression in a group of patients without symptoms of coronary artery disease, and significant regression of lesions in women.

The STARS (6) drug intervention arm was able to demonstrate that diet plus cholestyramine was more effective than diet alone in reducing LDL but that the frequency of disease progression and regression were remarkably similar in the two treatment groups. On the basis of their findings, Watts et al. recommended use of a lipid-lowering diet and, if necessary, appropriate drug treatment in men with coronary artery disease and mildly raised serum cholesterol levels.

The Monitored Atherosclerosis Regression Study (13) (MARS) published in 1993 ushered in the era of the HMG CoA reductase inhibitors as monotherapy to more powerfully reduce LDL levels (Table 6). Treatment with lovastatin plus diet slowed the rate of progression and increased the frequency of regression in coronary artery lesions, especially in more severe lesions (Table 4). The investigators were also able to demonstrate internal consistency in this trial between computer and observer scores on quantitative coronary angiography.

The Stanford Coronary Risk Intervention Project (14) (SCRIP) was an intensive multifactorial risk reduction study carried out over 4 years in men and women with CAD but not necessarily elevated blood lipids. The risk factor reduction significantly decreased angiographically defined progression of coronary atherosclerosis (Table 4) as well as hospitalizations for clinical cardiac events (Table 5). Angiographic benefits occurred even in subjects with a relatively low risk profile.

The Canadian Coronary Atherosclerosis Intervention Trial (15) (CCAIT) published by Waters et al. found that lovastatin monotherapy resulted in less worsening of established coronary artery disease and perhaps more importantly,

Table 4 Observed Progression and Regression Rates

Trial	N		Progression		Regression	
	Treatment	Control	Treatment	Control	Treatment	Control
Lifestyle intervention						
Heidelberg	40	52	8	22	12	2
STARS-diet	26	24	4	11	10	1
Lifestyle	22	19	4	10	18	8
Total	88	95	16	43	40	11
Percent			18%	45%	45%	12%
Combination drug therapy						
HARP	40	39	13	15	5	6
SCRIP	119	127	60	63	24	13
SCOR	40	32	8	13	13	4
FATS	74	46	17	21	26	5
CLAS	80	82	31	50	13	2
Total	353	326	129	162	81	30
Percent			37%	50%	23%	9%
Resin trials						
STARS-drug	24	24	3	9	8	1
NHLBI	59	57	19	28	4	4
Total	83	81	22	37	12	5
Percent			27%	46%	14%	6%

Fibrate trials						
BECAIT	42	39	31	33	9	5
Percent			74%	85%	21%	13%
Statin trials						
LCAS	171	169	29	39	14	8
CIS	104	101	36	54	19	13
CARS	39	41	7	16	1	1
Post CABG	599	593	234	302	22	20
REGRESS	314	327	142	181	54	30
PLAC I	163	157	60	79	31	24
CCAIT	146	153	48	76	14	10
MAAS	178	167	41	54	33	20
MARS	123	124	29	41	23	12
Total	1837	1832	646	869	225	144
Percent			35%	47%	12%	8%
Surgical therapy						
POSCH	333	301	125	197	42	14
Percent			38%	65%	13%	5%
Calcium Channel Blocker Therapy						
Montreal	168	167	66	69	27	24
INTACT	173	175	43	39	21	30
Total	341	342	109	108	48	54
Percent			32%	32%	14%	16%

Table 5 Cardiovascular Event Rates

Trial	Definition of CV event			CV events (n)	
		Rx	Control	Rx	Control
Lifestyle intervention					
Heidelberg	CV death, MI, CABG, PTCA	56	57	5	4
STARS diet	CV death, MI, CABG, PTCA	27	28	3	10
Lifestyle	CV death, unstable angina	28	20	1	1
Total		111	105	9	15
Percent				8%	14%
Combination drug therapy					
HARP	CV death, MI, CABG, PTCA, U/A, CHF	44	47	6	10
SCRIP	CV death, MI, CABG, PTCA	145	155	20	34
SCOR	CV death, MI, CABG, PTCA	40	32	0	1
FATS	CV death, MI, CABG, PTCA	94	56	5	10
CLAS	CV death, MI	94	94	14	18
Total		417	384	45	73
Percent				11%	19%
Resin trials					
STARS drug	CV death, MI, CABG, PTCA	26	28	1	10
NHLBI	CV death, MI	59	57	8	12
Total		85	85	9	22
Percent				11%	26%

Fibrate trials					
LOCAT	Death, MI, CABG, PTCA	197	198	7	7
BECAIT	CV death, MI, CABG, PTCA	47	45	3	11
Total		244	243	10	18
Percent				4%	7%
Statin trials					
LCAS	Death, MI, CABG, PTCA, U/A, CVA	214	215	31	41
CARS	CV death, MI, CABG, PTCA	45	45	1	3
CIS	CV death, MI, CABG, PTCA	129	125	15	19
REGRESS	CV death, MI, CABG, PTCA	450	434	54	86
PLAC I	CV death, MI, CABG, PTCA	206	202	55	81
CCAIT	CV death, MI, unstable angina	165	166	14	18
MAAS	CV death, MI, CABG, PTCA	193	188	40	51
MARS	CV death, MI, CABG, PTCA, U/A	123	124	22	31
Total		1525	1499	232	330
Percent				15%	22%
Surgical therapy					
POSCH	CV death, MI	421	417	82	125
Percent				19%	30%
Calcium channel blocker therapy					
Montreal	Death, MI, unstable angina	192	191	30	25
INTACT	CV death, MI, CABG, PTCA, U/A	214	211	60	46
Total		406	402	90	71
Percent				22%	18%

Table 6 Lipid Lowering Achieved

Trial	% Reduction in total cholesterol	% Reduction in LDL
Lifestyle intervention		
Heidelberg	5	7
STARS	12	13
Lifestyle	19	32
Combination drug therapy		
HARP	28	41
SCRIP	16	19
SCOR	22	27
FATS - LC	30	38
FATS - NC	19	25
CLAS	22	38
Resin trials		
STARS	23	30
NHLBI	15	21
Fibrate trials		
LOCAT	11	10
BECAIT[a]	9	2
Statin trials		
LCAS	14	21
CIS	29	35
CARS	14	22
Post CABG	25	40
REGRESS	20	29
PLAC I	21	29
CCAIT	20	27
MAAS	23	31
MARS	30	42
Surgical therapy		
POSCH	18	32

[a]BECAIT achieved a 31% reduction in triglyceride levels.
% Reduction = % reduction in intervention group -% reduction in controls.

in fewer new lesions (16% lovastatin vs. 32% placebo, $P = .001$). The beneficial effect of treatment was most pronounced in the more numerous, milder lesions and in patients whose baseline total or LDL cholesterol levels were above the group median. No significant reduction in clinical events was noted in this 2–year study.

The Multicentre Anti-Atheroma Study (16) (MAAS) published in 1994 assessed the effect of HMG-CoA reductase inhibitor monotherapy on diffuse and focal coronary atheroma in patients with heart disease and moderately elevated cholesterol levels. Treatment effects were observed regardless of diameter stenosis at baseline. Angiographic progression occurred more, regression was less frequent, and significantly more new lesions and new total occlusions developed in the placebo group. Again, there was no significant difference in clinical outcome.

Sacks et al., for the Harvard Atherosclerosis Reversibility Project (17) (HARP), investigated whether lipid-lowering therapy could have a benefit in normocholesterolemic patients with coronary artery disease. The two groups differed significantly in plasma lipid levels by the end of the study, but there was no angiographically measurable benefit on the coronary arteries (Table 4).

A small, single-blind trial published in 1997 by the Coronary Artery Regression Study Group (18) (CARS) investigated lipid lowering with a statin in normocholesterolemic Japanese patients and showed delayed progression of atherosclerosis in the treatment group vs. controls (18% vs. 39%, $P < .05$). The reasons for the discrepancy between this study and HARP were unclear but were postulated to be related to differences in patient selection and race.

The Regression Growth Evaluation Statin Study (19) (REGRESS) was a statin intervention study carried out on a broad range of coronary patients. A small but significant difference was found favoring the treatment group (pravastatin 40 mg daily) for both angiographic outcomes and clinical event rates.

The Pravastatin Limitation of Atherosclerosis in the Coronary Arteries Study (20) (PLAC I) was designed to evaluate the effect of pravastatin on progression of coronary atherosclerosis in patients with coronary artery disease and mild to moderate hyperlipidemia. Progression of atherosclerosis was reduced by 40% for minimal vessel diameter, particularly in lesions <50% at baseline ($P = .04$). There were also fewer new lesions in the treatment group. Similar rates of regression were found. Examples of regression are shown in Figure 1. Again, an unexpected finding in this study was the significant reduction in clinical event rates (Table 5).

In the Post Coronary Artery Bypass Graft Trial Investigators (21) (Post-CABG) published in 1997, 1351 coronary bypass patients were randomized to moderate LDL cholesterol lowering or aggressive lowering with a statin agent. The two-by-two factorial design also tested the hypothesis that low-dose anticoagulation would delay the late progression of atherosclerosis and occlusion in saphenous vein grafts. The analysis did not show a benefit of low-dose anticoagulation with warfarin. Patients in the aggressive cholesterol-lowering group had a 31% reduction in the mean per-patient percentage of grafts showing progression of atherosclerosis compared with the moderate treatment group (27% vs. 39%, $P <.001$) . No conclusions could be drawn regarding regression of atherosclerosis. The study was not powered to find a difference in clinical events but there was a trend in favor of aggressive therapy with drugs. Again, it was postu-

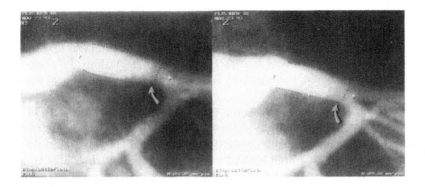

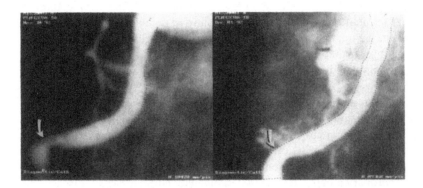

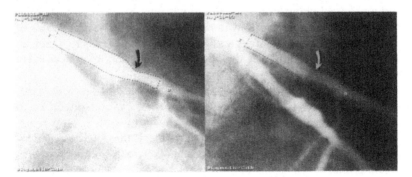

Figure 1   Figures are all obtained from analyses in the PLAC I study. The left-hand panel is the baseline frame and the right-hand panel is the follow-up film after 3 years of therapy with pravastatin 40 mg daily. (*Top*) Mild distal left main lesion which is absent on follow-up examination. (*Middle*) Moderate right coronary artery lesion that is markedly improved after therapy. (*Bottom*) Moderately stenosed mid-left anterior descending lesion that is markedly improved on the final angiogram.

lated that the removal of lipids from the lipid-rich layers of the arteries increases the stability of plaques, thus preventing their disruption and slowing the progression of atherosclerosis and occlusion, but without necessarily reducing plaque size (22).

The Lipoprotein and Coronary Atherosclerosis Study (22) (LCAS) published results in 1997 (24). The investigators were able to show coronary regression (14.6%) in the actively treated patients with moderate hypercholesterolemia. They also found that HDL-C response was a predictor of the future course of the disease. Those with HDL-C of <0.91 mmol/L had a threefold greater risk of progression.

The Bezafibrate Coronary Atherosclerosis Intervention Trial (25) (BECAIT) was a small but important trial of fibrate monotherapy in young, dyslipidemic male survivors of MI. The primary endpoint, minimum lumen diameter, was significantly larger at the end of treatment with bezafibrate than at the end of placebo treatment. The clinical effects were independent of the effect on LDL levels. There was also a significantly lower rate of coronary events among treatment patients. A much larger study (n = 3122) will be concluded in 1998 (after minimum 5 years of treatment) testing the effect of bezafibrate on a more general population.

The Lopid Coronary Angiography Trial (26) (LOCAT) studied 395 post coronary bypass men with HDL cholesterol level <1.1 and LDL cholesterol <4.5 who were randomized to gemfibrozil vs. placebo. After a time interval of 32 months there was a significant reduction in the number of new lesions in the bypass grafts in the treatment group. When all types of native coronary segments were taken into account, there was significantly less progression in the gemifibrozil group than the placebo. In this trial, patients were not classified as "progressors" or "regressors" therefore the results could not be included in this meta-analysis.

Another fibrate trial is being conducted on men and women with type II diabetes by the Diabetes Atherosclerosis Intervention Study (27) (DAIS). Four hundred eighteen individuals have been randomized in order to determine by quantitative angiography whether long-term correction of the dyslipoproteinemia of diabetes with fenofibrate results in evidence of decreased progression or regression of preexisting coronary atherosclerosis. The results of this 3-year trial should be available in late 1999 or early 2000.

The Multicenter Coronary Intervention Study (28) (CIS) from Germany reported a significant retardation of coronary artery disease in a study population of young men with severe coronary artery disease over a treatment period of 2.3 years. CIS was the first study to show a significant correlation between the therapeutically achieved LDL levels and the extent of change in minimum lumen diameter in the intervention group. No correlation could be found in the control group, suggesting that progression-inhibiting effects of LDL cholesterol lowering can only be expected beyond a certain threshold value.

The International Nifedipine Trial on Antiatherosclerotic Therapy (29) (IN-TACT) showed a suppression of new lesion formation with nifedipine. There were problems with the validity of this trial in that there were more previous MIs in the placebo group and more of the treatment patients were taking aspirin, tending to bias toward a treatment effect. An unexpected finding limiting the usefulness of this intervention was the higher death rate in the treatment group.

The Montreal Heart Institute Study (30) published by Waters in 1990 was a randomized study of nicardipine vs. placebo in 383 patients with established coronary artery disease. Nicardipine had no overall effect on the incidence of progression and regression of established coronary lesions. In a retrospective analysis, the incidence of progression was shown to be reduced in the patients with minimal lesions (lesions < 20%). As in INTACT, however, the clinical event rate was higher in the treatment arm.

The Quinapril Ischemic Event Trial (31) (QUIET) is the first prospective trial to investigate the long-term antiatherosclerotic effects of ACE inhibition. Normotensive, nonhyperlipidemic subjects (n = 1750) with normal left ventricular systolic function were randomly assigned to treatment or placebo at percutaneous transluminal coronary angioplasty (PTCA). The primary endpoint is time to first cardiac ischemic event. The primary QCA endpoint will be per-patient categorical designation as progressor or nonprogressor. Preliminary results, however, indicate that with an intention-to-treat analysis, neither aggregated clinical events, including angioplasty and bypass surgery, nor angiographic severity were different between the two arms of the study.

IV. LIPID LOWERING

The amount of lipid lowering achieved in each of the studies (Table 6) can be attributed to the difference in intensity and form of therapy. The earlier trials used less powerful medications but intensified the dietary differences between the treatment and control groups. Later trials using the more powerful statin agents provided dietary control and, often, a resin drug for the control participants because of the ethical problems of not providing this therapy in high risk groups.

A. Methods of Meta-Analysis

In order to assess trends and to synthesize the results of disparate trials, we looked closely at the reported trial results with respect to the main angiographic and clinical event endpoints. Odds ratios were calculated comparing progression and regression as dichotomous responses, excluding mixed or no-change responses (Fig. 2). Similarly, odds ratios were calculated for clinical events. Tests of homogeneity were performed and were not significant; i.e., it may be assumed that the different trials in each category estimate a common odds ratio even though definitions of progression and regression, and of clinical events differed somewhat among the trials. The significance of the calculated pooled odd ratios as well as 95% confidence intervals (CI) were calculated. Paired comparisons between combined odds ratios for different trial groups were carried out using Bonferroni's correction for multiple comparisons.

We also attempted to detect linkages between the angiographic and event findings. We did this using two approaches. First, numbers needed to treat (NNT) were calculated from the meta-analysis by first calculating the combined risk reduction, by using weights equal to the reciprocals of the variances of the individual rate differences, and then computing NNT with its standard error (SE) along with 95% CIs using our previously described method (32). Secondly, we undertook weighted regression analyses between relative reductions in cardiovascular events and the rates of regression and progression reported in each trial.

V. ANGIOGRAPHIC ENDPOINTS

Meta-analysis of the data provided by the trials attests to a statistically significant angiographic treatment effect (Fig. 2). The odds ratio for angiographic coronary regression vs. progression for the combined lifestyle trials was 10.7 (95%

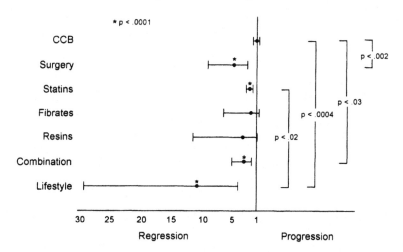

Figure 2 Odds ratios were calculated using frequency of regression vs. progression and ignoring mixed-change or no-change categories. Comb = combination drug therapies. CCB = calcium channel blockers.

CI 4.0–29, $P < .0001$). The odds ratio for the two combined resin trials was 3.2, but was not statistically significant. The odds ratio of regression vs. progression for the combination therapy trials was 3.0 (95% CI 1.8–5.1, $P < .0001$). The eight statin trials yielded a combined odds ratio of 2.1 (95% CI 1.6–2.7, $P < .0001$). The single fibrate trial yielded an odds ratio of 1.92 (NS), and the single surgical trial yielded an odds ratio of 4.7 (95% CI 2.5–9.0, $P < .0001$). The calcium channel blocker trials did not demonstrate a treatment effect (odds ratio = 1.01). The only potentially important and interesting finding in paired comparisons was a statistically significant difference between the lifestyle and the statin trials ($P = .02$). This may be a reflection of the difference between modifications of multiple risk elements in the lifestyle trials as compared to modification solely of lipids.

VI. CARDIOVASCULAR EVENT RATES

The cardiovascular event rates, which are defined differently in each trial (Table 5), include cardiovascular death, nonfatal MI, and emergency coronary artery bypass or percutaneous coronary angioplasty for unstable angina. Despite differences in explicit definitions, the general results are surprisingly similar between the treatment and the control groups, and tests of heterogeneity were negative. This yields a valid opportunity to compare general treatment effect sizes on major cardiac events even though specific syndromes could not be analyzed.

In contrast to calcium channel blocker therapy, all other interventions showed at least a favorable trend in reducing cardiovascular events (Fig. 3). The pooled odds ratios with 95% CI were 0.57 (0.23, 1.46) for the lifestyle trials (NS), 0.41 (0.17, 1.00) for the resin trials ($P = .049$); 0.54 (0.36, 0.81) for the combination drug trials ($P = .0031$); 0.67 (0.57, 0.80) for the statin trials ($P < .0001$); 0.55 (0.24, 1.28) for the fibrate trial (NS); 0.57 (0.41, 0.78) for the surgical trial ($P = .0005$) and 1.33 (0.94, 1.89) for the calcium channel blocker trials (NS). A limitation of this meta-analysis is that the reported rates in the various trials were not all standardized on a "per-year" basis (see trial duration in Table 2). The combination, surgery, and statin trial results were significantly different from the calcium channel blocker trials ($P < .009$) whereas no specific statistical differences could be detected among the lifestyle and lipid-lowering trials.

VII. RELATIONSHIP BETWEEN ANGIOGRAPHIC AND CLINICAL EVENTS

Long-term clinical follow-up of two of the angiographic trials has provided evidence that angiographically defined progression is predictive of future coronary events.

ODDS RATIOS:
Cardiovascular Event Rates

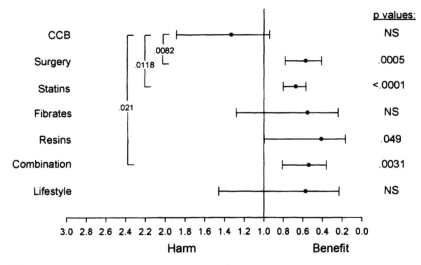

Figure 3 Odds ratios for cardiovascular events are shown.

The POSCH study was sufficiently powered to detect a difference in clinical event rates. The investigators examined 1866 film pairs over a mean follow-up period of 9.7 years (9,10). Angiographic progression was associated with a two-fold increase in risk for the combined endpoint of cardiac mortality and confirmed nonfatal MI compared to a score indicating no change or regression between baseline and 3 years ($P < .001$). The 13-year follow-up of the POSCH cohort was published in 1995 (35). An overall mortality rate of 10% occurred at 6.7 years in the control group and 9.4 years in the intervention group, for a gain in disease-free interval of 2.7 years in the intervention group ($P = .032$). A coronary heart disease (CHD) mortality rate of 8% occurred at 7.2 years in the control group and 11 years in the intervention group, for a gain of 3.8 years ($P = .046$). Twenty percent of patients demonstrated the combined endpoint of CHD mortality and confirmed nonfatal MI at 5.9 years in the control group and 11.4 years in the intervention group, for a gain of 5.5 years ($P < .001$). The intervention group gained 7 years of freedom from coronary bypass surgery, heart transplant, or angioplasty ($P < .001$).

Long-term follow-up of the CLAS cohort was published in 1996 (34). During follow-up of an average of 7 years after the 2-year angiogram 34% of the subjects (22 drug vs. 33 placebo) had one or more clinical coronary events. Coronary artery lesion progression was significantly related to any coronary event

(2.1<RR<2.2, P <.001). Change in all angiographic endpoints—minimum lumen diameter, percent diameter stenosis, and global change score—was predictive of future coronary events. For every 10% stenosis increase in average percent diameter stenosis over the 2-year interval, the risk for a clinical coronary event doubled over an average of 7 years of follow-up. Interestingly, progression of mild/moderate lesions (<50%) more accurately predicted future events than did progression of severe lesions (>50%). CLAS results also demonstrated that formation of new lesions in bypass grafts is a precursor of poor clinical outcome.

The Montreal Heart Institute study confirmed the validity of coronary progression as a strong independent predictor of future coronary events (35). An analysis of the data revealed that 16 of the 19 cardiac deaths occurred in progressors, a relative risk of 7.3 (95% CI 2.2–24.7, P <.001). Among the 40 patients with cardiac death or myocardial infarction, 25 were progressors; the relative risk (RR) was 2.3 (1.3–4.2, P = .009). The RR for any cardiac event was 1.7 (1.3–2.3, P <.001). These analyses suggest that angiographic endpoints may be useful surrogates for hard clinical events. This should not be misconstrued, however, as indicating the mechanism by which clinical events are prevented. This issue is discussed in detail in other chapters which outline the plethora of effects of these interventions on mechanisms that are more closely linked to clinical outcome (e.g., plaque stabilization and improvement of endothelial function).

Thus, the ability to halt the progression of atherosclerosis and to reverse it in some cases has been eclipsed, justifiably so, by the powerful impact of aggressive lipid lowering in preventing clinical events and improving overall prognosis. It should be pointed out, however, that the deemphasis of the angiographic findings has been achieved to some extent by comparing *absolute* changes in the angiogram, measured in fractions of millimeters or in a few percent diameter stenosis units, to *relative* changes in clinical events. It is understandable that to a clinician, a 50% decrease in events will impart greater impact than a 0.4 mm mean change in minimum lumen diameter. Conversely, however, one might compare a 50% decrease in rate of progression to an absolute decrease in events of only a handful of patients per trial. Conveyed this way, the impact to the practitioner of the angiographic results may erroneously deemphasize the interpretation of the clinical outcome information.

To avoid this, we chose to compare the clinical impact of the angiographic and clinical event outcomes using a common measure of the workload faced by a practitioner who wishes to apply the results of these trials to daily practice, i.e., NNT (number needed to treat). This parameter gives an indication of the treatment effort required on average to achieve a specified outcome during a specified period of therapy. Table 7 shows the results of these calculations. The average NNT to induce a beneficial angiographic result is between 4 and 9 over the course of 2 to 3 years of therapy, whereas more patients must be treated this way

Table 7 Number Needed to Treat (NNT) Data

	Prevention of cardiovascular events			Prevention of angiographic progression		
	NNT	Upper 95% CI	Lower[a] 95% CI	NNT	Upper 95% CI	Lower 95% CI
Lifestyle	43	175	1	4*	5	2
Combination	17	28	6	8	13	3
Resin	6	11	2	5	9	1
Fibrate	93	402	1	9	24	1
Statin	28	43	13	8**	10	6
Surgery	10	15	4	4***	5	3

$*P = .0009.$
$**P < .0001.$
$***P = .0003.$
[a]All lower CI truncated at 1.

to prevent an event. While this result is intuitively obvious, it serves to reiterate the efficiency with which angiographic trials have assisted in unraveling this therapeutic modality.

Figure 4 shows a direct linear relationship between the reported frequency of progression in the treatment and control groups and the observed cardiovascular event rates. An inverse relationship, best described by a quadratic curve, is noted between the reported frequency of regression and the observed cardiovascular event rates in Figure 5. Both analyses are limited, however, because of the variable definitions for progression and regression. To overcome this, the ratio of frequency of progression to regression was calculated. This emphasizes the relative angiographic findings *within* each trial, thereby minimizing any impact of the differing definitions of progression and regression *among* trials. Using this approach, there is a strong linear relationship shown in Figure 6. Thus, as the rate of progression/regression falls as a result of therapy, so too does the cardiovascular event rate.

VIII. CONCLUSIONS

This systematic review and meta-analysis attempts to summarize and put into perspective the burgeoning group of "regression trials" using lipid-lowering therapy. These trials show a consistent increase in the frequency of detection of

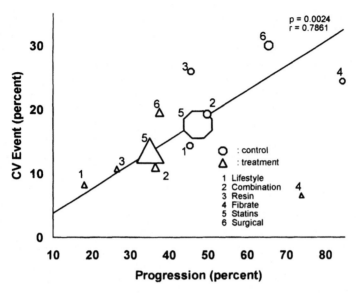

Figure 4 Weighted regression between cardiovascular event rates and the rate of detection of progression. Symbols are drawn in proportion to the square root of the sample size; i.e., the area of the symbol reflects sample size.

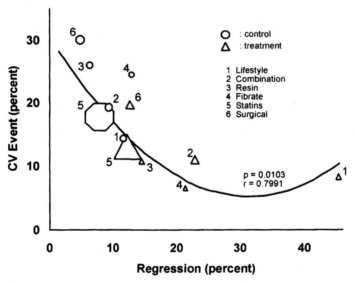

Figure 5 Weighted regression between cardiovascular event rates and the rate of detection of regression. Symbols are scaled as in Figure 4.

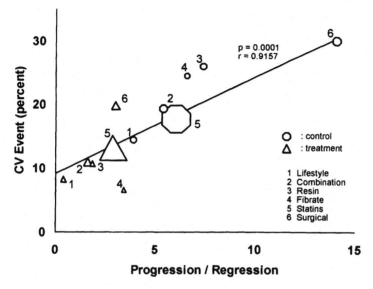

Figure 6 Weighted regression between cardiovascular event rates and the ratio of angiographic frequencies of progression to regression. Symbols are scaled as in Figure 4.

angiographic regression in comparison to angiographic progression in patients treated with diverse interventions that mainly lower lipids. These surrogate changes can be detected in fewer patients than can reductions in clinical events as emphasized by the NNT analyses. Finally, the differing interventions yield rather striking and consistent relationships between improved angiographic outcomes and clinical prognosis thereby validating the early and original premise underlying the planning of these studies, i.e., that angiographic changes can serve as surrogate endpoints. Currently, however, we appreciate that the processes by which salutary clinical outcomes are achieved are linked much more closely to important functional mechanisms such as plaque stabilization and improved endothelial function, than to the morphologic changes seen in epicardial coronary segments.

REFERENCES

1. Scoblionko DP, Brown G, Mitten S, et al. A new digital electronic caliper for measurement of coronary arterial stenosis: comparison with visual estimates and computer-assisted measurement. Am J Cardiol 1984; 53:689–693.
2. Gussenhoven EJ, Geselschap JH, van Lankeren W, Posthuma DJ, van der Lugt A.

Remodeling of atherosclerotic coronary arteries assessed with intravascular ultrasound in vitro. Am J Cardiol 1997; 79(5):699–702.

3. Waters D, Lesperance J, Craven TE, Hudon G, Gillam LD. Advantages and limitations of serial coronary arteriography for the assessment of progression and regression of coronary atherosclerosis: implications for clinical trials. Circulation 1993; 87(3):II38–II47.

4. Kane JP, Malloy MJ, Ports TA, Phillips NR, Diehl JC, Havel RJ. Regression of coronary atherosclerosis during treatment of familial hypercholesterolemia with combined drug regimens. JAMA 1990; 264(23):3007–3012.

5. Ornish D, Brown S, Scherwitz L, et al. Can lifestyle changes reverse coronary heart disease? The Lifestyle Heart Trial. Lancet 1990; 336:129–133.

6. Watts GF, Lewis B, Brunt JNH, et al. Effects on coronary artery disease of lipid-lowering diet, or diet plus cholestyramine, in the St. Thomas' Atherosclerosis Regression Study (STARS). Lancet 1992; 339:563–569.

7. Schuler G, Hambrecht R, Schlierf G, et al. Regular physical exercise and low-fat diet: effects on progression of coronary artery disease. Circulation 1992; 86(1):1–11.

8. Brensike JF, Levy RI, Kelsey SF, et al. Effects of therapy with cholestyramine on progression of coronary artery disease: results of the NHLBI type II coronary intervention study. Circulation 1984; 69(2):313–324.

9. Blankenhorn DH, Nessim SA, Johnson RI, Sanmarco ME, Azen SP, Cashin-Hemphill L. Beneficial effects of combined colestipol-niacin therapy on coronary atherosclerosis and coronary venous bypass grafts. JAMA 1987; 257(23):3233–3240.

10. Buchwald H, Varco RL, Matts JP, et al. Effect of partial ileal bypass surgery on mortality and morbidity from coronary heart disease in patients with hypercholesterolemia; report of the program on the surgical control of the hyperlipidemias (POSCH). N Engl J Med 1990; 323(14):946–955.

11. Buchwald H, Matts JP, Fitch LL, et al. Changes in sequential coronary arteriograms and subsequent coronary events. JAMA 1992; 268(11):1429–1433.

12. Brown G, Albers JJ, Fisher LD, et al. Regression of coronary artery disease as a result of intensive lipid-lowering therapy in men with high levels of apolipoprotein B. N Engl J Med 1990;323(19):1289–1298.

13. Blankenhorn DH, Azen SP, Kramsch DM, et al. Coronary angiographic changes with lovastatin therapy. Ann Intern Med 1993;119:969–976.

14. Haskell WL, Alderman EL, Fair JM, et al. Effects of intensive multiple risk factor reduction on coronary atherosclerosis and clinical cardiac events in men and women with coronary artery disease. Circulation 1994; 89(3):975–990.

15. Waters D, Higginson L, Gladstone P, et al. Effects of monotherapy with an HMG-CoA reductase inhibitor on the progression of coronary atherosclerosis as assessed by serial quantitative arteriography. Circulation 1994; 89(3):959–968.

16. MAAS Investigators. Effects of simvastatin on coronary atheroma: the Multicentre Anti-Atheroma Study (MAAS). Lancet 1994; 344:633–638.

17. Sacks FM, Pasternak RC, Gibson CM, Rosner B, Stone PH, Harvard Atherosclerosis Reversibility Project (HARP) Group. Effect on coronary atherosclerosis of

decrease in plasma cholesterol concentrations in normocholesterolemic patients. Lancet 1994; 344:1182–1186.

18. Tamura A, Mikuriya Y, Nasu M, Coronary Artery Regression Study (CARS) Group. Effect of pravastatin on progression of coronary atherosclerosis in patients with serum total cholesterol levels from 160 to 220 mg/dl and angiographically documented coronary artery disease. Am J Cardiol 1997; 79:893–896.

19. Jukema JW, Bruschke AVG, van Boven AJ, et al. Effects of lipid lowering by pravastatin on progression and regression of coronary artery disease in symptomatic men with normal to moderately elevated serum cholesterol levels. The Regression Growth Evaluation Study (REGRESS). Circulation 1995; 91(10):2528–2540.

20. Pitt B, Mancini GBJ, Ellis SG, et al. Pravastatin limitation of atherosclerosis in the coronary arteries (PLAC I): reduction in atherosclerosis progression and clinical events. J Am Coll Cardiol 1995; 26(5):1133–1139.

21. Post Coronary Artery Bypass Graft Trial Investigators. The effect of aggressive lowering of low density lipoprotein cholesterol levels and low dose anticoagulation on obstructive changes in saphenous vein coronary artery bypass grafts. N Engl J Med 1997; 336(3):153–162.

22. Fuster V, Vorchheimer DA. Editorial: Prevention of atherosclerosis in coronary artery bypass grafts. N Engl J Med 1997; 336(3):212–213.

23. West MS, Herd A, Ballantyne CM, et al. The lipoprotein and coronary atherosclerosis study (LCAS): design, methods and baseline data of a trial of fluvastatin in patients without severe hypercholesterolemia. Control Clin Trials 1996; 17:550–583.

24. Herd JA, Ballantyne CM, Farmer JA, et al. Effects of fluvastatin on coronary atherosclerosis in patients with mild to moderate cholesterol elevations (lipoprotein and coronary atherosclerosis study [LCAS]. Am J Cardiol 1997; 80:278–280.

25. Ericsson CG, Hamsten A, Nilsson J, Grip L, Svane B, de Faire U. Angiographic assessment of effects of bezafibrate on progression of coronary artery disease in young male postinfarction patients. Lancet 1996; 347:849–853.

26. Frick MH, Syvanne M, Nieminen MS, Kauma H, Taskinen MR, Lipid Coronary Angiography Trial (LOCAT) Study Group. Prevention of the angiographic progression of coronary and vein-graft atherosclerosis by genifibrozil after coronary bypass surgery in men with low levels of HDL cholesterol. Circulation 1997; 96:2137–2143.

27. Steiner G, DAIS Project Group. The Diabetes Atherosclerosis Intervention Study (DAIS): a study conducted in cooperation with the World Health Organization. Diabetologia 1996; 39:1655–1661.

28. Bestehorn HP, Rensing UFE, Roskamm H, et al. The effect of simvastatin on progression of coronary artery disease; the multicenter coronary intervention study (CIS). Eur Heart J 1997; 18:226–234.

29. Lichtlen PR, Hugenholtz PG, Rafflenbeul W, et al. Retardation of angiographic progression of coronary artery disease by nifedipine. Lancet 1990; 335:1109–1113.

30. Waters D, Lesperance J, Francetich M, et al. A controlled clinical trial to assess the effect of a calcium channel blocker on the progression of coronary atherosclerosis. Circulation 1990; 82(6):1940–1953.

31. Lees RS, Pitt B, Chan RC, et al. Baseline clinical and angiographic data in the Quinapril Ischemic Event (QUIET) Trial. Am J Cardiol 1996; 78(9):1011–1016.

32. Schulzer M, Mancini GB, 'Unqualified success' and 'unmitigated failure': number needed to treat related concepts for assessing treatment efficacy in the presence of treatment-induced adverse events. Int J Epidemiol 1996; 25(4):704–712.

33. Buchwald H, Campos CT, Boen JR, Nguyen PA, Williams SE, POSCH Group. Disease-free intervals after partial ileal bypass in patients with coronary heart disease and hypercholesterolemia. J Am Coll Cardiol 1995; 26(2):351–357.

34. Azen S, Mack W, Cashin-Hemphill L, et al. Progression of coronary artery disease predicts clinical coronary events. Circulation 1996; 93(1):34–41.

35. Waters D, Craven T, Lesperance J. Prognostic significance of progression of coronary atherosclerosis. Circulation 1993; 87(4):1067–1075.

9

Biochemical Correlates of Plaque Progression and Coronary Events

Melissa Ferraro-Borgida
University of Connecticut School of Medicine, Farmington, and Hartford Hospital, Hartford, Connecticut

David Waters
San Francisco General Hospital, San Francisco, California

I. INTRODUCTION

The relationship between serum cholesterol levels and the risk of cardiovascular disease has now been understood for many years. Despite this, the formulation of guidelines to diagnose and treat cholesterol levels is extremely complicated and has engendered considerable controversy. Reasons for this include the substantial overlap in serum cholesterol levels between subjects with and without atherosclerosis and the epidemiological observations that the risk gradient appears to extend across a very wide range of cholesterol levels. Among the 361,662 men screened in the Multiple Risk Factor Intervention Trial (MRFIT), the risk of coronary heart disease death decreased in a curvilinear fashion from 300 to 140 mg/dL (1). Among middle-aged factory workers in Shanghai, where only 7% of deaths were attributed to coronary heart disease (CHD) and where the mean serum cholesterol level was only 162 mg/dL, cholesterol levels and coronary mortality still strongly correlated (2).

Controversy also exists as to which cholesterol parameter should be used in guidelines. Total cholesterol is easy to measure, but knowledge of both LDL and HDL cholesterol provide a much more accurate assessment of risk. Cholesterol levels should not be viewed in isolation; the NCEP guidelines stratify treatment according to whether atherosclerosis is already present, and in its absence, whether other traditional risk factors are present or absent (1). Newly identified risk factors such as blood levels of homocysteine or lipoprotein(a) have not been incorporated into any guidelines.

Coincident with the decline in age-adjusted coronary mortality in the United States over the past three decades, total cholesterol levels of the adult population have declined (3). Furthermore, cholesterol levels of patients with documented coronary disease have also declined (4). Clinical trials have clearly demonstrated that angiographic progression of atherosclerosis and a reduction in coronary events accrue from cholesterol lowering in patients with coronary disease at baseline. Newer drugs make it possible to lower LDL cholesterol by up to 60% (5), well beyond the reductions obtained in recent clinical trials. That we can, does not mean that we should. The epidemiological association between low blood cholesterol levels and increased noncoronary mortality (6) has not translated into an increased noncoronary mortality in trials with HMG-CoA reductase inhibitors. However, benefit from more profound LDL cholesterol lowering has not yet been documented.

The purpose of this chapter is to assess the data from clinical trials relevant to the question of how low cholesterol levels should be lowered in patients with known coronary atherosclerosis. The chapter will also discuss two newer, potentially modifiable risk factors: homocysteine and lipoprotein(a).

II. RELATION OF CHOLESTEROL LOWERING TO CORONARY ANGIOGRAPHIC CHANGES

As reviewed in the previous chapter, 16 randomized controlled trials have examined the effect of various forms of cholesterol lowering on coronary arterial dimensions as assessed by serial coronary arteriography. In three of these trials cholesterol lowering was part of a more global coronary risk reduction treatment. Visual assessment was employed in the earlier studies to compare coronary dimensions, and computer-based quantitative measurement systems have been used in more recent trials. Uniform criteria for measuring and reporting angiographic data have not been developed, complicating comparisons among trials (7).

Nevertheless, despite major differences in patient populations, treatment regimens, and study durations, the results of these trials were overwhelmingly similar: cholesterol lowering slowed progression of coronary atherosclerosis and prevented the development of new coronary lesions (8). In most of the trials, regression occurred in more lesions and more patients in the actively treated groups. Although individually these trials were not powered to show a significant reduction in coronary events, in the aggregate total mortality was reduced by approximately one-quarter and myocardial infarction by approximately one-half (8).

Thompson et al. analyzed the first 11 trials that used quantitative angiography to assess the effect of lipid-lowering therapy on coronary atherosclerosis

(9). Linear regression analysis was performed on the 20 control and active treatment groups, weighted for their size. The mean change in percent diameter stenosis correlated well with the percent reduction in LDL cholesterol ($r = .74$, $P < .0005$). According to this relationship, a reduction of 44% in LDL cholesterol levels would be sufficient to arrest completely the progression of coronary atherosclerosis. In these trials the angiographic results did not correlate well with LDL cholesterol levels on treatment. The results of the six coronary angiographic trials (10–15) that compared an HMG-CoA reductase inhibitor to placebo yielded remarkably similar results, as depicted in Figure 1. Minimum lumen diameter narrowed by a mean of 0.03 to 0.05 mm per year in the control groups, and LDL cholesterol levels were reduced by 24% to 38% with active treatment. In all six trials the rate of progression was markedly reduced in the active treatment groups, to 0.014 to 0.03 mm per year.

It has been argued that small improvements such as this in arterial dimensions are out of proportion to the event reduction observed with cholesterol low-

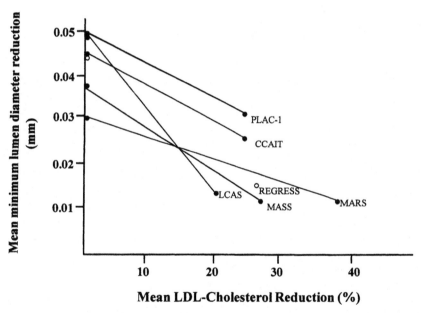

Figure 1 Relationship between LDL cholesterol reduction and coronary angiographic changes for the six randomized trials comparing an HMG-CoA reductase inhibitor to placebo (10–15). The mean minimum lumen diameter decrease per year is displayed on the y axis. The changes in mean LDL cholesterol levels in the control groups were <5% for each study, and for convenience are plotted as zero. LDL cholesterol was reduced from 24% to 38% in the active treatment groups. In each trial, the rate of progression, expressed as reduction of minimum lumen diameter per year, was slowed. The results of these six trials are remarkably consistent.

ering (16). Those who make such arguments usually compare absolute changes in arterial diameter to relative reductions in events. Most coronary lesions do not change over the short course of an angiographic trial, so the age change tends to be very small. However, coronary events can usually be linked to a lesion that has progressed (17). An example of a coronary lesion that progressed during an angiographic trial is illustrated in Figure 2.

An example of a coronary lesion that regressed during an angiographic trial is illustrated in Figure 3. The amount of regression that can be obtained with cholesterol lowering appears to be quite limited: in the colestipol/niacin treatment arm of the Familial Atherosclerosis Treatment Study (FATS), where LDL cholesterol levels were lowered by 32% and HDL cholesterol levels raised by 43%, only 39% of patients had regression of one or more lesions by ≥ 10 percentage points (18). During a second 2.5-year treatment period for a subset of FATS patients, the mean LDL cholesterol level was reduced by 52% to a mean of 101 mg/dL, but few additional coronary lesions regressed (19).

A.   Relation of Coronary Angiographic Change to Coronary Events

Clinical trials that use coronary angiographic measurements instead of coronary events as endpoints have the obvious advantage of being smaller, shorter, and thus much less expensive. However, for the results of such trials to have clinical relevance, it must be demonstrated that coronary angiographic changes are closely related to important clinical outcomes. This link has been clearly documented in each of the three angiographic trials where patients were subsequently followed for several additional years.

In the Program on the Surgical Control of the Hyperlipidemias (POSCH), Buchwald et al. compared the outcome over 6.7 years of follow-up between patients who had progressed and those who had not during the first 3 years of the trial (20). Total mortality, coronary mortality, and the combined endpoint of coronary mortality or myocardial infarction (MI) were all significantly more frequent during follow-up among progressors.

In a trial where the intervention had no influence on the rate of progression or regression, 141 of 335 (42%) patients had progression of at least one coronary lesion in 2 years (21). During a subsequent mean follow-up of nearly 4 years, 19 of the patients experienced cardiac death and 21 had nonfatal MIs. The relative risk of cardiac death for progressors was 7.3 (95% confidence interval (CI) 2.2–24.7) and for cardiac death or myocardial infarction was 2.3 (95% CI 1.3–4.2). Progression was as strong a predictor of future coronary events in this population as ejection fraction or number of diseased vessels, the predictors routinely used in clinical practice. The association between angiographic progression and a higher subsequent coronary event rate was also demonstrated in the

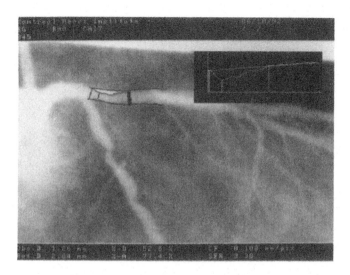

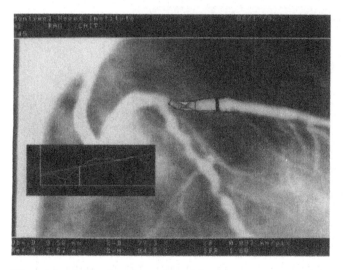

Figure 2 Example of progression of coronary atherosclerosis. At baseline (*top*) the proximal left anterior descending coronary artery contains a diameter narrowing measured at 52.5% with a minimum lumen diameter of 1.26 mm. After 2 years (*bottom*) the lesion has progressed to a narrowing of 77.3% and a minimum lumen diameter of 0.58 mm. Conservative cut points for a significant lesion change are 20% points for diameter stenosis and 0.4 mm for minimum lumen diameter.

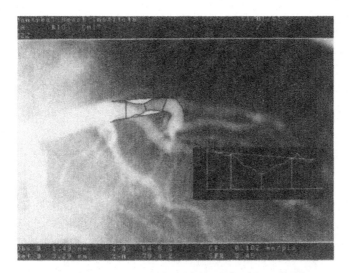

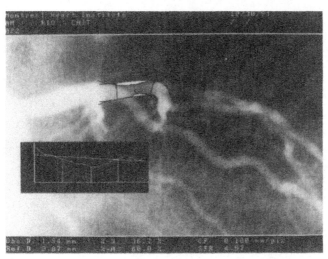

Figure 3 Example of regression of coronary atherosclerosis. At baseline (*top*) the proximal left anterior descending coronary artery contains a diameter narrowing measured at 54.6% with a minimum lumen diameter of 1.49 mm. After 2 years (*bottom*) the lesion has improved to a narrowing of 36.7% and a minimum lumen diameter of 1.94 mm.

long-term follow-up of the patients in the Cholesterol Lowering Atherosclerosis Study (CLAS) (22). Taken together, these three studies provide strong support for the use of angiographic endpoints as surrogates for coronary events in cholesterol lowering trials.

B. Relation of Cholesterol Lowering to Coronary Events

Eight large, randomized clinical trials (23–30) provide evidence that relates cholesterol lowering to coronary event reduction. Four of these trials enrolled mainly or entirely patients without known coronary disease: the Lipid Research Clinics Coronary Primary Prevention Trial (23), the Helsinki Heart Study (24), the West of Scotland Coronary Prevention Study (WOSCOPS) (26), and the Air Force/ Texas Coronary Atherosclerosis Prevention Study (AFCAPS/TexCAPS) (29). The three large secondary prevention trials are the Scandinavian Simvastatin Survival Study (4S) (25), the Cholesterol and Recurrent Events (CARE) trial (27), and the Long-term Intervention with Pravastatin in Ischemic Disease (LIPID) trial (30). Although the Post Coronary Artery Bypass Graft Trial was smaller and had angiographic endpoints, it is worthwhile to include it because its aggressively treated group attained lower LDL cholesterol levels than groups in any of the other trials.

Reduction in LDL cholesterol levels and coronary event rates for the treatment groups are summarized in Table 1 for the primary prevention trials and in Table 2 for the secondary prevention trials. Excluding the Helsinki Heart Study, the reduction in mean LDL cholesterol in these trials averaged 27% and the reduction in coronary events averaged 26%. This 1:1 relationship did not hold in the Helsinki Heart Study, where gemfibrozil reduced cardiac events by 34% even

Table 1 Percent LDL Reduction and Percent Coronary Event Reduction in Primary Prevention Trials

Study	Number of subjects	Treatment	% LDL reduction	% reduction in coronary events
LRC-CPP (23)	3,806	Cholestyramine	20%	19%
Helsinki Heart Study (24)	4,801	Gemfibrozil	10%	34%
WOSCOPS (26)	6,595	Lovastatin	26%	30%
AFCAPS/TexCAPS (29)	5,608	Lovastatin	25%	37%

Table 2 Percent LDL Reduction and Percent Coronary Event Reduction in Secondary Prevention Trials

Study	Number of subjects	Treatment	% LDL reduction	% reduction in coronary events
4S (25)	4,444	Simvastatin	35%	34%
CARE (27)	4,159	Pravastatin	32%	24%
LIPID (30)	9,014	Pravastatin	25%*	29%

*relative to placebo group

though mean LDL cholesterol levels were reduced by only 10%. This discrepancy can be attributed to the substantial increase in HDL cholesterol induced by gemfibrozil. The positive results from the Helsinki trial should be tempered by the observation that total mortality was actually slightly higher in gemfibrozil-treated patients than in controls.

In Figure 4, the LDL cholesterol values on treatment are plotted against the cardiac event rates for the treatment and control groups in these eight trials. There are minor limitations associated with presenting the data in this manner. Event rates in patients with coronary disease are higher, so secondary prevention trials are clustered near the top of the figure and primary prevention trials near the bottom. Event rates will be higher in longer trials, and the data are not corrected for trial length. In some of these trials, particularly 4S, the reduction

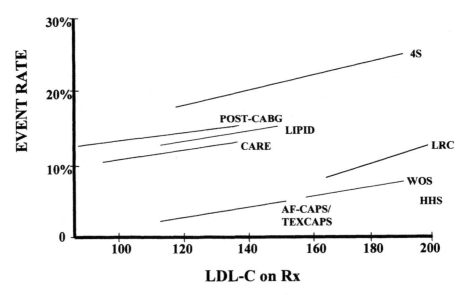

Figure 4   Cardiac event rates related to LDL cholesterol levels on treatment for the control and active treatment groups in the eight major trials (23—30). In each case the right-sided point represents the placebo or less aggressively treated group, and the left-sided point represents the aggressively treated group. The mean duration of patient follow-up varied from 4 to 7 years, and the definition of a cardiac event was not uniform for all the trials. The four primary prevention trials are located at the bottom of the graph (where the risk is lower) and the four secondary prevention trials are situated at the top of the graph. Note that the lines representing the risk reduction for each trial are relatively parallel. The benefit of cholesterol reduction is present across the entire range of LDL cholesterol represented. The benefit of the interventions is roughly similar, dependent on the degree of LDL cholesterol reduction.

in risk appeared to increase with the duration of treatment (31). The effect of this would be to make the benefit greater (and the slope of the line steeper) for longer trials.

In spite of these caveats, the striking feature of this graph is that the lines joining the treatment and control groups for each trial are nearly parallel. This finding suggests that if the initial risk of a group and the amount of LDL cholesterol lowering are known, the absolute risk on therapy can be estimated. It also implies that the benefit seen in these trials is derived from LDL cholesterol lowering. The data do not support claims that one HMG-CoA reductase inhibitor or another has special event-reducing properties independent of its lipid effects.

Epidemiological data such as MRFIT (1) suggest that the curve relating serum cholesterol to coronary events is steeper at higher values and flattens out slightly below 200 mg/dL. The number of data points generated by the treatment and control groups in these trials is obviously much too small to determine whether a similar curvilinear relationship applies to treated patients as well. In any case, cholesterol lowering clearly reduces the event rate across the range from 100 to 200 mg/dL. Lowering LDL cholesterol levels below 100 mg/dL may produce additional benefit, but this hypothesis has not yet been tested in a clinical trial.

The relationship between baseline cholesterol levels and treatment effect was assessed in 4S by dividing the patients into quartiles based upon their LDL cholesterol levels at study entry (32). The reduction in risk with treatment was nearly identical in each quartile, varying between 32% and 36%. Although the absolute risk reduction was less in the lowest quartile because the event rates were lower, the difference was still statistically significant (relative risk 0.65; 95% CI 0.50–0.85). In CARE (27), the investigators divided the patients into quintiles based upon their LDL cholesterol at study entry and presented the results with the four highest quintiles combined into two groups and the lowest quintile separate. The reduction in risk was 35% in the two highest quintiles, where LDL cholesterol was >150 mg/dL, and was 26% in the next two quintiles, where baseline LDL cholesterol was 125 to 150 mg/dL. In the lowest quintile, where pretreatment LDL cholesterol was <125 mg/dL, no benefit was seen. (The quintile contained only 851 patients and thus is greatly underpowered to detect a significant difference.) The authors noted that although this finding could not be considered definitive, it suggested that an LDL cholesterol level of approximately 125 mg/dL *"may be an approximate lower boundary for a clinically important influence of the LDL cholesterol level on coronary heart disease"* (27).

The CARE investigators noted that this conclusion is consistent with the findings of their small angiographic trial, which did not show benefit with LDL cholesterol lowering in this range (33). But a much larger angiographic trial published later documented considerable angiographic benefit in this category of patient (15). The Post Coronary Artery Bypass Graft Trial showed benefit, albeit in bypass grafts, with LDL cholesterol reduction to a mean of 93 mg/dL, compared to 136 mg/dL in the moderately treated group (28).

In the absence of more conclusive data, it seems reasonable to continue to adhere to the LDL cholesterol target of 100 mg/dL for patients with known atherosclerosis, as recommended by the National Cholesterol Education Panel (1). This chapter has focused on LDL cholesterol; however, HDL cholesterol and triglyceride levels are also appropriate targets for treatment when they are abnormal. Other conventional risk factors—diabetes, hypertension, and smoking in particular—are as important as cholesterol in controlling coronary risk.

III. HOMOCYSTEINE AS A RISK FACTOR

Despite the clear association between LDL and HDL cholesterol levels and coronary disease, many patients with "normal" cholesterol levels and no other obvious risk factors experience MI. It has been estimated that the classical risk factors only account for approximately half of all cases of coronary disease.

Homocysteine, a nonlipid, biochemical marker for atherosclerosis, has recently become recognized as an important risk factor. That elevated blood homocysteine levels caused atherosclerosis was first proposed almost 30 years ago by McCully (33). He had performed autopsies on two children with homocystinuria, a rare autosomal-recessive disorder characterized by very elevated blood homocysteine levels and homocystinuria. Patients with this disease suffer from mental retardation, skeletal abnormalities, and lens dislocations, and they have a strong tendency toward arterial and venous thromboses. McCully postulated that milder elevations of blood homocysteine levels could cause atherosclerosis in a much broader population than the few patients with the severe inherited metabolic disorder. Several years later, a controlled study demonstrated alterations in methionine metabolism and elevated homocysteine levels among patients with premature vascular disease (34). Since then, a large number of retrospective studies including thousands of patients, and large, prospective studies have confirmed this association. Several biological mechanisms by which homocysteine may promote thrombosis, endothelial damage, and atherosclerotic plaque formation have been demonstrated in vitro and in vivo. Supplementation with a combination of folic acid, pyridoxine, and riboflavin reduces homocysteine levels. Further prospective studies are needed to determine whether this will result in a reduction in progression of atherosclerosis and/or coronary events.

A. Biochemistry of Homocysteine

Homocysteine is a sulfur-containing amino acid formed during the metabolism of methionine, an essential amino acid derived from dietary protein. The metabolism of homocysteine is directed toward either excretion or conservation, through the transsulfuration and remethylation pathways. In the transsulfuration pathway, vitamin B_6 acts as a cofactor for the enzymes cystathionine β-synthase and γcystathionase, which sequentially convert homocysteine to sulfate and water. These end products are excreted in the urine. In the remethylation pathway, homocysteine is recycled back to methionine by either methionine synthase or betaine-homocysteine methlytransferase. Methionine synthase requires both vitamin B_{12} as a cofactor and methyltetrahydrofolate as a cosubstrate. When excess methionine is present, the transsulfuration pathway is favored by upregulation of the enzyme cystathionine β-synthase. With methionine deficiency, conservation of methionine is achieved by upregulation of the remethylation cycle (35). Thus, methionine metabolism and homocysteine levels are dependent upon the availability of folate, vitamin B_6, and vitamin B_{12} as well as the presence and activity of the enzymes involved in these pathways. Elevations in plasma homocysteine can be caused by factors that interfere with either excretion, remethylation, or transsulfuration. Elevated serum creatinine levels tend to increase homocysteine levels, such that mild to moderate hyperhomocysteinemia is commonly present among patients with end-stage renal disease (36). The increase in homocysteine levels seen with aging may also be related to creatinine levels (37). Deficiencies of folate, vitamin B_6 and vitamin B_{12} interfere with homocysteine metabolism and can result in elevated levels (38–40). Hyperhomocysteinemia may also be seen in disease states that result in low folate levels, such as psoriasis and cancer, as well as with the use of drugs that interfere with folate metabolism or the availability of B vitamins. Homocysteine levels increase after cardiac transplantation (41). Women have lower homocysteine levels than men until after menopause, when gender differences become less. Wouters et al. reported that homocysteine levels after methionine loading correlated negatively with 17β-estradiol levels (42). This mechanism may be one of the ways that estrogen protects against cardiovascular disease.

B. Epidemiological Studies of Homocysteine

McCully's observations were not evaluated in a controlled study until 1976, when Wilcken et al. (34) reported that patients with angiographically proven, premature coronary atherosclerosis had higher homocysteine levels after methionine loading than did controls. Although a follow-up study in 1983 (43) failed to confirm their findings, many other investigators were able to demonstrate a positive correlation between homocysteine levels and coronary disease, as well as cerebrovascular and peripheral vascular disease (44,45). Among a study population <55 years of age, homocysteine levels were elevated after methionine loading in no controls but in 42% of patients with cerebrovascular disease, 28% of those with peripheral vascular disease, and 30% of those with coronary disease (45).

The European Concerted Action Project (46) was the first large-scale study to examine the relationship between elevated homocysteine levels and vascular disease. In this case control study of subjects <60 years of age, those with plasma homocysteine levels in the top fifth percentile had a two-fold increase in vascular disease risk, compared to the rest of the study population. This was similar to the risk attributable to hypercholesterolemia or smoking.

A thermolabile form of the remethylation enzyme methylenetetrahydrofolate reductase (MTHFR) has been identified and found to be associated with elevated homocysteine levels (47) and in some studies, increased coronary risk (47,48). However, among participants in the Physicians' Health Study, homozygosity for the gene for this variant was associated with high homocysteine levels ($P < .01$) but not with risk of MI (49). In a study of young women, elevated plasma homocysteine and low folate levels were risk factors for MI, but homozygosity for the thermolabile variant of MTHFR was not (50).

The relationship between plasma homocysteine and coronary events was prospectively examined in the Physicians' Health Study, a placebo-controlled 2 × 2 factorial trial of aspirin and β-carotene in male physicians (51). The 271 participants who suffered MIs during the study had higher homocysteine levels than did controls matched for age and smoking status ($P = .03$). The adjusted relative risk for the top fifth percentile of homocysteine levels versus the rest of the study population was 3.4 (95% CI 1.3–8.8). Notably, a threshold of increased risk was seen with homocysteine levels >15.8 nmol/mL, the 95th percentile level for controls.

Other prospective studies have demonstrated a graded "dose-response" relationship between homocysteine and coronary events. In a series from Norway of patients with angiographically defined coronary artery disease and homocysteine levels at baseline (52), a strong relationship of homocysteine to subsequent overall mortality and cardiovascular mortality was seen. After adjustment for age, sex, left ventricular ejection fraction (LVEF), serum creatinine, total cholesterol, number of involved coronary arteries, and mortality ratios (compared to a reference category of homocysteine <9 μmol/L) were 2.3 for levels of 9–14.9 μmol/L, 2.5 for levels of 15–19.9 μmol/L, and 7.8 for levels ≥ 20 μmol/L (P for trend = .01). Homocysteine was the strongest modifiable predictor of cardiovascular mortality in this study.

Such a relationship has not been found in all studies. For example, in a Finnish population followed for 9 years, homocysteine levels did not predict subsequent MI or stroke (53). Homocysteine levels were low in both cases and controls (approximately 10 μmol/L), suggesting either a true feature of the population or a difference in measurement technique. In a meta-analysis of 27 studies relating homocysteine to vascular disease (54), 14 of the 17 studies that examined the relationship between coronary disease and homocysteine found a positive association. The authors calculated a summary odds ratio of 1.7 (95%

CI 1.5–1.9) based upon the 15 studies that provided sufficient data. They calculated an odds ratio for coronary disease of each 5 μmol/L increment of total homocysteine as 1.6 (95% CI 1.4–1.7) for men and 1.8 (95% CI 1.3–1.9) for women. They attributed 10% of the risk of coronary disease to homocysteine in this analysis (54).

C. How Homocysteine May Induce Atherosclerosis

The mechanisms by which homocysteine stimulates the progression of atherosclerosis are incompletely understood, but prothrombotic effects and an impairment of endothelial function have been reported. Endothelial cells in culture exposed to homocysteine for 3 hours lose their ability to produce endothelium-derived relaxing factor (EDRF) (55). Homocysteine has been reported to stimulate smooth muscle cell proliferation and decrease endothelial cell DNA synthesis (56). Endothelial cells in culture were lysed when exposed to homocysteine that had been oxidized by the addition of copper (57). Homocysteine also potentiates the auto-oxidation of LDL cholesterol (58).

Studies examining the effects of homocysteine on platelets or on prostacyclin production have had contradictory results (59–64). Reduced serum antithrombin activity was seen in patients with hyperhomocysteinemia (65). Pyridoxine and folate supplements normalized antithrombin activity in such patients in one study, despite persistent elevation of homocysteine levels (66). In vivo, a procoagulant activity of homocysteine has been demonstrated by a variety of mechanisms (35).

The infusion of L-homocysteine into baboons has been shown to cause patchy desquamation of vascular endothelium and a decrease in platelet survival times (59,60,66). In patients with hyperhomocysteinemia, endothelium-dependent, but not endothelium-independent, flow-mediated vasodilation is impaired in brachial arteries (67). Children with homocystinuria also have abnormal endothelial function, but their parents, who obligate heterozygotes, do not (68).

Taken together, these data from cell cultures, experimental animals, and human subjects establish multiple mechanisms by which homocysteine may induce atherosclerosis and its complications. This information complements the epidemiological findings and produces a solid body of evidence incriminating homocysteine as an important risk factor.

D. Treatment of Hyperhomocysteinemia

Hyperhomocysteinemia has been clearly linked to folate deficiency, and several studies have demonstrated that folic acid supplementation will effectively reduce homocysteine levels in patients with normal folate, B_{12}, and B_6 levels at baseline (54). Doses of 650 to 10,000 μg of folic acid were effective in these studies; for example, in men with homocysteine levels >16.3 µmol/L, 650 µg/day reduced homocysteine levels by 42% in one study (69).

Homocysteine levels are inversely related to folate intake, and a dose of at least 400 µg/day is required to maintain a low homocysteine level (70). In a recent editorial (71) Stampfer and Malinow contended that folate consumption is inadequate in a large proportion of the population. The U.S. recommended daily allowance of folate was reduced a few years ago from 400 to 200 µg/day, but recent insights into the prevention of neural tube defects with folate supplementation during pregnancy have prompted a reevaluation of this dosage.

In patients with coronary disease, effective reductions of homocysteine levels have been demonstrated with various dosages and combinations of folic acid and vitamins B_{12} and B_6 (72,73). Cobalamin supplementation alone has been shown to reduce homocysteine levels only in patients with B_{12} deficiency (74). Pyridoxine reduces homocysteine levels after methionine loading but is ineffective when used alone for reduction of baseline homocysteine levels (75). The optimal dosages of folate, cobalamin, and pyridoxine necessary to reduce homocysteine levels optimally are as yet unclear and require further study.

Another potential approach to the treatment of hyperhomocysteinemia is estrogen therapy. As mentioned above, homocysteine levels rise after menopause and correlate inversely with 17β-estradiol levels (42). Replacement therapy with continuous micronized 17β-estradiol and cyclic desogestrone resulted in a >10% decrease in homocysteine levels in postmenopausal women in a prospective study (76). Only women with high baseline homocysteine levels responded to therapy, and the effect plateaued after 6 months of treatment.

Although hyperhomocysteinemia is amenable to safe and inexpensive vitamin therapy, the value of lowering homocysteine levels in patients with vascular disease is as yet unknown. No trials have yet examined the effects of homocysteine reduction on either the angiographic progression of coronary disease or on the incidence of clinical events. The potential impact upon coronary disease of treatment with innocuous and inexpensive vitamins is potentially great, and represents an exciting area of clinical research.

IV. LIPOPROTEIN(A)

Although a lipoprotein similar to LDL cholesterol was first discovered in 1963, the association between lipoprotein(a) [Lp(a)] and coronary disease was not appreciated until the early 1970s. Intense investigation into the molecular biology of Lp(a) has revealed that it is similar in composition to LDL cholesterol but with an additional apolipoprotein: apo(a). There is a striking structural homology between apo(a) and plasminogen, implying that Lp(a) may be prothrombotic, while its presence within intracoronary plaques and foam cells indicates that it is likely also more directly atherogenic. High Lp(a) levels have clearly been linked to an increased incidence of coronary disease; however, whether lowering Lp(a) levels reduces risk has not yet been tested. Lp(a) levels are relatively resistant to diet and to currently available drug therapies, with the exception of estrogen and high doses of niacin, making this a difficult risk factor to modify.

A. Structure and Function of Lp(a)

Like LDL cholesterol, Lp(a) is composed of cholesterol, phospholipid, and apolipoprotein [(apo)B-100]. Its distinguishing feature is the addition of apo(a), a heavily glycosylated apolipoprotein which is linked to (apo)B-100 by disulfide bond. Like plasminogen, apo(a) consists of a kringle-containing domain and a serine protease domain. Kringle-4 of plasminogen is homologous to the highly repeated domain of apo(a). The genes for plasminogen and apo(a) are adjacent to each other on chromosome 6q26–27 (78,79). The number of kringle-4 repeats of apo(a) is genetically determined and accounts for the significant size heterogeneity of Lp(a) in plasma. Persons with fewer kringle-4 repeats appear to be at greater risk for coronary disease, suggesting that smaller forms of Lp(a) may be more atherogenic (80,81).

The remarkable structural similarity between plasminogen and Lp(a) has intrigued researchers because of the potential role for this lipoprotein as a link between thrombosis and atherosclerosis. Lp(a) binds fibrin and fibrinogen competitively with plasminogen in vitro and inhibits fibrinolysis (82–84). It also competes with plasminogen for binding sites on endothelium, on platelets, and on macrophages (83). Thrombosis stimulated by intracoronary plaque rupture is considered to play a crucial role in the progression of coronary atherosclerosis, and by inhibiting fibrinolysis, Lp(a) may favor this process.

The role of Lp(a) in active thrombus development was assessed in a study of patients with unstable angina who were randomized to receive either intravenous t-PA or placebo (85). Serum Lp(a) levels fell 48% from baseline within 12 hours in the t-PA group, but not in the placebo group, and began to return to baseline by 72 hours. The acute reduction in serum Lp(a) was considered due to a plasmin-mediated increase in Lp(a) binding to fibrin, fibrinogen, or other intravascular receptors (9). The finding of Lp(a) in the walls of arteries and venous bypass grafts suggests that it also plays a more direct role in atherosclerosis (86). Elevated serum levels of Lp(a) have been found to correlate with the amount of arterial deposition of both apo(a) and apo-B (87). Lp(a) promotes the formation of oxygen-derived free radicals in monocytes (88) and makes LDL cholesterol more susceptible to oxidative modification, which, of course, increases its atherogenicity (89).

Many risk factors for coronary disease impair endothelial function (90–92). Endothelial dysfunction often predates the appearance of symptomatic coronary disease by decades. In two angiographic studies where endothelial function in coronary arteries was assessed with acetylcholine (93,94), elevated serum Lp(a) levels were shown to correlate with impaired endothelium-dependent vasodilation. Taken together, the available data indicate that Lp(a) plays an important role in the evolution of atherosclerosis and its thrombotic complications.

B. Measurement of Lp(a)

Early studies identified Lp(a) qualitatively on gel electrophoresis as a band with prebeta mobility. The presence of this band was correlated with an increased risk of coronary disease in many case control and prospective studies. Quantitative assays (ELISA) revealed that prebeta positivity correlated with elevated Lp(a) levels of >20–30 mg/dL (95,96). In several case control studies and in five prospective studies, Lp(a) levels above the cutpoint of 30 mg/dL were independent markers of coronary risk (95–99).

Lp(a) levels are inversely related to the size of apo(a), and the evidence suggests that apo(a) phenotypes are genetically predetermined (100). Serum concentrations of Lp(a) remain relatively constant over a person's lifetime (101), although transient increases have been reported after acute MI and surgical procedures (102). Menopause was thought to be associated with increased serum levels of Lp(a), but this relationship was not seen in the Framingham Offspring Study (103,104). Oral estradiol and progesterone reduce Lp(a) in postmenopausal women (105). Higher Lp(a) levels have been reported with diabetic nephropathy and end-stage renal failure (106).

The pattern of distribution of Lp(a) levels differs among ethnic groups, with a normal distribution seen among African-Americans and a distribution that is skewed to the left in white Americans and Asian Indians (106). The studies linking Lp(a) to an increase in coronary risk contain an overrepresentation of white subjects, so that generalizations of the findings to other ethnic groups should be done with caution.

C. Lp(a) and Coronary Risk

Lp(a) levels have been measured in baseline blood samples and correlated with subsequent coronary events in 10 prospective trials. Seven of these studies (95,99,107,108) reported a positive correlation, while three others (109–111) did not.

In a small prospective case control study in Sweden (99), men who went on to experience coronary events during a 6-year follow-up had higher serum Lp(a) levels than controls (278 vs. 173 mg/dL; 95% CI 18–192 mg/dL). Men in the highest quintile of Lp(a) concentration (>365 mg/dL) experienced more than twice the incidence of coronary events than men in the other four quintiles. Lp(a) was independently associated with coronary events by multiple logistic regression analysis. Among 1332 Icelandic men (107), apolipoprotein(a) was a significant independent risk factor for the development of coronary disease (odds ratio 1.22) during an 8.6-year follow-up. In the Gottingen Risk Incidence and Prevalence Study (97), Lp(a) level was an independent risk factor for myocardial infarction, but had less predictive power than HDL or LDL cholesterol level, family history of myocardial infarction, or fibrinogen level. In the British United Provident Association Study (108), apo-B was the strongest independent predictor of coronary risk but apo(a) was weakly predictive. The Lipid Research Clinics Primary Prevention Trial (98) enrolled 3806 middle-aged men with hypercholesterolemia and followed them for 7.4 years. Baseline Lp(a) levels were higher among those who developed coronary events than among controls (23.7 vs. 19.5 mg/dL, $P < .02$), and by multiple logistic regression analysis, the Lp(a) risk was independent of age, body mass index, cigarette smoking, blood pressure, and LDL and HDL cholesterol levels ($P < .01$).

Lp(a) has been studied in both men and women from Framingham. Among 3103 women without coronary disease, 434 had elevated Lp(a) levels detected as a sinking prebeta band on gel electrophoresis (96). The population attributable risk for myocardial infarction of elevated levels was 17% for Lp(a), 19% for total cholesterol, and 19% for HDL cholesterol. Among 2192 men in the Framingham study (95), the relative risk of coronary disease before age 55 for elevated Lp(a) was 1.9 (95% CI 1.2–2.9). This was similar to the relative risk for a total cholesterol ≥240 mg/dL and for an HDL cholesterol ≤ 35 mg/dL, both of which were 1.8.

Three prospective studies failed to show a positive correlation between Lp(a) levels and coronary risk. Two of them, the Helsinki Heart Study (109) and the study of Colemen et al. (110), lacked statistical power and may have suffered because frozen serum samples were not stored under cold enough conditions to prevent degradation of Lp(a). Lp(a) levels were not predictive of MI in the Physicians' Health Study either (111), perhaps because half of the study population took aspirin, which may have attenuated the prothrombotic effects of Lp(a).

D. Lp(a) and Angiographic Coronary Disease

Few data are available on the impact of Lp(a) on the progression of coronary atherosclerosis. The relative resistance of Lp(a) levels to cholesterol-lowering drugs makes it difficult to conduct a randomized prospective trial of Lp(a) reduction. However, two small studies have reported a positive association between Lp(a) levels and coronary angiographic progression (112,113) while two others found no such correlation (114,115). Positive correlations between high Lp(a) levels and bypass graft disease (116) and restenosis after angioplasty (117,118) have also been reported. These findings should be interpreted cautiously because they are derived from studies where Lp(a) was not the main focus of investigation; often it was only one of many measurements made at baseline. Under these circumstances, a strong bias exists for positive but not negative results to be published.

In one of these studies (112), 21 of 79 patients with coronary disease had progression of at least one stenosis when angiography was repeated after a mean interval of only 66 days. Lp(a) levels were higher among progressors than nonprogressors (median 66 vs. 13 mg/dL, $P = .01$). Two-thirds of the rapid progressors and one-third of the nonprogressors had Lp(a) levels >25 mg/dL ($P = .007$) (38).

In a prospective study designed to evaluate the impact of risk factors on coronary progression, serum Lp(a) levels were measured in 85 Japanese coronary patients. Repeat angiography after 3 years revealed progression in 48 and no progression in 37. Lp(a) ($P = .0002$), smoking ($P = .002$), serum HDL ($P = .003$), and serum LDL cholesterol ($P = .01$) predicted progression, but only Lp(a) ($P = .003$) and HDL cholesterol ($P = .03$) were independent predictors (113).

In contrast to these two studies, the Familial Hypercholesterolemia Regression Study (114) showed no association between Lp(a) levels and angiographic outcomes. The only lipoprotein variable found to correlate significantly with angiographic change was apo-B. In an angiographic trial assessing the effects of exercise and a low-fat diet in men with coronary disease (115), no correlation between Lp(a) levels and angiographic progression was found in the subgroup of 38 patients who had Lp(a) measurements at baseline.

Lp(a) levels were reported to be associated with the severity of saphenous vein bypass graft stenosis in one study ($r = .24$, $P = .002$) (116). The pathogenesis of venous bypass graft disease involves both accelerated atherogenesis and thrombosis, processes that higher Lp(a) levels would be expected to facilitate.

Restenosis due mainly to myointimal hyperplasia occurs within 6 months after 30% to 40% of coronary angioplasties. The injury induced by angioplasty results in thrombus formation and platelet activation, with subsequent release of growth factors that stimulate smooth muscle cell proliferation and deposition of extracellular matrix (117,118). Lp(a) may play a role in this process by promoting thrombogenesis and platelet activation. Among 69 patients undergoing coronary angiography within 10 months after angioplasty, Lp(a) was found to be the only significant independent predictor of restenosis in one study (119). The odds ratio for restenosis for the highest quintile of Lp(a) level was 11 (95% CI 9–13, $P = .033$).

In another study (120), Lp(a) levels were measured before and 1 day, 3 days, and 4 months after angioplasty in 138 patients. An acute decrease in Lp(a) levels was seen the first day after angioplasty in both patients who would and would not subsequently develop restenosis; however, the decrease was significantly greater among the restenosis group, and persisted to 3 days but not to 4 months. Baseline Lp(a) levels were also significantly higher among the restenosis group.

Taken together, these angiographic studies are not as consistent or as numerous as the studies linking elevated Lp(a) levels to coronary events. Nevertheless, because coronary events are the consequence of angiographic progression, a relationship between Lp(a) levels and angiographic change certainly exists. Further studies are needed to define this relationship more accurately.

E. Treatment of Lp(a)

Many of the treatments for hypercholesterolemia have no effect on Lp(a) levels. As reviewed by Stein and Rosenson (106), the HMG-CoA reductase inhibitors do not lower Lp(a) levels, and actually raised them significantly in some clinical trials. Among the fibrates, bezafibrate, but not clofibrate or gemfibrozil, has been shown to reduce Lp(a) levels effectively. In addition, high doses of niacin (3–4 g/day) have proven to be efficacious at Lp(a) reduction. Neomycin, either alone or in combination with niacin, has also been successful. Angiotensin-converting enzyme (ACE) inhibitors reduce Lp(a) levels in patients with nephrotic syndrome. N-acetylcysteine, by reducing the disulfide bridge between apo(a) and apo-B, reduces Lp(a) concentrations also.

Lp(a) levels are influenced by hormonal status. Tamoxifen, estrogen, and combined estrogen-progesterone therapy each produce significant reductions in Lp(a) levels in postmenopausal women (121,122). The effects of oral contraceptives have been variable (123,124). In a study of estrogen replacement in women after surgical menopause, oral estradiol was found to be effective but transdermal administration was not (125). Lp(a) levels fell significantly in elderly men with prostatic carcinoma receiving exogenous estrogens (126). Anabolic steroids are also effective, but their serious side effects preclude their use for this purpose (55–57).

In three angiographic trials of cholesterol lowering, no additional benefit of lowering Lp(a) was realized if LDL cholesterol was adequately reduced (114,115,128). This evidence raises the possibility that Lp(a) may be highly pathogenic only in the presence of elevated LDL cholesterol. Furthermore, it is possible that antiplatelet therapy may mitigate some of the risk associated with high Lp(a) levels (111). Whether Lp(a) reduction will prove to be a therapeutic goal with documented benefit, either generally or in selected populations, is at present speculative. A better understanding of this particle is required, in terms of both its role in the pathogenesis of vascular disease and its relationship with other risk factors. Although only meager support can be mustered for treating elevated Lp(a) levels, patients with this abnormality are at high risk, and their other risk factors, particularly LDL cholesterol, must be managed aggressively.

REFERENCES

1. National Cholesterol Education Program Expert Panel, Second report of the Expert Panel on Detection, Evaluation, and Treatment of High Blood Cholesterol in Adults (Adult Treatment Panel II). *Circulation* 1994;89:1329–1444.
2. Chen Z, Peto R, Collins R, et al. Serum cholesterol concentration and coronary heart disease in population with low cholesterol concentrations. *BMJ* 1991;303:276–282.
3. Johnson CL, Rifkind BM, Sempos CT, et al. Declining serum total cholesterol levels among US adults. *JAMA* 1993;269:3002–3008.
4. Rubins HB, Robins SJ, Collins D, et al. Distribution of lipids in 8,500 men with coronary artery disease. *Am J Cardiol* 1995;75:1196–1201.
5. Nawrocki JW, Weiss SR, Davidson MH et al. Reduction of LDL cholesterol by 25% to 60% in patients with primary hypercholesterolemia by atorvastatin, a new HMG-CoA reductase inhibtor. *Arterioscler Thromb Vasc Biol* 1995;15:678–682.
6. Jacobs DR Jr, Blackburn H, Higgins M, et al. Report of the Conference on Low Blood Cholesterol: Mortality Associations. *Circulation* 1992:86:1046–1060.
7. Waters D, Lespérance J, Craven TE, et al. Advantages and limitations of serial coronary arteriography for the assessment of progression and regression of coronary atherosclerosis. Implications for clinical trials. *Circulation* 1993;87(suppl II):II-38–II-47.
8. Waters D. Lessons from coronary atherosclerosis "regression" trials. In: Abrams J, Pasternak B (eds.) *Risk Factors for Coronary Artery Disease. A New Paradigm for the Prevention and Treatment of Coronary Heart Disease; Endothelial Function, Plaque Behavior, Regression Trials. Cardiovascular Clinics.* Philadelphia:W.B. Saunders, 1996;31–50.
9. Thompson GR, Hollyer J, Waters DD. Percentage change rather than plasma level of LDL-cholesterol determines therapeutic response in coronary heart disease. *Curr Opin Lipidol* 1995;6:386–388.
10. Blankenhorn DH, Azen SP, Kramsch DM, et al. Coronary angiographic changes with lovastatin therapy: the Monitored Atherosclerosis Regression Study (MARS). *Ann Intern Med* 1993;119:969–976.
11. Waters D, Higginson L, Gladstone P, et al. Effect of monotherapy with an HMG-CoA reductase inhibitor on the progression of coronary atherosclerosis as assessed by serial quantitative arteriography: the Canadian Coronary Atherosclerosis Intervention Trial. *Circulation* 1994;89:959–968.
12. MAAS Investigators. Effect of simvastatin on coronary atheroma: the Multicentre Anti-Atheroma Study (MAAS). *Lancet* 1994;344:633–638.
13. Pitt B, Mancini GJB, Ellis SG, et al. Pravastatin Limitation of Atherosclerosis in the Coronary Arteries (PLAC I): reduction in atherosclerosis progression and clinical events. *J Am Coll Cardiol* 1995;26:1133–1139.
14. Jukema JW, Bruschke AVG, van Boven AJ, et al. Effects of lipid lowering by pravastatin on progression and regression of coronary artery disease in symptomatic men with normal to moderately elevated serum cholesterol levels. The Regression Growth Evaluation Statin Study (REGRESS). *Circulation* 1995;91:2528–2540.

15. Herd JA, Ballantyne CM, Farmer JA, et al. Effects of fluvastatin on coronary atherosclerosis in patients with mild to moderate cholesterol elevations (Lipoprotein and Coronary Atherosclerosis Study [LCAS]). *Am J Cardiol* 1997;80:278–286.
16. Topol EJ, Nissen SE. Our preoccupation with coronary luminology. The dissociation between clinical and angiographic findings in ischemic heart disease. *Circulation* 1995;92:2333–2342.
17. Brown BG, Zhao X-Q, Sacco DE, et al. Lipid lowering and plaque regression. New insights into prevention of plaque disruption and clinical events in coronary disease. *Circulation* 1993;87:1781–1791.
18. Brown G, Albers JJ, Fisher LD, et al. Regression of coronary artery disease as a result of intensive lipid-lowering therapy in men with high levels of apolipoprotein B. *N Engl J Med* 1990;323:1289–1298.
19. Stewart BF, Brown BG, Zhao X-Q, et al. Coronary atherosclerosis regression is less pronounced during a second course of comparably effective lipid-lowering therapy. *Circulation* 1993;88(suppl I):I-363. Abstract.
20. Buchwald H, Matts JP, Fitch LL, et al. Changes in sequential coronary arteriograms and subsequent coronary events. *JAMA* 1992;268:1429–1433.
21. Waters D, Craven T, Lespérance J. Prognostic significance of progression of coronary atherosclerosis. *Circulation* 1993;87:1067–1075.
22. Azen SP, Mack WJ, Cashin-Hemphill L, et al. Progression of coronary artery disease predicts clinical coronary events. Long-term follow-up from Cholesterol Lowering Atherosclerosis Study. *Circulation* 1996;93:34–41.
23. Lipid Research Clinics Program. The Lipid Research Clinics Coronary Primary Prevention Trial results. I. Reduction in incidence of coronary heart disease. *JAMA* 1984;251:351–363.
24. Frick MH, Elo O, Haapa K, et al. Helsinki Heart Study: primary-prevention trial with gemfibrozil in middle-aged men with dyslipidemia. Safety of treatment, changes in risk factors, and incidence of coronary heart disease. *N Engl J Med* 1987;317:1237–1245.
25. Scandinavian Simvastatin Survival Study Group. Randomised trial of cholesterol lowering in 4444 patients with coronary heart disease: the Scandinavian Simvastatin Survival Study (4S). *Lancet* 1994;344:1383–1389.
26. Shepherd J, Cobbe SM, Ford I, et al. Prevention of coronary heart disease with pravastatin in men with hypercholesterolemia. *N Engl J Med* 1995;333:1301–1307.
27. Sacks FM, Pfeffer MA, Moye LA et al. The effect of pravastatin on coronary events after myocardial infarction in patients with average cholesterol levels. *N Engl J Med* 1996;335:1001–1009.
28. Post Coronary Artery Bypass Graft Trial Investigators. The effect of aggressive lowering of low-density lipoprotein cholesterol levels and low-dose anticoagulation on obstructive changes in saphenous-vein coronary-artery bypass grafts. *N Engl J Med* 1997;336:153–162.
29. Downs, JR, Clearfield M, Weis S, et al. Primary prevention of acute coronary events with Lovastatin in men and women with average cholesterol levels: Results of AFCAPS/TexCAPS. *JAMA* 1998; 279:1615–1622.
30. LIPID study group. Prevention of cardiovascular events and death with pravastatin

in patients with coronary heart disease and a broad range of initial cholesterol levels. NEJM 1998; 339:1349–1357.

31. Yusuf S, Ananad S. Cost of prevention. The case for lipid lowering. *Circulation* 1996;93:1774–1776.

32. Scandinavian Simvastatin Survival Study Group. Baseline serum cholesterol and treatment effect in the Scandinavian Simvastatin Survival Study (4S). *Lancet* 1995;345:1274–1275.

33. McCully K. Vascular pathology of homocysteinemia: implications for the pathogenesis of arteriosclerosis. *Am J Pathol* 1969;56:111–128.

34. Wilcken DEL, Wilcken B. The pathogenesis of coronary artery disease: a possible role for methionine metabolism. *J Clin Invest* 1976;57:1079–1082.

35. Mayer EL, Jacobsen DW, Robinson K. Homocysteine and coronary atherosclerosis. *J Am Coll Cardiol* 1996;27:517–527.

36. Wilcken DEL, Gupta VJ. Sulfur containing amino acids in chronic renal failure with particular reference to homocystine and cysteine-homocysteine mixed disulfide. *Eur J Clin Invest* 1979;9:301–307.

37. Anderson A, Brattstrom L, Israelsson B, et al. Plasma homocysteine before and after methionine loading with regard to age, gender, and menopausal status. *Eur J Clin Invest* 1992;22:79–87.

38. Brattstrom L, Israelsson B, Lingarde F, Hultberg B. Higher total plasma homocysteine in vitamin B_{12} deficiency than in heterozygosity for homocystinuria due to cystathionine B-synthase deficiency. *Metabolism* 1988;37:175–178.

39. Kang S-S, Wong PWK, Norusis M. Homocysteinemia due to folate deficiency. *Metabolism* 1987;36:458–462.

40. Miller JW, Ribaya-Mercado JD, Russell RM, et al. Effect of vitamin B-6 deficiency on fasting plasma homocysteine concentrations. *Am J Clin Nutr* 1992; 55:1154–1160.

41. Berger PB, Jones JD, Olson LJ, et al. Increase in total plasma homocysteine concentration after cardiac transplantation. *Mayo Clin Proc* 1995;70:125–131.

42. Wouters MGAJ, Moorrees TEC, van der Mooren MJ, et al. Plasma homocysteine and menopausal status. *Eur J Clin Invest* 1995;25:801–805.

43. Wilcken DEL, Reddy SG, Gupta VJ. Homocysteinemia, ischemic heart disease and the carrier state for homocystinuria. *Metabolism* 1983;32:363–370.

44. Boers GHJ, Smals AGH Trijbels FJM, et al. Heterozygosity for homocystinuria in premature peripheral and cerebral occlusive arterial disease. *N Engl J Med* 1985;313:709–715.

45. Clarke R, Daly L, Robinson K, et al. Hyperhomocysteinemia: an independent risk factor for vascular disease. *N Engl J Med* 1991;324:1149–1155.

46. Graham IM, Daly LE, Refsum HM, et al. Plasma homocysteine as a risk factor for vascular disease: the European Concerted Action Project. *JAMA* 1997;277:1775–1781.

47. Kang S-S, Wong PWK, Zhou J, et al. Thermolabile methylenetetrahydrofolate reductase in patients with coronary artery disease. *Metabolism* 1988;37:611–613.

48. Kang S-S, Passen EL, Ruggie N, et al. Thermolabile defect of methylene-tetrahydrfolate reductase in coronary artery disease. *Circulation* 1993-88:1463–1469.

49. Ma J, Stampfer MJ, Hennekens CH, et al. Methylenetetrahydrofolate reductase polymorphism, plasma folate, homocysteine, and risk of myocardial infarction in US physicians. *Circulation* 1996;94:2410–2416.
50. Schwartz SM, Siscovick DS, Malinow MR, et al. Myocardial infarction in young women in relation to plasma total homocysteine, folate, and a common variant in the methylenetetrahydrofolate reductase gene. *Circulation* 1997;96:412–417.
51. Stampfer MJ, Malinow MR, Willett WC, et al. A prospective study of plasma homocyst(e)ine and risk of myocardial infarction in US physicians. *JAMA* 1992; 268:877–881.
52. NyGard O, Nordrehaug JE, Refsum H, et al. Plasma homocysteine levels and mortality in patients with coronary artery disease. *N Engl J Med* 1997;337:230–236.
53. Alfthan G, Pekkanen J, Jauhianen M, et al. Relation of serum homocysteine and lipoprotein (a) concentrations to atherosclerotic disease in a prospective Finnish population based study. *Atherosclerosis* 1994; 106:9–19.
54. Bousney CJ, Beresford SAA, Omenn GS, Motulsky AG. A quantitative assessment of plasma homocysteine as a risk factor for vascular disease: probable benefits of increasing folic acid intakes. *JAMA* 1995;274:1049–1057.
55. Stamler JS, Osborne JA, Jaraki O, et al. Adverse vascular effects of homocysteine are modulated by endothelium-derived relaxing factor and related oxides of nitrogen. *J Clin Invest* 1993;91:308–318.
56. Tsai J-C, Perrella MA, Yoshizumi M, et al, Promotion of vascular smooth muscle cell growth by homocysteine: a link to atherosclerosis. *Proc Natl Acad Sci USA* 1994; 91:6369–6373.
57. Starkebaum G, Harlan JM. Endothelial cell injury due to copper-catalyzed hydrogen peroxide generation from homocysteine. *J Clin Invest* 1986;77:1370–1379.
58. Heinecke JW, Rosen K Suzuki LA, Chait A. The role of sulfur-containing amino acids in superoxide production and modification of low density lipoprotein by arterial smooth muscle cells. *J Biol Chem* 1987;262:10098–10103.
59. Harker LA, Slichter SJ, Scott CR, Ross R. Homocysteinemia. Vascular injury and arterial thrombosis. *N Engl J Med* 1974;291:537–543.
60. Harker LA, Ross R, Slichter SJ, Scott CR. Homocystine-induced arteriosclerosis. The role of endothelial cell injury and platelet response in its genesis. *J Clin Invest* 1976;58:731–741.
61. Uhlemann ER, TenPas JH, Lucky AW, et al. Platelet survival and morphology in homocystinuria due to cystathionine synthase deficiency. *N Engl J Med* 1976;295:1283–1286.
62. Hill-Zobel RL, Pyeritz RE, Scheffel U, et al. Kinetics and distribution of [111]indium- labeled platelets in patients with homocystinuria. *N Engl J Med* 1982;307:781–786.
63. Panganamala RV, Karpen CW, Merola AJ. Peroxide mediated effects of homocysteine on arterial prostacyclin synthesis. *Prostaglandins Leukotrienes Med* 1986;22:349–356.
64. Wang J, Dudman NPB, Wilcken DEL. Effects of homocysteine and related compounds on prostacyclin production by cultured human vascular endothelial cells. *Thromb Haemostas* 1993;70:1047–1052.

65. Palareti G, Coccheri S. Lowered antithrombin III activity and other clotting changes in homocystinuria: effects of a pyridoxine-folate regimen. *Haemostasis* 1989;19(suppl):24–28.

66. Harker LA, Harlan JM, Ross R. Effects of sulfinpyrazone on homocysteine-induced endothelial cell injury and arteriosclerosis in baboons. *Circ Res* 1983;53:731–739.

67. Tawakol A, Omland T, Gerhard M, et al. Hyperhomocyst(e)inemia is associated with impaired endothelium-dependent vasodilation in humans. *Circulation* 1997;95:1119–1121.

68. Celermajer DS, Sorensen K, Ryalls M, et al. Impaired endathelial function occurs in the systemic arteries of children with homozygous homocystinuria but not in their heterozygous parents. *J Am Coll Cardiol* 1993;22:854–858.

69. Ubbink JB, Vermaak WJH, van der Merwe, et al. Vitamin requirements for the treatment of hyperhomocysteinemia in humans. *J Nutr* 1994;124:1927–1933.

70. Selhub J, Jacques PF, Wilson PW, et al. Vitamin status and intake as primary determinants of homocysteinemia in an elderly population. *JAMA* 1993;270:2693–2698.

71. Stampfer MJ, Malinow MR. Can lowering homocysteine levels reduce cardiovascular risk? *N Engl J Med* 1995;332:328–329.

72. Dudman NPB, Wilcken DEL, Wang J, et al. Disordered methionine/homocysteine metabolism in premature vascular disease. Its occurrence, cofactor therapy, and enzymology. *Arterioscler Thromb* 1993;13:1253–1260.

73. Landgren F, Israelsson B, Lindgren A, et al. Plasma homocysteine in acute myocardial infarction: homocysteine-lowering effect of folic acid. *J Intern Med* 1995;237:381–388.

74. Naurath HJ, Joosten E, Reizler R, et al. Effects of vitamin B_{12}, folate, and vitamin B_6 supplements in elderly people with normal serum vitamin concentrations. *Lancet* 1995;346:85–89.

75. Brattstrom L, Israelsson B, Norrving B, et al. Impaired homocysteine metabolism in early-onset cerebral and peripheral occlusive arterial disease: effects of pyridoxine and folic acid treatment. *Atherosclerosis* 1990;81:51–60.

76. Van der Mooren MJ, Wouters MGAJ, Blom HJ, et al. Hormone replacement therapy may reduce high serum homocysteine in postmenopausal women. *Eur J Clin Invest* 1994;24:733–736.

77. Berg K. A new serum type system in man: the Lp system. *Acta Pathol Microbiol Scand* 1963;59:369–382

78. Drayna DT, Hegele RA, Hass PE, et al. Linkage between lipoprotein (a) and a DNA polymorphism in the plasminogen. *Genomics* 1988;3:230–236.

79. Gavish D, Azrolan N, Breslow JL. Plasma Lp(a) concentration is inversely correlated with ratio of kringle IV/kringle V encoding domains in the apo(a) gene. *J Clin Invest* 1989;84:4171–4174.

80. Kronenberg F, Trenkwalder E, Dieplinger H, Utermann G. Lipoprotein(a) in stored plasma samples and the ravages of time: why epidemiological studies might fail. *Arterioscler Thromb Vasc Biol* 1996; 16:1568–1572.

81. Kraft HG, Lingenhel A, Kochl S, et al. Apolipoprotein (a) kringle IV repeat num-

ber predicts risk for coronary heart disease. *Arterioscler Thromb Vasc Biol* 1996; 16:713–719.

82. Harpel PC, Gordon BR, Parker TS. Plasmin catalyzes binding of lipoprotein(a) to immobilized fibrin and fibrinogen. *Proc Natl Acad Sci USA* 1989;86:3847–3851.

83. Hajjar KA, Gavish D, Breslow JL, Nachman RL. Lipoprotein (a) modulation of endothelial cell surface fibrinolysis and its potential role in atherosclerosis. *Nature* 1989;339:303–305.

84. Gonzales-Gronow M, Edelberg JM, Pizzo SM. Further characterization of the cellular plasminogen binding sites: evidence that plasminogen 2 and lipoprotein (a) compete for the same site. *Biochemistry* 1989;28:2374–2377.

85. Hegele RA, Freeman MR, Langer A, et al. Acute reduction of lipoprotein (a) by tissue-type plasminogen activator. *Circulation* 1992;85:2034–2038.

86. Cushing GL, Gambatz JW, Nava ML, et al. Quantification and localization of apolipoprotein (a) and B in coronary artery bypass vein grafts resected at reoperation. *Arteriosclerosis* 1989;9:593–603.

87. Rath M, Niendorf A, Reblin T, et al. Detection and quantification of lipoprotein (a) in the arterial wall of 107 coronary bypass patients. *Arteriosclerosis* 1989;9:597–592.

88. Hansen PR, Kharazmi A, Jauhianen M, Enholm C. Induction of oxygen free radical generation in human monocytes by lipoprotein (a). *Eur J Clin Invest* 1994;24:497–499.

89. Naruszewicz M, Selinger E, Davignan J. Oxidative modification of lipoprotein (a) and the effect of B-carotene. *Metabolism* 1992;41:1215–1224.

90. Celermajer DS, Sorenson KE, Georgakpoulos D, et al. Cigarette smoking is associated with dose-related and potentially reversible impairment of endothelium-dependent dilatation in healthy young adults. *Circulation* 1993;88:2149.

91. Johnstone MT, Creager SJ, Scales KM, et al. Impaired endothelium-dependent vasodilatation in patients with insulin-dependent diabetes mellitus. *Circulation* 1993;88:2150.

92. Panza JA, Quyyumi AA, Brush JE, et al. Abnormal endothelium-dependent vascular relaxation in patients with essential hypertension. *N Engl J Med* 1990;323:22–27.

93. Tsurumi Y, Nagashima H, Ichikawa K-I, et al. Influence of plasma lipoprotein (a) levels on coronary vasomotor response to acetylcholine. *J Am Coll Cardiol* 1995;26:1242–1250.

94. Schachinger V, Halle M, Minners J, et al. Lipoprotein (a) selectively impairs receptor- mediated endothelial vasodilator function of the human coronary circulation. *J Am Coll Cardiol* 1997;30:927–934.

95. Bostom AG, Cupples LA, Jenner JL, et al. Elevated plasma lipoprotein(a) and coronary heart disease in men aged 55 years and younger: a prospective study. *JAMA* 1996;27:544–546.

96. Bostom AG, Gagnon DR, Cupples LA, et al. A prospective investigation of elevated lipoprotein (a) detected by electrophoresis and cardiovascular disease in women: the Framingham Heart Study. *Circulation* 1994;90:1688–1695.

97. Cremer P, Nagel D, Labrot B, et al. Lipoprotein Lp(a) as a predictor of myocar-

dial infarction in comparison to fibrinogen, LDL cholesterol and other risk factors: results from the prospective Gottingen Risk Incidence and Prevalence Study (GRIPS). *Eur J Clin Invest* 1994;24:444–453.

98. Schaefer EJ, Lamon-Fava S, Jenner JL, et al. Lipoprotein (a) levels and risk of coronary heart disease in men. *JAMA* 1994;271:999–1003.

99. Rosengren A, Wilhelmsen L, Eriksson E, et al. Lipoprotein (a) and coronary heart disease: a prospective case-control study in a general population sample of middle-aged men. *BMJ* 1990;301:1248–1251.

100. Utermann G, Menzel HJ, Kraft HG, et al. Lp(a) glycoprotein phenotypes: inheritance and relation to Lp(a)-lipoprotein concentrations in plasma. *J Clin Invest* 1987;80:458–465.

101. Albers JJ, Marcovina SM, Lodge MS. The unique lipoprotein (a): properties and immunochemical measurement. *Clin Chem* 1990;36:2019–2026.

102. Maeda S, Abe A, Selshima M, et al. Transient changes of serum lipoprotein (a) as an acute phase protein. *Atherosclerosis* 1989;78:145–150.

103. Schriewer H, Assman G, Shulte H. Lp(a) and risk factors for coronary heart disease. *J Clin Chem Clin Biochem* 1984;22:591–596.

104. Jenner JL, Ordovas JM, Laman-Fava S, et al. Effects of age, sex, and menopausal status on plasma lipoprotein (a) levels: the Framingham Offspring Study. *Circulation* 1993;87:1135–1141.

105. Haines C, Chung T, Chang A, et al. Effect of oral estradiol on Lp(a) and other lipoproteins in postmenopausal women. *Arch Intern Med* 1996; 156:866–872.

106. Stein JH, Rosenson RS. Lipoprotein Lp(a) excess and coronary heart disease. *Arch Intern Med* 1997;157:1170–1176.

107. Sigurdsson G, Baldursdottir A, Sigvaldson H, et al. Predictive value of apolipoproteins in a prospective survey of coronary artery disease in men. *Am J Cardiol* 1992;69:1251–1254.

108. Wald NJ, Law M, Watt HC, et al. Apolipoproteins and ischaemic heart disease: implications for screening. *Lancet* 1994;343:75–79.

109. Jauhiainen M, Koskinen P, Ehnholm C, et al, Lipoprotein (a) and coronary heart disease risk: a nested case-control study of the Helsinki Heart Study participants. *Atherosclerosis* 1991;89:59–67.

110. Colemen MP, Key TJA, Wang DY, et al. A prospective study of obesity, lipids, apolipoproteins and ischaemic heart disease in women. *Atherosclerosis* 1992;92:177–185.

111. Ridker PM, Hennekens CH, Meir MJ, Stampfer MJ. A prospective study of lipoprotein (a) and the risk of myocardial infarction. *JAMA* 1993;270:2195–2199.

112. Terres W, Tatsis E, Pfalzer B, et al. Rapid angiographic progression of coronary artery disease in patients with elevated lipoprotein (a). *Circulation* 1995;91:948–950.

113. Tamura A, Watanabe T, Mikuriya Y, Nasu M. Serum lipoprotein (a) concentrations are related to coronary disease progression without new myocardial infarction. *Br Heart J* 1995;74:365–369.

114. Thompson GR, Maher VMG, Matthews S, et al. Familial Hypercholesterolemia Regression Study: a randomised trial of low-density-lipoprotein apheresis. *Lancet* 1995;345:811–816.

115. Marburger C, Hambrecht R, Niebauer J, et al. Association between lipoprotein (a) and progression of coronary artery disease in middle-aged men. *Am J Cardiol* 1994;73:742–746.

116. Hoff HF, Beck GJ, Skibinski CI, et al. Serum Lp(a) level as a predictor of vein graft stenosis after coronary artery bypass surgery in patients. *Circulation* 1988;77:1238–1244.

117. Casscells W. Migration of smooth muscle and endothelial cells: critical events in restenosis. *Circulation* 1992;86:723–729.

118. Popma JJ, Califf RM, Topol EJ. Clinical trials of restenosis after coronary angioplasty. *Circulation* 1991;84:1426–1436.

119. Hearn JA, Donohue BC, Ba'albaki H, et al. Usefulness of senun lipoprotein (a) as a predictor of restenosis after percutaneous transluminal coronary angioplasty. *Am J Cardiol* 1992;69:736–739.

120. Horie H, Takahashi M, Izumi M, et al. Association of an acute reduction in lipoprotein (a) with coronary artery restenosis after percutaneous transluminal coronary angioplasty. *Circulation* 1997;96:166–173.

121. Shewmon DA, Stock JL, Rosen CJ, et al. Tamoxifen and estrogen lower circulating lipoprotein(a) concentrations in healthy postmenopausal women. *Arterioscler Thromb* 1994;14:1586–1593.

122. Soma MR, Meschia M, Bruschi F, et al. Hormonal agents used in lowering lipoprotein (a). *Chem Phys Lipids* 1994;67-68:345–350.

123. Kuhl H, Marz W, Jung-Hoffman C, et al. Effect on lipid metabolism of a biphasic desogestrel-containing oral contraceptive: divergent changes in apolipoprotein B and E and transitory decrease in Lp(a) levels. *Contraception* 1992;47:69–83.

124. Deplanque B, Beaumont V, Lemort N, Beaumont JL. Lp(a) levels and antiestrogen antibodies in women with and without thrombosis in the course of oral contraception. *Atherosclerosis* 1993;100:183–188.

125. Lobo RA, Notelovitz M, Bernstein L, et al. Lp(a) lipoprotein: relationship to cardiovascular disease risk factors, exercise, and estrogen, *Am J Obstet Gynecol* 1992; 166:1182–1190.

126. Hiraga T, Harada K, Kobayashi T, Murase T. Reduction of serum lipoprotein (a) using estrogen in a man with familial hypercholesterolemia. *JAMA* 1992;267:2328.

127. Crook D, Sidhu M, Seed M, et al. Lp(a) levels in women given danazol, an impeded androgen. *Atherosclerosis* 1992;92:41–47.

128. Maher VMG, Brown BG. Lipoprotein (a) and coronary heart disease. *Curr Opin Lipidol* 1995; 6:229–235.

10

Lipid Therapy to Stabilize the Vulnerable Atherosclerotic Plaque

New Insights Into the Prevention of Cardiovascular Events

B. Greg Brown and Xue-Qiao Zhao
University of Washington School of Medicine, Seattle, Washington

I. INTRODUCTION

Recent large clinical trials detailed elsewhere in this volume have confirmed that lipid therapy in selected patients reduces substantially the risk of clinical cardiovascular events without causing other clinical morbidity. Earlier trials examining the effect of lipid-altering therapy on the coronary artery had already provided compelling evidence for clinical benefit, and identified a spectrum of dynamic changes in the coronary stenosis that appear to explain these benefits and provide targets for more effective disease prevention. These advances in understanding have followed technical advancements in arterial and thrombus imaging, in assessment of vasomotion, and in techniques of vessel wall and plaque histomorphometry. In particular, the quantitative arteriographic finding that intensive lipid therapy dramatically reduces the frequency of episodes of major plaque instability has been an observation central to understanding the mechanism of therapy benefit. This finding has caused us to reexamine a number of important related observations, and has fostered a new generation of studies probing plaque structure and biology for further insight.

This chapter reviews evidence supporting the idea that processes stabilizing the vulnerable lipid-rich atherosclerotic plaque are central to the prevention of clinical instability in patients with occult or with manifest coronary atherosclerosis.

II. PATHOGENIC MECHANISMS IN DYSLIPIDEMIA

This chapter focuses briefly on several clinically important aspects of plaque biology that are adversely affected in dyslipidemia, and on current understanding of the effects of lipid-altering therapy on these atherogenic processes.

A. Plaque Lipid Accumulation

LDL and more recently Lp(a) have been localized in the intimal extracellular space. Intimal lipid also accumulates intracellularly in subendothelial monocyte-derived macrophages (1,2). Such "foam cell" formation is thought to occur by unregulated scavenger receptor uptake of modified LDL (3,4) and possibly of Lp(a) (5,6) by macrophages. Smooth muscle cells may also become foam cells. Foam cells are abundant in precursor fatty streak lesions (7), in the cap of early fibrous plaques, and in the shoulders, cap, and basilar neovascular complex of advanced plaques (8).

By expert consensus (9) there are six histologic classifications of athero-sclerotic plaque, with three variants of end-stage severe obstructive disease. In the first four stages, intimal monocytes accumulate and are transformed to mac-rophages and then to foam cells. Extracellular lipid droplets form and coalesce as a core lipid region overlaid by a thickening fibrous cap of smooth muscle cells and their synthesized collagen. Lipid may enter the core region of the fibrous plaque by transmural flux of its more mobile forms (lipoprotein particles, perifibrous droplets, and vesicles (10–12), or it may be deposited there as a re-sult of foam cell necrosis (7). In the core, lipids coalesce into lower-energy phases dictated by the local cholesterol, phospholipid, and cholesteryl ester concentra-tions (13). Cholesteryl ester droplets and vesicles and cholesterol monohydrate crystals are the dominant core lipids (8,13). Thus the evolution of the plaque is dominated in its early stages by intimal lipid accumulation and a monocytic in-flammatory response with foam cell formation and subsequent necrosis creating coalescent intimal pools of extracellular lipid. Plaque formation may proceed gradually to severe obstruction (Fig. 1A), or the plaque may be transformed sud-denly by erosion or rupture of its cap, with plaque hemorrhage and/or thrombo-sis, to become the culprit lesion for an acute ischemic event (Fig. 1B).

Roberts and co-workers (14–17) have studied plaque composition as it re-lates to certain aspects of the spectrum of clinical atherosclerosis. Briefly, by quantitative morphometry of the histologic section, they have determined the proportion of intimal area which is contributed by each of the principal plaque components: (1) dense fibrous tissue; (2) cellular fibrous tissue; (3) calcific de-posits; (4) inflammatory infiltrates; (5) extracellular core lipid (as "pultaceous debris"); and (6) foam cell lipid (14). They separated serially sampled arterial sections into four ranges of lumen area reduction. They found, on average, that

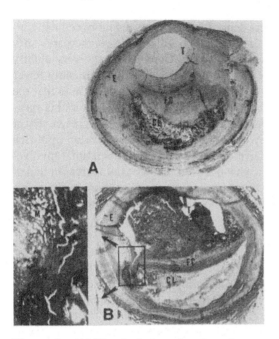

Figure 1   (A) Histological section through a structurally stable coronary plaque in a patient with vasospastic angina. Morphological features include: E, internal elastic lamina; FC, a thick fibrous cap composed largely of collagen and smooth muscle cells (SMC); CL, core lipid, here largely crystalline; T, a small tag of thrombus. (B) Section through a structurally unstable coronary plaque in a patient dying from myocardial infarction. The lumen, only moderately narrowed by the plaque, is acutely occluded by thrombus (T). There are many features in common with the section in (A). In the unstable plaque, core lipid (some dislodged by sectioning artifact) comprises a much larger fraction of the plaque. The fibrous cap is much thinner than in (A), is fissured (or vented) at its left shoulder, permitting a small pocket of hemorrhage (H) in the plaque. This fissure, the associated hemorrhagic pocket, and the plaque shoulder, here rich in lipid-laden macrophages (M) (round, bright spots), are shown at increased magnification in the inset. Also at higher magnification (not shown), the fibrous cap has few SMC but many M. (From Ref. 28.)

early intimal involvement is almost entirely fibrocellular, but at the stage of severe arterial obstruction, the cellular contribution has declined to about 25% of total intimal area while dense fibrous tissue occupies about 50% (Fig. 1A). Foam cells appear in some numbers when intimal involvement is moderate; their fraction increases to 10% of area in the more severe stages, then declines in the most severe. Calcific deposits and extracellular lipid become relatively abundant in the more severe stages; each increases progressively to contribute, on average, about 10% of intimal area in the most severe. Younger women (<40 y.o.) with CAD were found to have significantly less dense fibrous tissue and more cellular connective tissue and lipid-rich foam cells than their older male counterparts, suggesting a greater potential for reversibility (15). Conversely, the very elderly showed a tendency toward a more fibrotic disease and had significantly fewer foam cells than younger men and women (16). It is important to note that the above reports describe *average plaque morphology*. Some plaques in this spectrum have considerably greater lipid deposits and, as seen below, greater risk of plaque disruption. It is unclear why certain arterial segments develop lipid-rich plaques while adjacent segments have the more stable fibrous intimal involvement. Since Gimbrone et al. (18–20) have shown that areas of increased leukocyte adhesion molecule activity overlay areas of plaque formation, it is possible that this aspect of endothelial biology is a localizing factor for atherogenesis.

B. Regression and Progression of Coronary Lumen Obstruction

The recanalization of a previously obstructed coronary artery and the improvement of ordinary lesions are occasionally observed in a later arteriogram. Thus the critical question is not "does *regression* happen in patients?" (it does), but, "can it be promoted with sufficiently great *magnitude* and *frequency* to favorably alter the clinical course of the disease?" The emerging evidence regarding this question is extremely encouraging.

1. Experimental Observations

That atherosclerosis can regress with lipid lowering has been convincingly shown in studies in cholesterol-fed primates (21–24). Such "atherogenic" diets increase coronary artery collagen (3×), elastin (4×), and cholesterol (7×, mostly esterified). Cholesterol gradually accumulates in foam cells, and deep in the intimal lipid pool as ester droplets and monohydrate crystals. When the animals are changed to a vegetarian "regression" diet, serum cholesterol returns to normal (140 mg/dL), and the arterial lipid and connective tissue changes partially regress over 20 to 40 months. Collagen content does not decrease much from its peak value (-20%), but elastin (-50%), and cholesterol (-60%, mostly esterified) do; and the plaques shrink (22,24).

2. Evidence of Regression in Patients

Regression in patients has been defined indirectly in terms of arteriographic lumen improvement and may occur in several ways. Plaque lipid may be depleted (25) as, to a lesser extent, may its connective tissues (23). Lysis of fully occlusive thrombi or mural thrombi is commonly seen. Remodeling of the underlying vascular architecture or relaxation of excess vasomotor tone may improve lumen size independently of changes in plaque size (260. The importance of endothelial dysfunction as a basis for abnormal vasomotor and thrombogenic states, and the relationship(s) of therapy to functional recovery are becoming clarified in many of these processes (27).

Beginning in 1984, a series of randomized clinical arteriographic trials has documented the magnitude of, frequency of, and conditions under which regression can occur in patients. These have been summarized in detail (28), and are illustrated in Figures 2, 3, and 4. Of interest, in only about half of these trials was there an entry requirement for even modest hyperlipidemia. Despite the het-

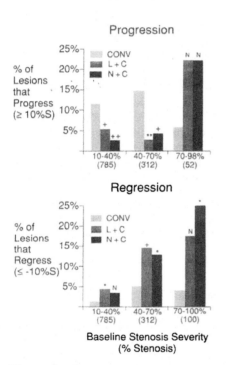

Figure 2 Frequency of definite lesion change in FATS, expressed as the percentage of lesions that decrease in severity (regress) by a measured 10% stenosis or more. Lesions from 120 patients are subgrouped into 785 mild stenoses (10→40%S), 312 moderate stenoses (40→70%S), and 52 severe stenoses (70→98%S). Also, 48 lesions, initially totally occluded, are added to the severe lesion regression analysis because these may "regress" by recanalization. In general, change is relatively infrequent. The more severe the lesion at baseline, the more likely its change. Intensive lipid-lowering therapy increases regression frequency at all levels of severity. χ^2 statistical comparisons versus control group (CONV) frequency: *$P < .05$; +$P < .02$; **$P < .005$; ++$P < .001$; N, not significant). (From Ref. 28.)

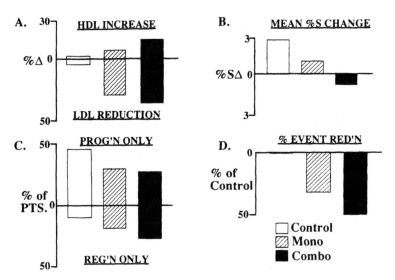

Figure 3 Graphical summary of the results of 17 angiographic trials, 8 comparing combination therapy with a control regimen and 9 comparing single-drug monotherapy with a control regimen. (See Ref. 29 for details.) (A) Mean response of LDL and HDL cholesterol among control groups □, monotherapy groups, ▨, and combination therapy groups ■. (B) Mean change in mean proximal stenosis severity, using the same treatment groups as in panel A. (C) Frequency of patient in these treatment groups with at least one lesion showing progression by at least 10%, or with at least one lesion progressing by that amount. (D) Reduction relative to control in the frequency of patients experiencing at least one of the following cardiovascular events: death, nonfatal MI, revascularization for refractory ischemia.

erogeneity among these studies in clinical presentation, lipid entry requirements, treatment regimens, and methods for arteriographic analysis, their results are surprisingly consistent (see Fig. 3). The control group in each study had minimal (<10%) LDL reduction in virtually all cases and no change in HDL. The LDL reductions were substantial in all treated groups; HDL also rose when niacin was used. Each study demonstrated an arterial treatment benefit. As a generalization from all these studies (Fig. 3C), less than one-twelfth of the *control* group patients were judged to have any improvement in arterial obstruction ("regression") during the study period. By contrast, more than one-fourth of *treated* patients were improved (a three- to fourfold increase) (Fig. 3C). Furthermore, average estimates of coronary disease severity, per patient, worsened (progressed) by about 3% stenosis among the controls, while improving (regressing) by 1% to 2% stenosis among the treated patients. In nearly every study the frequency of clinical car-

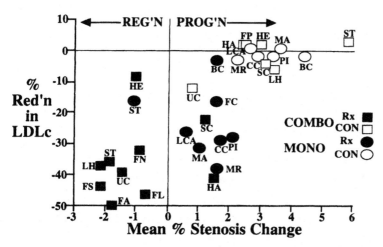

Figure 4 Data from reported angiographic trials relating the percentage of LDL lowering, from baseline, with the average change in coronary percent stenosis, per patient, in each treatment group. Comparisons are made between the therapy (Rx) and control (CON) groups for studies using drug combinations (COMBO) and single drugs (MONO). Codes: LH, lifestyle; FN, FL, FC, FP, FATS (niacin, lovastatin, colestipol, placebo); ST, STARS; SC, SCRIP; HE, Heidelberg; HA, HARP; MR, MARS; CC, CCAIT; PI, PLAC I; MA, MAAS; BC, BCCAIT; LCA, LCAS; FS, FA, FHRS. (See Ref. 29 for details on individual studies.)

diovascular events was decreased substantially with therapy, although owing to their small sizes, the reductions achieved statistical significance in only half of these trials.

3. Regression Among Lesions

As seen in Figure 2, the likelihood that a given lesion will regress was surprisingly low. In FATS, only 5% of control group and 12% of treated group lesions were seen to improve by the criterion amount of $\leq-10\%S$, which we consider "definite" regression with our method. Thus very few lesions undergo natural, or spontaneous, regression. This number can be increased significantly by lipid-lowering therapy; nevertheless a large majority of lesions are unaltered by therapies that can be characterized as "intensive" and which result in marked alterations in the lipid and lipoprotein profile. Paradoxically, these regimens are commonly associated with substantial reductions in clinical event rate. We will return to this apparent paradox below.

4. Retarding Progression of CAD

Data supporting the idea that lipid-lowering therapy can effectively retard progression of atherosclerotic arterial obstruction are detailed in Refs. 28 and 29 and Figures 2–4. Again, despite the diversity of these trials, the evidence for reduced disease progression with therapy is surprisingly consistent. Over one-half of the control group, but only one-fourth of treated patients, were judged to have worsening arterial obstruction during the study periods.

C. Plaque Disruption in Acute Ischemic Syndromes

Acute ischemic syndromes are most commonly precipitated when sites of mild or moderate atherosclerotic narrowing become disruptively transformed into severely obstructive culprit lesions. This process is illustrated in Figures 1B and 5. Evolving insights into three major aspects of atherosclerosis have greatly altered our understanding of the precipitation of plaque events leading to acute clinical events.

First, *mild* and *moderate* coronary lesions (<70% stenosis) may progress abruptly to severe obstruction, with resulting unstable angina, myocardial infarction, or death. In fact, a majority of clinical events occur under these circumstances. When the lesion precipitating an MI has, by chance, been seen on a recent angiogram, its preinfarct severity averages 50% stenosis, and its morphology will not usually suggest that it is destined to soon become occluded (30–33). Severe (≥70%) lesions are less likely to cause clinical events because they are only a small fraction of a patient's total number of lesions and because the majority of occlusions of severe stenoses occur without symptoms (34).

A second insight was originally brought into focus by Constantinides (35) but has received renewed attention (36-43). It is that, for the great majority of ischemic coronary events, a "culprit" lesion can be identified with variations of the following disruptive morphologic features at histologic examination: (1) a fissure, tear, vent, or erosion of the fibrous cap overlying the core lipid pool; (2) thrombus adherent at the site of the fissure; (3) bleeding into the core lipid region; and (4) severe arterial obstruction by the composite mass of expanded plaque and thrombus. Angiographic examples of disrupted human coronary plaques are provided in Figure 5.

A third insight is that there are features of plaque structure and lipid composition that predict the risk of fissuring (39–47). Fissures of the arterial intima rarely occur in the absence of atheroma. The greater the plaque core lipid content, the greater the likelihood of its disruption (40,41). Fissuring occurs most commonly at the shoulder of an eccentric lipid-rich plaque, a location of high macrophage density (39) and also of high circumferential stress (40,41). The

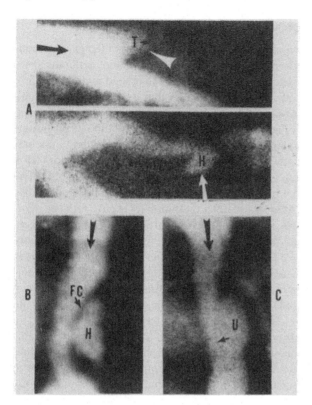

Figure 5 Highly magnified arteriographic images of structurally unstable plaques causing unstable angina or myocardial infarction. (A) This LAD is acutely occluded; 24 h after t-PAIV, the thrombotic component (T) of the obstruction is lysed, revealing a pocket of contrast (H) protruding beyond the lumen boundaries into presumed plaque and fed through a narrow-necked fissure. This appears to be the arteriographic counterpart of the hemorrhagic pocket (H) in Fig. 1B. (B) Angiographically visualized plaque hemorrhage. Bleeding into the lipid-rich core of the plaque appears to have formed a hemorrhagic pocket (H), which has driven the thin fibrous cap (FC) into the lumen, progressively obstructing it. The opposite wall has been remodeled, curving outward to preserve lumen size in the face of the expanding plaque. On cine, contrast enters this pocket from the lumen via small breaks or channels at its upstream shoulder and exits via a mid-FC vent. (C) A large ulcer (U) is seen after t-PA for unstable angina. This image appears to have been created by full-length erosion or eruption of the fibrous cap, of which only a thin arteriographic vestige remains (arrow). (From Ref. 28.)

perception of plaque disruption as a "passive" process related to the softness and size of the core lipid pool and the strength of the fibrous cap is being refined as our understanding of the "active" macrophage inflammatory mechanisms evolves (43–51).

D. Enhanced Thrombogenesis

Thrombus formation at sites of plaque rupture is the final step in the process leading to the acute ischemic syndromes. Although this review cannot detail the spectrum of prothrombotic processes, the reader is referred to related reviews (52,53).
There are some intriguing links to hyperlipidemia that will be discussed here
briefly. Risk of arterial thrombosis is determined by a complex interplay among
factors leading to disruption of atherosclerotic plaque, to activation of prothrom-
bin and fibrinogen, and to impaired fibrinolysis. The last is dependent on a balance between the cellular synthesis and release of plasminogen activators (tPA
and UPA) and of plasminogen activator inhibitors, principally PAI-1 (54). The
procoagulant state associated with elevated fibrinogen levels has been epidemio-
logically linked to clinical CAD (55). Platelet inhibition by aspirin reduces myocardial infarction risk by 44% among men (56). Furthermore, platelet aggre-
gability is increased in hypercholesterolemia (57) and cigarette smoking, possibly
related to inhibition of platelet NO release (58). Lp(a), an LDL particle complexed
to a protein with high sequence homology to plasminogen (59), has been associated with increased risk of myocardial infarction (60), is experimentally shown
to interfere with fibrinolysis (61), and is linked with the stuttering course of acute
MI (62).

Cells of the vessel wall may also contribute to a procoagulant state. Endothelial dysfunction may impair the cellular production or release of tPA. Macrophages and smooth muscle cells colocalize with thrombogenic tissue factor in
atherectomy specimens from patient with unstable (but not with stable) angina
(63).

E. Endothelium-Dependent Processes

At the critical interface between flowing blood and the vessel wall, the endothelial monolayer mediates important short- and long-term homeostatic responses.
In certain atherogenic states, specific forms of endothelial dysfunction have been
clearly documented. Once the endothelium is better understood, we are likely to
find that its generalized dysfunction in atherogenic states broadly diminishes
vascular homeostatic capacity. The following is a brief review of our understanding of endothelial dysfunction.

1. Vasodilatory Dysfunction

The epicardial coronary arteries and the coronary microvascular bed normally
dilate in response to increased flow demands. One mechanism is the flow-dependent endothelial production of EDRF (64), which is nitric oxide (NO) (65).
Endothelium-dependent vascular relaxation is impaired by most risk factors in-

cluding hypercholesterolemia (66), diabetes mellitus (67), estrogen deficiency (68), low HDL (69), elevated Lp(a) (70), small, dense LDL particles (71), and hypertension (72). Absent this vasodilatory mechanism, the circulating vasoconstrictors dominate vascular response at loci of such endothelial dysfunction. This appears to account for the apparently paradoxical epicardial coronary vasoconstriction effects of isometric (73) and aerobic (74) exercise in patients with CAD. Vasodilatory dysfunction is experimentally reversed by reducing dietary cholesterol (27), despite persistence of intimal thickening; and vascular smooth muscle responsiveness to direct dilators is largely unaltered by atherosclerosis. Vasodilatory dysfunction is therefore felt to be a direct effect of these risk factors on the local availability of NO, due either to inhibition of constitutive NO synthesis (25) or to quenching of released NO by superoxide ion (76). In recent years, evidence has accumulated that therapy of these risk factors reverses the associated vasodilatory dysfunction (68, 76–82). Indeed, the effects of estrogen and of LDL apheresis are almost immediate (68,81). Restoration of flow-mediated vasodilatation could raise the angina threshold, perhaps accounting for diminished ambulatory ischemia within 4 to 6 months after initiating lovastatin therapy (83). However, evidence that lipid-lowering therapy with statins diminishes angina or ischemia is inconclusive. The 4S patients did not self-report reduced angina with therapy (84), although treated patients experienced angina progression significantly less than controls. Conversely, among medically treated (not revascularized) patients in the REGRESS trial, pravastatin therapy improved perfusion rate (hyperemic transit time) by 51%, compared to an 18% worsening of this index among controls. Treated patients self-reported a significant (nearly 1 grade) relative improvement in angina severity and had an increase in treadmill exercise time that correlated well with the perfusion index (85).

2. Healing and Remodeling

We have yet to fully decipher the processes involved in the homeostatic regulation of arterial lumen size. Evidence indicates that flow and fluid shear stress at the endothelial interface play important roles. By developmental regulation of vessel size in the branching epicardial coronary tree, basal blood flow velocity is held relatively constant at about 10 to 20 cm/sec in branches ranging from large to small (86). This implies a relatively constant basal shear stress. Shear stress is teleologically and experimentally an appropriate regulatory variable for determining arterial size. In conditions of high basal flow, such as A-V fistulae, coronary segments can grow to four times normal diameter, stabilizing size at diameters which result in subturbulent Reynolds numbers ($Re = 4\rho/\pi \, \mu D < 2000$) (87). This capacity for flow-dependent remodeling appears to require a functioning endothelium (88). Indeed, the molecule biology of the transduction of shear stress to regulate endothelial gene expression of vasoactive products has been characterized in some detail by Gimbrone et al. (89–91). Glagov et al. (26) and Clarkson et al. (92) have shown for humans and for nonhuman primates that overall vessel size increases in concert with the growth of intimal atherosclerosis to maintain a constant lumen size. This adaptive remodeling fails when plaque area approaches 40% of the area circumscribed by the internal elastica.

Flow-dependent remodeling, while incompletely understood, almost certainly stands as another example of vascular homeostasis dependent on normal endothelial function. States of endothelial dysfunction are likely to interfere with this process.

3. Role of Endothelial Adhesion Molecules in Plaque Lipid Accumulation

Monocytes in the circulation are precursors for intimal macrophages. They enter the intima after binding to endothelial cell membrane adhesion molecules (18,19), discussed in further detail above. These adhesion molecules are seen in high-density overlying areas of increased intimal lipid, apparently induced by cytokines elaborated in the intimal cells in atherogenic conditions (20,93).

F. Timing of Effects: Mechanistic Implications

We have described certain pathologic mechanisms that are favorably altered by lipid therapy. The relative contribution of each mechanism to the overall clinical benefit is unclear. However, the time courses of certain of these mechanisms provide clues.

1. Lag Time to Reduced Clinical Events

In virtually all large clinical studies, there is a delay between the onset of therapy and the divergence of major coronary event rate (coronary death or nonfatal MI) between the treated and the control groups. Lag time varies between 7 months in WOSCOPS (94) and 26 months in CARE (95). Based on our unpublished review of these data, the variable most predictive of a short lag time is the percent reduction in major coronary events at study's end. A less important determinant is the baseline disease severity; lag time averages 13 months in the trials of healthy individuals, compared to 17 months in trials among those with clinically established coronary disease ($P = NS$). The degrees of LDL reduction or of HDL elevation with therapy do not appear to be important independent determinants of lag time. Nor do patient age or baseline LDL or HDL levels. It should be cautioned that the above conclusions are based on our preliminary analysis of only nine blinded, randomized, placebo-controlled trials.

2. Time Course of Diminished Ischemia

We have only a limited number of studies of the effect of lipid therapy on signs of ischemia. Pravastatin has significantly improved regional perfusion rate and reduced self-reported angina between an initial and final study done 2 years later (85). Self-reported angina was improved from baseline to the 1-year assessment in the treatment-unblinded Lifestyle Heart Trial (96); and quantitative PET studies documented improved regional perfusion (97). Lovastatin dramatically decreased the frequency of ambulatory episodes of spontaneous ST-segment depression as assessed at baseline and at 4 to 6 months of treatment (83). Thus, reduced ischemic may occur as early as 4 months after therapy onset. This is consistent with the clinical experience of many, but not all (84), practitioners of lipid therapy. We believe that angina, when present initially, decreases in a clinically important amount in one-third to one-half of patients, and usually does so in the first 6 months. More controlled studies are definitely needed in this area.

3. Time Course of Plaque Lipid Depletion

Regression of coronary stenosis severity, which may well be an indication of plaque lipid depletion, is reported to occur within 1 year of initiation of an intensive lifestyle-and-diet regimen (96). Other studies have reported improved stenosis severity when assessed at 2 to 3 years after treatment onset. In experimental studies, intimal foam cells induced by cholesterol feeding virtually completely disappear within 6 months of return to a normal diet with normalization (< 150 mg/dL) of plasma cholesterol. Arterial cholesterol content, greatly increased during cholesterol feeding, gradually decreases by 60% (mostly in the ester fraction) over 2 years. Cholesterol monohydrate, in crystalline form, tends to persist. Within the first 6 months of normal diet the lipid content of the plaque core region actually increases. This is thought to reflect the deposition in the core of the cholesterol content of disintegrating foam cells.

4. Time Course of Recovery of Endothelial Function

Lipid therapy has rather quickly improved the vascular response to vasodilatory stimuli such as increased flow or direct acetylcholine infusion. Endothelial dysfunction has been shown to occur within several hours of ingestion of a meal rich in saturated fat (97). Intravenous replacement of estrogen in postmenopausal women immediately corrects the endothelial dysfunction associated with estrogen deficiency (68) and LDL apheresis changes forearm vascular responsiveness within hours (81). Simvastatin has progressively improved brachial artery vasomotor function over a period of 3 months in middle-aged normolipidemic men; significant improvement occurred at 2 weeks, at which point cholesterol was lowered by 20% (99). Lovastatin alone (79), pravastatin (78), and lovastatin plus probucol (80) have been found to restore coronary vasomotor function when restudied at 6 months after therapy onset.

Thus the recovery of endothelial function with lipid therapy occurs well before plaque regression. Indeed, it seems likely that recovery of endothelial function may occur in parallel with the therapeutic normalization of circulating or vessel wall atherogenic factors such as LDL, oxidized LDL, low HDL, blood pressure, or cigarette agents. In the case of hyperlipidemia, endothelial dysfunction has been related to superoxide production by intimal lipids; superoxide quenches vasodilatory nitric oxide as it is produced by the endothelium (76).

G. Working Hypotheses: Mechanisms of Benefit

The time courses of the various direct vascular effects and clinical benefit of lipid therapy may be used to support the following preliminary hypotheses.

1.   Early (several months) recovery of endothelial vasodilatory function is likely to explain the relatively early improvement in angina and spontaneous ambulatory ischemia. These benefits occur at a time when the disappearance of macrophage foam cells and the regression of significant stenoses are probably not complete.

2.   Plaque stabilization, as manifest by a reduced frequency of the acute coronary syndromes, is likely more directly due to plaque lipid (cholesterol ester) depletion from the core lipid regions or from foam cells. Acute coronary events begin to decrease in many cases after 7 to 15 months of therapy, a time course best correlated with the experimentally observed decrease in intimal inflammatory macrophage and T-cells, and with the depletion of core lipid volume. Both intimal lipid-laden macrophage density and core lipid content predict vulnerability of the plaque to disruption, as discussed below.

3.   It is unclear, at present, whether the intensity of lipid therapy or underlying patient characteristics are importantly related to the timing of clinical benefits.

4.   If recovery of the endothelial tPA/PAI-1 balance parallels the recovery of endothelial vasodilatory function, one could argue, based on such temporal considerations, that endothelial thromboresistance is not of major importance to the therapy benefit.

III. LIPID DEPLETION FOR PLAQUE AND CLINICAL
STABILITY: A HYPOTHESIS THAT FITS THE FACTS

Therapy directed at reduction of plasma LDL would be expected to reduce the likelihood of plaque disruption because of the experimentally documented favorable effects of LDL lowering on macrophage foam cell density, core lipid pool size, and intimal LDL concentration (25). As a clinical consequence, LDL-lowering therapy ought to decrease the frequency of abrupt progression to clinical events. *Indeed, this has been the case.* Analysis of 13 coronary events among 146 FATS patients reveals that disruption of mild or moderate coronary lesions was >90% less likely among those treated with intensive cholesterol-lowering therapy than among those treated with conventional therapy (28,100). The following specific observations support this hypothesis.

A. Observations

1. Of unstable clinical episodes (e.g., myocardial infarction), 60% to 90% are due to disruption and thrombosis of lipid-rich plaques (35,37,39,42–44).
2. Plaque lipid may be depleted by normalizing elevated plasma cholesterol (21–25, 101).

   a. Lipid-laden macrophages (foam cells) disappear within 6 months (23–25).
   b. Core lipid volume increases during the first 6 months of therapy, but has begun to diminish after 12 months (25).
   c. After 2 years, 60% of plaque cholesterol is depleted (22)
   d. Plaque shrinkage is primarily due to cholesteryl ester depletion (22,25).

3. Plaque instability is predicted largely by its lipid-related features (37).

   a. Area of the core lipid region as a percent of total plaque area (42,47).
   b. Foam cell number in fibrous cap and shoulder regions (35,37,49,42–44).
   c. Foam cell stromalysin, one of a family of inflammatory metalloproteinases that can weaken the fibrous cap (45,46,59,51).

4. Only about 15% of human coronary lesions are lipid-rich plaques (>50% core lipid, by volume) (74).
5. Only about 12% of all coronary lesions visibly regress (10% stenosis change) with intensive lipid lowering (28).
6. Clinical benefits from intensive lipid lowering are associated with a 10-fold reduction in the frequency of abrupt progression of mild or moderate coronary lesions to become severe lesions (28,100).

B. Unifying Hypothesis

Angiographic coronary stenosis regression, seen in 12% of all coronary lesions during intensive lipid-lowering therapy, reflects depletion of cholesteryl esters selectively from the "vulnerable" subgroup of lipid- and foam cell-rich lesions,

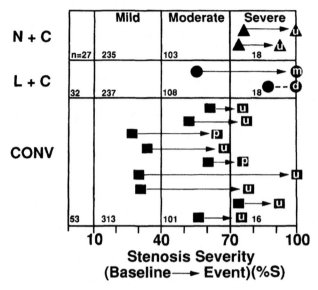

Figure 6 Lesion changes associated with the 13 coronary events as measured from 1316 lesions in 120 FATS patients. Among lesions exposed to intensive lipid-lowering therapy, only one of 683 mild or moderate lesions, at baseline, among 74 such patients progressed to a clinical event (see Fig. 5 for definitions), while eight of 414 such lesions among 46 conventionally treated patients did so (per patient or per lesion, $P < .004$). By this standard, severe lesions did not appear to benefit from therapy. N, niacin; C, colestipol; L, lovastatin; CONV, conventional therapy; U, unstable angina event; M, myocardial infarction; D, death; P, progressive angina; %S, percentage diameter stenosis. The number in each panel represents the number of lesions at risk, at baseline, in each subgroup. (From Ref. 28.)

which comprise about 15% of all visible coronary lesions. Such lipid depletion typically reduces stenosis severity by 10% to 20%, but more importantly, stabilizes the plaque in terms of its mechanical strength, inflammatory activity, and endothelial functional integrity. The plaque fibrous cap is strengthened by favorable geometric changes and by a marked reduction in the number of intimal inflammatory cells (macrophages and T- lymphocytes) which secrete proteolytic enzymes and tissue factor. Plaque stabilization by these mechanisms appears to explain the substantial reduction of clinical events associated with intensive lipid lowering.

IV. SUMMARY

The consensus of evidence from angiographic trials demonstrates both coronary artery and clinical benefits from lipid lowering, using any of a variety of treatment regimens. The findings of decreased arterial disease progression and increased regression have been convincing but, at best, modest in their magnitude. For example, among those treated intensively in FATS, the mean improvement in proximal stenosis severity was <1%S per patient, and only 12% of all lesions showed convincing regression. In view of these modest arterial benefits, the associated reductions in major cardiovascular events have been surprisingly great (24% to 35% in three recent large trials and 50% or more in angiographic trials using combination therapies). This apparent paradox appears to be explained in terms of the process of *plaque disruption*, which, with hemorrhage and thrombosis, triggers most clinical coronary events. Morphologic features of the plaque at high risk for disruption include a large accumulation of core lipid and a high density of lipid-laden macrophages in its thinned fibrous cap. Histochemical risk factors include matrix-degrading proteinases and procoagulant tissue factors, colocalized with regions of macrophage inflammatory activity. Lesions with these characteristics comprise only 10% to 20% of the overall lesion population but account for 80% to 90% of the acute clinical events.

The reduction in clinical events observed in these trials appears to be best explained by favorable effects of lipid-lowering therapy on these "high-risk" features of plaque morphology and biology, which are the lipid and foam cell content of the plaque, the strength of its fibrous cap, and the functional integrity of its endothelial surface. The composite of data presented here supports the hypothesis that lipid-lowering therapy selectively lipid-depletes (regresses) that relatively small but dangerous subgroup of fatty lesions containing a large lipid core and dense clusters of intimal macrophages. By doing so, these lesions are effectively *stabilized* and clinical event rate is accordingly decreased.

ACKNOWLEDGMENTS

The efforts of Brad Sousa in preparing this manuscript are greatly appreciated. This chapter has been extensively modified from an article published previously (27).

Supported in part by NIH grants (P01 HL 30086, R01 HL 19451, and R01 49546) from the National Heart, Lung, and Blood Institute; a grant from the John Locke Jr. Charitable Trust; the University of Washington Clinical Research Center (NIH #RR-37); and a grant (DK-35816) to the University of Washington Clinical Nutrition Research Unit from the National Institute of Diabetes, Digestive and Kidney Disorders.

REFERENCES

1. Gerrity RG. The role of monocyte in atherogenesis. I. Transition of bloodborne monocytes into foam cells in fatty lesions. Am J Pathol 1981; 103:181–190.

2. Ross R. The pathogenesis of atherosclerosis-an update. N Engl J Med 1986; 314:488–500.
3. Steinberg D, Parthasarathy S, Carew TE, Khoo JC, Witztum JL. Beyond cholesterol: modifications of low-density lipoprotein that increase its atherogenicity. N Engl J Med 1989; 320:915–924.
4. Berliner JA, Territo MC, Sevanian A, Ramin S, Kim JA, Barnshad B, Esterson M, Fogelman AM: Minimally modified LDL stimulates monocyte endothelial interactions. J Clin Invest 1990; 85:1260–1266.
5. Yamaguchi J, Hoff MF. Apolipoprotein B accumulation and development of foam cell lesions in coronary arteries of hypercholesterolemic swine. Lab Invest 1984; 51:325–332.
6. Krempler F, Kostner GM, Roscher A, Bolzano K, Sandhofer F. The interaction of human apoB containing lipoproteins with mouse peritoneal macrophages: a comparison of Lp(a) with LDL. J Lipid Res 1984; 25:283–287.
7. Stary HC. Changes in the cells of atherosclerotic lesions as advanced lesions evolve in coronary arteries of children and young adults. In: *Pathobiology of the Human Atherosclerotic Plaque*. New York: Springer-Verlag, 1989;93–106.
8. Guyton JR, Klemp KF. The lipid-rich core region of human atherosclerotic fibrous plaques. Am J Pathol 1989; 34:705–717.
9. Stary HC, Chandler AB, Dinsmore RE, Fuster V, Glagov S, Insull W, Rosenfeld ME, Schwartz CJ, Wagner WD, Wissler RW. A definition of advanced types of atherosclerotic lesions and a histologic classification of atherosclerosis. Circulation 1995; 92:1512–1531.
10. Guyton JR, Bocan TMA. Human aortic fibrolipid lesions. Progenitor lesions for fibrous plaques, exhibiting early formation of the cholesterol-rich core. Am J Pathol 1985; 120:193–206.
11. Smith EB, Evans PH, Pownham MD. Lipid in the aortic intima: the correlation of morphological and chemical characteristics. J. Atheroscler Res 1967; 7:171–186.
12. Guyton JR, Bocan TM, Schifani TA. Quantitative ultrastructural analysis of perifibrous lipid and its association with elastin in nonatherosclertic human aorta. Arterio 1985; 5:644–652.
13. Small DM, Shipley GG. Physical-chemical basis of lipid deposition in atherosclerosis. Science 1974;185:222–229.
14. Kragel AH, Reddy SG, Wittes JT, Roberts WC. Morphometric analysis of the composition of atherosclerotic plaques in the four major epicardial coronary arteries in acute myocardial infarction and in sudden coronary death. Circulation 1989; 80:1747–1756.
15. Dollar AL, Kragel AH, Fernicola DJ, Waclaview MA, Roberts WC. Composition of atherosclerotic plaques in coronary arteries in women less than 40 years of age with fatal coronary artery disease and implications for plaque reversibility. Am J Cardiol 1991; 67:1223–1227.
16. Gertz SD, Malezadah S, Dollar, Kragel AH, Roberts WC. Composition of atherosclerotic plaques in the four major epicardial coronary arteries in patients greater than or equal to 90 years of age. Am J Cardiol 1991; 67:1228–1233.

17. Kragel AH, Roberts WC. Composition of atherosclerotic plaques in the coronary arteries in homozygous familial hypercholesterolemia. Am Heart J 1991; 121:210–211.
18. Luscinskas FW, Gimbrone MA Jr. Endothelial-dependent mechanisms in chronic inflammatory leukocyte recruitment. Annu Rev Med 1996;47:413–421.
19. De Canterina R, Libby P, Peng HB, Thannickal VJ, Rajavashisth TB, Gimbrone MA, Jr, Shin WS, Liao JK. Nitric oxide decreases cytokine-induced endothelial activation. Nitric oxide selectively reduces endothelial expression of adhesion molecules and proinflammatory cytokines. J Clin Invest 1995; 6:60–68.
20. Gerzten RE, Luscinskas FW, Ding HT, Dichek DA, Stoolman LM, Gimbrone MA, Jr, Rosenzweig A. Adhesion of memory lymphocytes to vascular cell adhesion molecule-1-transduced human vascular endothelial cells under simulated physiological flow conditions in vitro. Circ Res 1996; 79;1205–1215.
21. Wissler RW, Vesselinovitch D. Can atherosclerotic plaques regress? Anatomic and biochemical evidence from nonhuman animal models. Am J Cardiol 1990; 65:33–40.
22. Armstrong ML, Megan MB. Lipid depletion in atheromatous coronary arteries in rhesus monkeys after regression diets. Circ Res 1972; 30:675–680.
23. Clarkson TB, Bond MG, Bullock BC, Marzetta CA. A study of atherosclerosis regression in *Macaca mulatta*. IV. Changes in coronary arteries from animals with atherosclerosis induced for 19 months and then regressed for 24 or 48 months at plasma cholesterol concentrations of 300 or 200 mg/dl. Exp Mol Pathol 1981; 34:345–368.
24. Armstrong MC, Megan MB. Arterial fibrous protein in cynomolygous monkeys after atherogenic and regression diets. Circ Res 1975;36:256–261.
25. Small DM, Bond MG, Waugh D, Prack M, Sawyer JK. Physiochemical and histological changes in the arterial wall of nonhuman primates during progression and regression of atherosclerosis. J Clin Invest 1984; 73:1590–1605.
26. Glagov S, Weisenberg E, Zarins CK, Stankunavicius R, Kolettis GJ. Compensatory enlargement of human atherosclerotic coronary arteries. N Engl J Med 1987; 316:1371–1375.
27. Harrison DG, Armstrong ML, Freeman PC, Heistad DD. Restoration of endothelium-dependent relaxation by dietary treatment of atherosclerosis. J Clin Invest 1987; 80:808–811.
28. Brown BG, Qiao X-Q, Sacco DE, Albers JJ. Lipid lowering and plaque regression: new insights into plaque disruption and clinical events in coronary disease. Circulation 1993; 87:1781–1791.
29. Brown BG, FusterV. Impact of management in stabilization of coronary disease. In: *Atherosclerosis and Coronary Artery Disease*. Fuster V, Ross R, Topol EJ, eds. Philadelphia: Lipincott-Raven, 1996: 191–205.
30. Brown BG, Gallery CA, Badger RS, Kennedy JW, Mathey D, Bolson EL, Dodge HT: Incomplete lysis of thrombus in the moderate underlying atherosclerotic lesion during intracoronary infusion of streptokinase for acute myocardial infarction: Quantitative angiographic observations. Circulation 1986; 73:653–661.
31. Ambrose JA, Tannenbaum MA, Alexopoulos D, et al. Angiographic progression

of coronary artery disease and the development of myocardial infarction. J Am Coll Cardiol 1988; 12:56–62.

32. Little WC, Constantinescu M, Applegate RM, et al. Can coronary angiography predict the site of a subsequent myocardial infarction in patients with mild-to-moderate coronary artery disease? Circulation 1988; 78:1157–1166.

33. Little WC. Angiographic assessment of the culprit coronary artery lesion before acute myocardial infarction. Am J Cardiol 1990; 66:44G–47G.

34. Webster MWI, Cheesebro JH, Smith HC, et al. Myocardial infarction and coronary artery occlusion: a prospective 5-year angiographic study. J Am Coll Cardiol 1990; 15(suppl A):218A. Abstract.

35. Constantinides P. Plaque hemorrhages, their genesis and their role in supra-plaque thrombosis and atherogenesis. In: *Pathobiology of the Human Atherosclerotic Plaque*. Glagov S, Newman WP, Schaffer SA, eds. New York:Springer-Verlag, 1990:393–411.

36. Fuster V, Badimon L, Badimon JJ, Chesebro JH. The pathogenesis of coronary artery disease and the acute coronary syndromes. N Engl J Med 1992; 326:242–250, 310–318.

37. Tracey RE, Devaney K, Kissling G. Characteristics of the plaque under a coronary thrombus. Virchows Arch Pathol Anat 1985; 405:411–427.

38. Davies MJ, Krikler DM, Katz D. Atherosclerosis: inhibition or regression as therapeutic possibilities. B Heart J 1991; 65:302–310.

39. Lendon CL, Davies MJ, Born GVR, Richardson PD. Atherosclerotic plaque caps are locally weakened when macrophage density is increased. Athero 1991; 87:87–90.

40. Richardson PD Davies MJ, Born GVR. Influence of plaque configuration and stress distribution on fissuring of coronary atherosclerotic plaques. Lancet 1989; iii:941–944.

41. Loree HM, Kamm RD, Strongfellow RG, Lee RT. Effects of fibrous cap thickness on peak circumferential stress in model atherosclerotic vessels. Circ Res 1992; 71:850–858.

42. Davies MJ. Stability and instability: two faces of coronary atherosclerosis. The Paul Dudley White Lecture 1995. Circulation 1996; 94:2013–2020.

43. Van der Wal AC, Becker AE, Van der Loos CM, Das PK. Site of intimal rupture or erosion of thrombosed coronary atherosclerotic plaques is characterized by an inflammatory process irrespective of the dominant plaque morphology. Circulation 1994; 89:36–44.

44. Moreno PR, Falk E, Palacios IF, Newell JB, Fuster V, Fallon JT. Macrophage infiltration in acute coronary syndromes: implications for plaque rupture. Circulation 1994; 90:775–778.

45. Shah PK, Falk E, Badimon JJ, et al. Human monocyte-derived macrophages express collagenase and induce collagen breakdown in atherosclerotic fibrous caps: implication for plaque rupture. Circulation 1993; 88(suppl I):I–254. Abstract.

46. Henney AM, Wakeley PR, Davies MJ, Foster K, Hembrey R, Murphy G, Humphries S. Location of stromelysin gene in atherosclerotic plaques using in-site hybridization. Proc Natl Acad Sci USA 1991; 88:8154–8158.

47. Davies MJ, Richardson PD, Woolf N, Katz DR, Mann J. Risk of thrombosis in human atherosclerotic plaques: role of extracellular lipid, macrophages, and smooth muscle cell content. Br Heart J 1993; 69:377–381.
48. Libby P. Molecular bases of the acute coronary syndromes. Circulation 1995; 91:2844–2850.
49. Galis ZS, Sakhova GK, Lark MW, Libby P. Increased expression of matrix metalloproteinases and matrix-degrading activity in vulnerable regions of human atherosclerotic plaques. J Clin Invest 1994; 94:2493–2503.
50. Mach F, Schönbeck U, Sukhova GK, et al. Functional CD40 ligand is expressed on human vascular endothelial cells, smooth muscle cells, and macrophages: implications for CD40-CD40 ligand signalling in atherosclerosis. Proc Natl Acad Sci USA 1997; 94:1931–1936.
51. Malik N, Greenfield BW, Wahl AF, Kiener PA. Activation of human monocytes through CD40L induces matrix metalloproteinases. J Immunol 1996; 156:3952–3960.
52. Ridker PM, Vaughn DE, Stampfer MJ, Sacks FM, Hennekens CH. A cross section study of endogenous tissue plasminogen activator, total cholesterol, HDL cholesterol, and apolipoproteins A-I, A-II, B-100. Arterioscler Thromb 1993; 13:1587–1592.
53. Harker LA, Hanson SR, Kelly AB. Antithrombotic strategies targeting thrombin activities, thrombin receptors and thrombin generation. Thromb Haemostas 1997; 78:736–741.
54. Linjen HR, Collen D. Impaired fibrinolysis and the risk for coronary heart disease. (Editorial.) Circulation 1996; 94:2052–2054.
55. Benderly M, Graff E, Reicher-Reiss H, Behar S, Brunner D, Goldbourt U. Fibrinogen is a predictor of mortality in coronary heart disease patients. Arterioscler Thromb Vasc Biol 1996; 16:351–356.
56. Steering Committee of the Physician's Health Study Research Group. Final report on the aspirin component of the ongoing Physician's Health Study. N Engl J Med 1989;321:129–135.
57. Carvalho ACA, Colman RW, Less RS. Platelet function in hypercholesterolemia. N Engl J Med 1974; 290:434–438.
58. Ichiki K, Hisao I, Haramaki N, Ueno T, Imaigumi T. Long-term smoking impairs platelet-derived nitric oxide release. Circulation 1996; 94:3109–3114.
59. McLean JW, Tomlinson JE, Kuang WJ, Lawn RE. CDNA sequence of human apolipoprotein (a) is homologous to plasminogen. Nature 1987; 330:132–137.
60. Maher VMG, Brown BG. Lipoprotein (a) and coronary heart disease. Curr Opin Lipidiol 1995; 6:229–235.
61. Jaijar KA, Gavish D, Breslow JL, Nachman RL. Lipoprotein (a) modulation of endothelial cell surface fibrinolysis and its potential role in atherosclerosis. Nature 1989; 339:303–305.
62. Haider AW, Andreotti F, Thompson GR, Kluft C, Maseri A, Davies GJ. Serum lipoprotein (a) level is related to thrombin generation and spontaneous intermittent coronary occlusion in patients with acute myocardial infarction. Circulation 1996; 94:2072–2076.

63. Moreno P, Bernardi V, Lopez-Cuellar J, Marcia AM, Palacios I, Gold HK. Macrophages, smooth muscle cells, and tissue factor in unstable angina. Implications for cell-mediated thrombogenicity in acute coronary syndromes. Circulation 1996; 94:3090–3097.

64. Furchgott RF, Zawadski JV. The obligatory role of endothelial cells in the relaxation of arterial smooth muscle by acetylcholine. Nature 1989; 299:373–376.

65. Moncada S, Palmer RM, Higgs EA. Nitric oxide physiology, pathophysiology, and pharmacology. Pharmacol Rev 1991; 43:109–142.

66. Yokoyama I, Ohtake T, Mormomura S-I, Nishikawa J, Sasaki Y, Omata M. Reduced coronary flow reserve in hypercholesterolemic patients without overt coronary stenosis. Circulation 1996; 94:3232–3238.

67. Clarkson P, Celermajer DS, Donald AE, Sampson M, Sorensen KE, Adams M, Yue DK, Betteridge DJ, Deanfoeld JE. Impaired vascular reactivity in insulin-dependent diabetes mellitus is related to disease duration and LDL cholesterol levels. *J Am Coll Cardiol* 1996; 28:573–579.

68. Lieberman EH, Gerhard MD, Uchata A, Walsh BW, Selwyn AP, Gang P, Yeung AC, Creager MA. Estrogen improves endothelium-dependent, flow-mediated vasodilation in post-menopausal women. Ann Intern Med 1994; 121:936–941.

69. Zeiher AM, Schächinger V, Hohnloser SH, Saurbier B, Just H. Coronary atherosclerotic wall thickening and vascular reactivity in humans: elevated high-density lipoprotein levels ameliorate abnormal vasoconstriction in early atherosclerosis. Circulation 1994; 89:2525–2532.

70. Sorensen KE, Celermajer DS, Geogakopoulos D, Hatcher G, Betteridge DJ, Deanfield JE. Impairment of endothelium-dependent dilation is an early event in children with familial hypercholeste role mia and its relation to the lipoprotein (a) level. J Clin Invest 1994; 93:50–55.

71. Dyce MC, Anderson TJ, Yeung AC, Selwyn AP, Gang P. Indices of LDL particle size closely relate to endothelial function. Circulation 1993; 88(suppl I):I-466–I-471.

72. Panza JA, Quyzumi AA, Brush JE, Epstein SE. Abnormal endothelium-dependent vascular relaxation in patients with essential hypertension. N Engl J Med 1990; 323:22–27.

73. Brown BG, Lee AB, Bolson EL, Dodge HT. Reflex constriction of significant coronary stenosis as a mechanism contributing to ischemic ventricular dysfunction during isometric exercise. Circulation 1984; 70:18–24.

74. Hess OM, Bortone A, Eid K, Coronary vasomotor tone during static and dynamic exercise. Eur Heart J 1989; 10(suppl F):105–110.

75. Liao JK, Shin WS, Lee WY, Clark SL. Oxidized low density lipoprotein decreases the expression of endothelial nitric oxide syntheses. J Biol Chem 1995; 270:319–324.

76. Muegge A, Edwell JH, Peterson TE, Hofmeyer TG, Heistad DD, Harrison DD. Chronic treatment with polyethylene-glycolated superoxide dismutase partially restores endothelium-dependent vascular relaxations in cholesterol-fed rabbits. Circ Res 1991; 69:1293–1300.

77. Leung WH, Lau CP, Wong CK. Beneficial effect of cholesterol-lowering therapy

on coronary endothelium-dependent relaxation in hypercholesterolemic patients. Lancet 1993; 341:1496–1500.

78. Egoshira K, Hirooka Y, Kai H, et al. Reduction in serum cholesterol with pravastatin improved endothelium-dependent coronary vasomotion in patients with hypercholesterolemia. Circulation 1994; 89:2519–2524.

79. Treasure CB, Kleis JL, Weintroub WS, et al. Beneficial effects of cholesterol-lowering therapy on the coronary endothelium in patients with coronary artery disease. N Engl J Med 1995; 332:481–487.

80. Anderson TJ, Meredith IT, Yeung AC, Frei B, Selwyn AP, Gang P. The effect of cholesterol-lowering and antioxidant therapy on endothelium-dependent coronary vasomotion. N Engl J Med 1995; 332:488–493.

81. Jamori O, Matsuoka H, Itabe H, Wada Y, Kohno K. Imaizumi T, Single LDL apheresis improves endothelium-dependent vasodilation in hypercholesterolemic humans. Circulation 1997; 95:76–82.

82. Stroes ES, Koomons HA, de Bruin TW, Tabelink TJ. Vascular function in the forearm of hypercholesterolemic patients off and on lipid-lowering medication. Lancet 1995; 346:467–471.

83. Andrews TC, Raby K, Barry J, Naimi CL, Alfred E, Ganz P, Selwyn AP. Effect of cholesterol reduction on myocardial ischemia in patients with coronary disease. Circulation 1997; 95:324–328.

84. Kjekshus J, Pederson TR, Pyorala K, Olsson AG. Effect of simvastatin on ischemic signs and symptoms in the 4S. *J Am Coll Cardiol* 1997; 29:75A. Abstract.

85. Aengevaeren WR, Uijen GJ, Jukema JW, Bruschke AV, van der Werf J. Functional evaluation of lipid-lowering therapy by pravastatin in the Regression Growth Evaluation Statin Study (REGRESS). Circulation 1997; 96:429–435.

87. Jaffe RB, Glancy DC, Epstein SE, Brown BG, Morrow AG. Coronary arterial-right heart fistulae: long-term observations in seven patients. Circulation 1973; 47:133–143.

88. Langille, BL, O'Donnell F. Reductions in arterial diameter produced by chronic decreases in blood flow are endothelium-dependent. Science 1986; 231:405–407.

89. Topper JN, Cai J, Falb D, Gimbrone MA Jr. Identification of vascular endothelial genes differentially responsive to fluid mechanical stimuli: cyclo-oxygenase-2, manganese superoxide dismutase, and endothelial cell nitric oxide synthase are selectively up-regulated by steady laminar shear stress. Proc Natl Acad Sci USA 1996; 93:10417–10422.

90. Resnick Y, Yahav H, Khchigan LM, Collins T, Anderson KR, Dewey FC, Gimbrone MA, Jr. Endothelial gene regulation by laminar shear stress. Adv Exp Med Biol 1997:155–164.

91. Khachigian LM, Resnick N, Gimbrone MA, Jr., Collins T. Nuclear factor-kappa B interacts functionally with the platelet-derived growth factor B-chain shear-stress response element in vascular endothelial cells exposed to fluid shear stress. J Clin Invest 1995; 96:1169–1175.

92. Clarkson TB, Pritchard RW, Morgan TM, Petrick GS, Klein KP. Remodeling of coronary arteries in human and non-human primates. JAMA 1994; 271:317–318.

93. O'Brien KD, McDonald TO, Chait A, Allen MD, Alpers CE. Neovascular expression of E-selectin, intercellular adhesion molecule-1, and vascular cell adhesion molecule-1 in human atherosclerosis and their relation to intimal leukocyte content. Circulation 1996; 93:672–682.

94. Shepherd J, Cobbe SM, Ford I, Prevention of coronary heart disease with pravastatin in men with hypercholesterolemia. N Engl J Med 1995; 333;1301–1307.

95. Sacks FM, Pfeffer MA, Moye LA, Rouleau JL, Rutherford JD, Braunwald E. The effect of pravastatin on coronary events after myocardial infarction in patients with average cholesterol levels. N Engl J Med 1996; 335:1001–1009.

96. Ornish D. Can lifestyle changes reverse coronary heart disease? Lancet 1990; 336:129–133.

97. Gould KL, Ornish D, Scherwitz L, Brown S, Edens RP, Hess MJ. Changes in myocardial perfusion abnormalities by positron emission tomography after long-term, intense risk factor modification. JAMA 1995; 274:894–901.

98. Vogel RA, Corretti MC, Plotnick GD. Effect of a single high fat meal on endothelial function in healthy subjects. Am J Cardiol 1997; 79:350–354.

99. Vogel RA, Corretti MC, Plotnick GD. Changes in flow-mediated brachial artery vasoactivity with lowering of desireable cholesterol levels in healthy middle-aged men. Am J Cardiol 1996; 77:37–40.

100. Brown BG, Albers JJ, Fisher LD, et al. Regression of coronary artery disease as a result of intensive lipid-lowering therapy in men with high levels of apoliproptein B. N Engl J Med 1990; 323:1289–1298.

101. Zhao XQ, Chun Y, Huss-Frechette EH, Maravilla KR, Brown BG. Effects of intensive lipid-lowering therapy on the characteristics of carotid atherosclerotic plaques by MRI (abstract). JACE 1999; 33(Suppl A):A–253.

11

Cholesterol-Lowering Trials with Carotid Ultrasonographic Outcomes

Robert P. Byington
Wake Forest University School of Medicine, Winston-Salem, North Carolina

I. INTRODUCTION

B-mode ultrasonography has been employed since the mid-1980s as a method of measuring the progression of carotid atherosclerosis in clinical trials designed to test a variety of cardiovascular treatment modalities, including lipid–lowering regimens. This chapter will begin with a short historical background describing the rationale for using this imaging technique. This will be followed by a discussion of six trials that have used ultrasonography to test the effects of lipid-lowering treatments on carotid atherosclerosis.

II. BACKGROUND: EARLIER TRIALS WITH EVENT OUTCOME MEASURES

By the mid-1960s it had been accepted that elevated blood cholesterol was a risk factor for experiencing a clinical coronary heart disease (CHD) event. But whether this relationship between cholesterol and the clinical consequences of atherosclerotic disease was causal had yet to be demonstrated. In the 22 years between 1965 and 1987, a series of large, long-term clinical trials designed to test this "lipid hypothesis" were conducted in high-risk participants. Four of these trials may be considered classical and pivotal: the WHO Clofibrate trial, the Coronary Primary Prevention trial, the Helsinki Heart Study, and the Coronary Drug Project. The first three were primary prevention trials in men with elevated cholesterol levels, and the fourth was a secondary prevention trial in men with a prior myocardial infarction (MI) documented by an electrocardiogram. Each of the four trials had clinical events as the primary outcome measure. The emphasis on primary prevention partly reflected the assumption that the amount of atherosclerosis in individuals who had experienced a clinical event may be far beyond the benefits of lipid lowering. (Also, most of the secondary prevention trials at that time were focused on treating the ischemic, thrombotic, electrical, and mechanical problems experienced by coronary patients.)

The first of the large primary prevention trials testing the lipid hypothesis, the WHO Cooperative Trial on Primary Prevention of Ischaemic Heart Disease Using Clofibrate (WHO Clofibrate Trial), began following 10,627 men in 1965. The initial findings were published 13 years later, in 1978, after 5.3 years of follow-up (=56,323 person-years of follow-up) (1). A 9% reduction in total cholesterol was achieved with clofibrate. Disturbingly, although there was a 20% reduction in major coronary events associated with clofibrate treatment (primarily a reduction in nonfatal infarction (2–3)), there was a concomitant 25% increase in all-cause mortality.

The second large primary prevention trial was the Lipid Research Clinics Coronary Primary Prevention Trial (CPPT), which was begun in 1973 testing the lipid-lowering effects of the bile acid sequestrant cholestyramine in 3806 men followed for an average of 7.4 years (= 28,164 person-years of follow-up) (4). Similar to the results from the WHO Clofibrate Trial, the CPPT investigators reported a a 9% reduction in total cholesterol and a 19% reduction in major coronary events associated with the lipid lowering (5,6). Contrary to the results of the WHO Clofibrate Trial, however, CPPT reported that all-cause mortality was reduced by 7%.

The third large primary prevention trial was the Helsinki Heart Study of gemfibrozil, which was begun in 1981. In this trial, 4081 men were followed for 5 years (= 20,405 person-years of follow-up). Similar to the other trials, participants randomized to the lipid-lowering agent had a 9.7% reduction in total cholesterol, but a slightly larger reduction of 34% reduction in fatal and nonfatal coronary events (7).

The fourth pivotal trial testing the lipid hypothesis was the Coronary Drug Project (CDP), which was a secondary prevention trial begun in 1966 to assess the long-term efficacy of five lipid-altering regimens (estrogen [2 dosage levels], clofibrate, dextrothyroxine sodium, and niacin) in 8341 men with a documented MI followed for 6.2 years (8). The two estrogen regimens and dextrothyroxine were discontinued early because of adverse events, and clofibrate appeared to be ineffective. The niacin group, followed for an average of 6.2 years (= 24,000 person-years), demonstrated a modest reduction in nonfatal recurrent MI, but no effect on all-cause mortality (9), although in a posttrial follow-up of participants (mean follow-up of 15 years), all-cause mortality was 11% lower in the niacin group (10).

Taken together, and coupled with almost 20 smaller trials, these trials ultimately led to the acceptance of the lipid hypothesis. Other trials, and pooled analyses of trials, demonstrated that the increase in all-cause mortality in the WHO clofibrate trial was a function of the agent used to lower cholesterol (clofibrate), not the process of lipid lowering itself (11,12). Furthermore, the recently reported results from the Scandinavian Simvastatin Survival Study (studying 4444 coronary patients) (13) and the West of Scotland trial (studying 6595 high-risk men (14) have gone far in demonstrating that lipid lowering can, in both primary and secondary prevention settings, translate into improved overall survival.

Many questions remain unanswered. The four pivotal trials described above were conducted in middle-aged men with (usually) elevated lipid levels. Would women experience similar benefits? Would older individuals? Would individuals with less than severely elevated lipids? Smokers? Diabetics? Hypertensives? The list of population subgroups demanding analysis is almost endless. Also, the WHO Clofibrate Trial and the CDP remind us that drugs can have unforeseen adverse effects. Any new agent must therefore be tested for safety beyond the usual Phase II/III requirements. Related to this, the newer lipid-lowering agents have the ability to lower cholesterol to levels far lower than the earlier agents. Is lower actually better? Are there plateau effects? Given that all cells require some low-density lipoprotein (LDL cholesterol or LDL), is there a level at which LDL-lowering actually becomes dangerous?

Clinical trials of coronary disease with event outcomes have historically been long and have required many thousands of participants. The four pivotal trials described above, which gave us only preliminary information on the effects of lipid lowering and clinical events, followed a total of 26,855 men for up to 10 years, giving us over 150,000 person-years of follow-up. By the early 1980s it was recognized that event trials, although of "bottom-line" importance, were too long and too expensive to be used extensively to address the thousands of questions that were confronting clinicians and public health workers.

III. ANGIOGRAPHY AND ULTRASONOGRAPHY

In addition to the resource-hungry and time-consuming challenges inherent in the large clinical trials, an appreciation was developing during the 1980s that the large trials were not actually studying the disease coronary atherosclerosis itself, but rather the clinical consequences of the disease. The intriguing concept of actually studying atherosclerosis itself became an focus of attention.

The use of imaging techniques, which were becoming popularly employed in clinical settings, were found to be useful in clinical trials settings. There are many advantages to using an imaging outcome measure. From a statistical point of view, a trial with an imaging outcome measure would allow each participant to contribute an almost equivalent amount of information to the statistical test, in contrast to an event trial, in which individuals with an event contribute disproportionately more information. Therefore, instead of doing a trial with thousands of individuals followed for 5 to 10 years, one could do a trial with hundreds of individuals followed for 2 to 3 years. Furthermore, from a biological point of view, imaging techniques measure atherosclerosis (or its luminal consequences) at the anatomic level. We would therefore be testing the drug at the very level in which we would expect to see an effect.

The two primary imaging techniques currently used in cardiovascular clinical trials are angiography of the coronary arteries and B-mode ultrasonography of the carotid arteries. (Other techniques, such as magnetic resonance imaging, are only now being developed and employed in population studies and clinical trials.) Each technique has its advantages and disadvantages. Angiography is a popular diagnostic tool in cardiology and its results are easily understood by practicing physicians. In fact, as demonstrated in other chapters in this book, angiography is a popular tool in clinical trials. However, coronary angiography does have its disadvantages. It is an invasive technique and only warranted in individuals with documented clinical disease (e.g., clinical angina or myocardial infarction). Even among these people its use is generally restricted to one examination at baseline and a second at the end of follow-up. It is therefore not used in a primary prevention setting, or among individuals with early atherosclerosis. Also, from a biologic point of view, angiography does not directly measure atherosclerosis, but rather measures a consequence of atherosclerosis: the diminution of a coronary arterial lumen caused by the encroachment of atherosclerosis from the arterial wall. Furthermore, the mere absence of a focal angiographic lesion does not imply absence of atherosclerosis because of compensatory arterial dilatation in the early stages of the disease: as atherosclerosis develops within the wall of the vessel, the lumen caliber is maintained via vascular remodeling (15).

Carotid ultrasonography is the other popular imaging technique in both clinical and clinical trial settings. However, in the clinical setting, it is the lumen and the search for a stenotic lesion that are very often the objects of interest. In clinical trials, on the other hand, it is the measurement of the mural intimal-medial thickness (IMT) that is of interest. This is the hallmark and primary advantage of this imaging technique as a clinical trial methodology: the measurement of atherosclerosis itself in the arterial wall (16–18). Other advantages are that it is low-cost and noninvasive, allowing multiple measurements over the course of follow-up. These advantages permit savings in sample size and allow ascertainment of interim follow-up data on participants who eventually die or who are lost to follow-up before the scheduled termination of the trial.

The primary disadvantage of ultrasonography is that it is not currently available for use in measuring disease progression in the coronary arteries, which are too small and too mobile to permit accurate measurements at this time. However, atherosclerosis is a generalized disease and increasing carotid IMT, as a measure of atherosclerosis, is correlated with angiographic coronary disease (19,20), clinical atherosclerotic disease (21–23), and the traditional coronary disease risk factors (24,25).

IV. B-MODE ULTRASONOGRAPHY

A. Imaging and Measuring Intimal-Medial Thickness

The use of B-mode ultrasonography to image carotid atherosclerosis is based upon the pulse-echo SONAR principle in which the arrival time of echos from an ultrasound pulse source is measured to create a two-dimensional real-time display of echo-producing arterial structures. The ultrasound device required for the assessment of atherosclerosis typically uses short pulses of 7.5 to 10 MHz ultrasound, which provides an axial resolution of 0.1 to 0.2 mm (18,26).

Highly trained and certified sonographers and readers are required to perform and analyze reproducible examinations on trial participants (26). Standardized protocols are used that define the method of interrogation, the anatomic landmarks to be followed, and the identification of the intimal, medial, adventitial, and luminal boundaries. In the typical clinical trial, the sonographer in the clinic scans the neck and identifies and records on videotape the longitudinal B-mode images of the arterial wall boundaries. The reader at the Ultrasound Reading Center analyzes the videotape and records the IMTs in the trial-specified vessel walls.

Not all trials record the IMTs in the same arteries or in the same walls. For example, some investigators record the IMTs in only the far walls that are visualized on the tape because there is some evidence that the near-wall IMT may not truly represent the thickness of the histological intimal-medial layers in this wall but may contain some ultrasonic noise (27,28). Other investigators continue to feel comfortable measuring the near wall because the progression of its thickness (regardless of any possible artifactual component) has been shown to be indistinguishable from IMT progression in the far wall (29). Given this, the continued inclusion of near-wall data into the statistics would effectively lower the required sample size for a trial. Similarly, some investigators record IMTs only from the common carotid artery because this artery is the most linear, thus reducing the variability of the measurements. Other investigators, wishing to understand more about other vessels, include the bifurcation and/or internal carotid arteries into their analyses.

The extent of atherosclerosis can also be quantified in a number of ways. For example, some investigators only examine and measure the mean IMT from across the length of the far wall of the common carotid artery on the right side of the neck (30,31). Other investigators examine more of the carotid arterial tree. These investigators measure, record, and average the maximum thicknesses across 12 arterial walls (near and far walls of the common, bifurcation, and internal arteries on both side of the neck) (32–34). This latter measurement (which sometimes excludes the four walls in the internal artery (35) has the advantage of more broadly representing the total amount of disease in the individual.

     Further details and references on the methodology, reproducibility, and validity of B-mode ultrasonography in observational studies and in clinical trials may be found elsewhere (17,18,26,29).

B. Statistical Issues in the Analysis of Serial Ultrasonographic Data

The general analytic plan for the measurement of the progression of carotid disease using ultrasonography has usually included multiple cross-sectional examinations performed during the course of follow-up (for example, at annual or semiannual clinic visits). For any specific participant in the trial, a least-squares regression line could then be fitted through the IMTs at the cross-sectional time points. Disease progression (or regression) could then be estimated by the slope of the line. This is depicted in Figure 1 for a fictional trial participant. In this example, the participant had a scan at baseline and every 6 months thereafter for 3 years, with an examination missing at 18 months. Treatment group differences

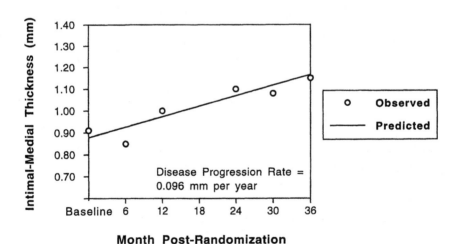

Figure 1 Hypothetical depiction of observed versus predicted intimal-medial thicknesses over time for a fictional participant.

in disease progression would compare the mean slopes between the two treatment groups (36).

It is characteristic that IMT measurements cannot be obtained from all walls at all visits. This issue may be approached statistically, given that methods are now available for estimating missing data (37). It is also characteristic that it is not always possible for a clinic to obtain one of the follow-up B-mode examinations for a participant. This challenge highlights the advantage of multiple B-mode examinations performed over follow-up: any available data contribute some information to the estimation of the progression/regression of disease. This is in contrast to the typical angiographic trial that only has measurements at baseline and at the end of the trial.

This situation is depicted in Figure 1. Here it is noted that the participant was not able to have an examination at 18 months. An estimate of the mean treatment group rate of change could still be made with the available data, with the caveat that when the mean treatment group rate of change is estimated, this participant's estimated rate will be weighted less.

V. CHOLESTEROL-LOWERING CLINICAL TRIALS WITH B-MODE ULTRASONOGRAPHIC OUTCOME MEASURES

There are six cholesterol-lowering clinical trials in the literature with B-mode ultrasonographic intimal-medial thickness outcome measures. These are, in the order of their initial publication: CLAS (Cholesterol Lowering Atherosclerosis Study) (30); ACAPS (Asymptomatic Carotid Artery Progression Study) (32); PLAC-II (Pravastatin, Lipids, and Atherosclerosis in the Carotid Arteries) (33); KAPS (Kuopio Atherosclerosis Prevention Study) (35); MARS (Monitored Atherosclerosis Regression Study) (31); and CAIUS (Carotid Atherosclerosis Italian Ultrasound Study) (34).

A. General Comparison of the Designs of the Six Trials

Table 1 describes the designs of each of the six trials. All were randomized and placebo-controlled and, with the exception of CLAS (38), all were double-masked. Three of the trials (CLAS, PLAC II, and MARS) were secondary prevention; two trials (ACAPS and CAIUS) were primary prevention; one trial (KAPS) obtained its study population from hyperlipidemics living in a free-living population and was thus a mixture of primary and secondary prevention. It should be noted that the distinction between primary and secondary prevention in the cases of ACAPS and CAIUS is blurred because both trials, although excluding participants with a history of a clinical coronary disease, required the

Table 1 Trial Designs and Entry Criteria

Trial	Type of prevention	General design	Years of follow-up	Total sample size	Age criteria (years)	Primary entry criteria	Lipid criteria
CLAS	Secondary	Randomized, placebo controlled	2–4	78 men (4 years); 46 men (with matching angiograms at 2 years)	40–59	History CABG, angiographically confirmed CAD, nonsmokers	Cholesterol 185–350 mg/dL
ACAPS	Primary	Randomized, placebo controlled, double-masked	3	919 men and women	40–79	Thickened carotid IMT (from B-mode)	Serum LDL 60–90th percentile
PLAC-II	Secondary	Randomized, placebo controlled, double-masked	3	151 men and women	50–75	Prior MI or angiographically confirmed CAD, and thickened carotid IMT (from B-mode)	Plasma LDL 60–90th percentile
KAPS	Primary/ Secondary	Randomized, placebo controlled, double-masked	3	424 men	44–65	Hyperlipidemia (free-living population)	Serum LDL > 155 mg/dL and total cholesterol < 290 mg/dL
MARS	Secondary	Randomized, placebo controlled, double-masked	2–4	188 men and women	37–67	Angiographically confirmed CAD	Plasma total cholesterol 190–295 mg/dL
CAIUS	Primary	Randomized, placebo controlled, double-masked	3	305 men and women	45–65	Thickened carotid IMT (from B-mode)	LDL > 150 to < 250 mg/dL and triglyceride < 250 mg/dL

CABG: coronary artery bypass graft; CAD: coronary artery disease; MI: myocardial infarction; IMT: intimal-medial thickness.

presence of a thickened IMT. This requirement connotes the presence of advanced atherosclerosis (39).

Five of the six trials were testing one of the new HMG CoA reductase inhibitors (or statins). ACAPS and MARS tested the effects of lovastatin; PLAC-II, KAPS, and CAIUS tested the effects of pravastatin. CLAS, the only trial not testing a statin, used colestipol and niacin in an attempt to reduce blood cholesterol by 40% (38). PLAC-II and ACAPS titrated their respective statins in attempts to achieve LDL reductions to the range of 90 to 110 mg/dL. Ultimately, about three-quarters of participants in both trials were on the higher dose of medication (40 mg/day).

In contrast to the classical, large, long-term event trials, these ultrasonographic trials were, as expected, smaller and of shorter duration. ACAPS was the largest trial with 919 participants and CLAS was the smallest with 78. CLAS and MARS followed participants for the longest period of time, 4 years. The other four trials followed participants for 3 years.

CLAS, KAPS, MARS, and CAIUS included participants with moderately to severely elevated lipid levels. ACAPS and PLAC-II included only participants with moderately elevated LDL and excluded participants with LDL values in the top 10% of their age/gender LDL distributions (41).

In contrast to the earlier, event trials that only studied men, four of the six ultrasonographic trials included women. Similar to those earlier trials, the ultrasonographic trials generally targeted middle-aged populations. However, ACAPS and PLAC-II did recruit participants who were older than 69 years. All trials, except CLAS, included smokers and nonsmokers; CLAS studied only nonsmokers.

Organizationally, the CLAS and MARS trials were subsets of larger angiographic trials (38,40,42). The KAPS population was derived from a larger observational study—the Kuopio Ischemic Heart Disease Risk Factor Study. CAIUS and PLAC-II shared the same Ultrasound Reading Center (in Winston-Salem, North Carolina) as did CLAS and MARS (in Los Angeles).

B. Specific Design Considerations

The Cholesterol Lowering Atherosclerosis Study (CLAS) was designed as an angiographic trial testing aggressive lipid-lowering treatment with colestipol and niacin (vs. placebo) in nonsmoking men who had had a coronary bypass and who had angiographically confirmed coronary disease (30,38,42). The B-mode portion of the trial was conducted in a subset of 78 participants, 46 of whom had a matching cervical angiogram. (An analysis of the femoral arteries was also conducted in a subset of participants.)

The Asymptomatic Carotid Artery Progression Study (ACAPS) was designed as an ultrasonographic trial testing lovastatin and/or warfarin therapies versus placebo in men and women, free of clinical vascular disease, with moderately elevated LDL cholesterol and evidence of intimal-medial thickening (32,43,44). One of the hallmarks of the trial was that it was conceived as a primary prevention trial of a lipid-lowering agent (lovastatin) given to people without severely elevated lipid levels. A factorial design was employed and participants were randomized to either lovastatin alone, warfarin alone, a lovastatin-warfarin combination, or double placebo. Because of an observed interaction with warfarin in the progression of IMT, only the B-mode results for the 461 participants not on warfarin are presented here, although the entire cohort of 919 participants was used in the analysis of events.

The Pravastatin, Lipids, and Atherosclerosis in the Carotid Arteries Trial (PLAC-II) was the first clinical trial specifically designed to use B-mode ultrasonography to evaluate the effect of any intervention on carotid atherosclerotic progression (33,45). All participants (men and women) were required to have either a documented history of a prior MI or angiographic evidence of coronary disease (>50% stenosis in at least one vessel). The trial was given the designation PLAC-II to differentiate it from its sister trial PLAC-I (Pravastatin Limitation of Atherosclerosis in the Coronary Arteries), which was an almost identically designed, concurrent trial with angiographic outcome measures (46).

The Kuopio Atherosclerosis Prevention Study (KAPS) was designed as an ultrasonographic trial testing whether pravastatin could slow the progression of carotid and/or femoral intimal-medial thickness in male hyperlipidemic patients derived from a free-living population in Finland (35). The primary outcome measure was the rate of carotid atherosclerotic progression (common + bifurcation), although the femorals were also examined. Although presented as a primary prevention trial, 7.6% of those randomized had had a prior myocardial infarction (35).

The Monitored Atherosclerosis Regression Study (MARS) was designed as an angiographic trial testing high-dose (80 mg/day) lovastatin in 270 men and women with angiographically defined coronary artery disease (31,40). The B-mode portion of the trial was conducted in a subset of 188 patients followed for 2 to 4 years.

The Carotid Artery Atherosclerosis Italian Ultrasound Study (CAIUS) was designed as a primary prevention, ultrasonographic trial testing whether a lipid-lowering therapy (pravastatin) could slow, halt, or reverse carotid atherosclerosis in 305 men and women with ultrasonographic evidence of early carotid atherosclerosis and moderately elevated LDL cholesterol (34).

C. B-mode Specifications in the Six Trials

Table 2 presents the B-mode specifications for the carotid examinations conducted in the six trials. Five of the six trials used a Biosound device; MARS used

Table 2 B-Mode Specifications

Trial	B-mode device	Summary measure of IMT	Carotid arteries that were interrogated	Walls examined	Side of neck
CLAS	Biosound, 9 MHz transducer	Average over length of single wall	Common carotid	Far wall only	Right only
ACAPS	Biosound phase 2, 10 MHz transducer	Average maximum thickness of 12 walls	Common carotid, bifurcation, internal carotid	Near and far walls	Right and left
PLAC II	Biosound 2000 II s.a., 8 MHz transducer	Average maximum thickness of 12 walls	Common carotid, bifurcation, internal carotid	Near and far walls	Right and left
KAPS	Biosound phase 2, 10 MHz transducer	Average maximum thickness of 4 walls	Common carotid, bifurcation	Far wall only	Right and left
MARS	Diasonics CV400, 7.5 MHz transducer	Average over length of single wall	Common carotid	Far wall only	Right only
CAIUS	Biosound 2000 II s.a., 8 MHz transducer	Average maximum thickness of 12 walls	Common carotid, bifurcation, internal carotid	Near and far walls	Right and left

IMT: intimal-medial thickness.

a Diasonics device. The transducer frequencies were generally similar, ranging from 7.5 to 10 MHz.

CLAS and MARS primarily examined only the far wall of the common carotid artery on the right side of the neck. KAPS also examined only the far wall, but this time in the common carotid and in the bifurcation, and on both sides of the neck. ACAPS, PLAC-II, and CAIUS examined both near and far walls of the common, bifurcation, and internal carotid arteries, on right and left sides of the neck. Given these examination specifications, it is noted that CLAS and MARS examined a single wall, KAPS examined four walls, and ACAPS, PLAC-II, and CAIUS examined 12 walls.

The summary measure of cross-sectional disease extent (= intimal-medial thickness) also differed slightly among the trials. CLAS and MARS averaged the IMT over the length of the far wall of the right common carotid artery. ACAPS, PLAC-II, and CAIUS averaged the maximum IMT measured across the 12 walls that were examined. KAPS averaged the maximum IMT measured across the four walls that were examined.

D. Lipid-Lowering Regimens/Responses in the Trials

In these trials, the level of LDL reduction was three to four times greater than the lipid reductions noted in the early event trials. ACAPS, PLAC-II, KAPS, and CAIUS experienced LDL reductions ranging from 23% to 28% (Table 3). CLAS and MARS experienced LDL reductions of 42% and 45%, respectively. It is noted that the dose of lovastatin used in MARS was two times the maximum dose used in ACAPS (80 vs. 40 mg/day).

E. Baseline Descriptions of the Six Trials

Table 4 presents the nonultrasonographic baseline characteristics of the six trials. There was only a 9-year difference in mean ages among the trials. As expected from the entry criteria (Table 1), ACAPS and PLAC-II had the highest mean ages (62 and 63 years, respectively). CLAS and CAIUS had the lowest mean ages (54 and 55 years, respectively). Whereas CLAS and KAPS excluded women, almost half of ACAPS and CAIUS were women. The KAPS participants had the highest means for blood pressure and LDL and the highest proportion of smokers (25%). As noted above, CLAS included only nonsmokers.

Table 5 presents the ultrasonographic baseline characteristics for five of the six trials. Recalling that CLAS and MARS used the average IMT in the right far wall of the common carotid, it is not surprising that their mean measurements are less than the average *maximum* IMTs noted for ACAPS, PLAC-II, and CAIUS. Note the baseline difference between the treatment groups for MARS ($P < .001$).

Table 3 Lipid Responses to Treatment

| | | Lipid-altering treatment | | |
| | | | Response to Treatment (mg/dL) | |
Trial	Control regimen	Regimen	LDL	HDL
CLAS	Placebo + diet	30.1 g/d colestipol + 4.2 g/d niacin + diet	−42.2% (for 46 men)	+33.3%
ACAPS	Placebo + diet	20 to 40 mg/day lovastatin + diet	−27.8%	+5.0%
PLAC-II	Placebo + diet	20 to 40 mg/day pravastatin + diet	−28.2%	+3.8%
KAPS	Placebo	40 mg/day pravastatin	−27.4%	.0%
MARS	Placebo + diet	80 mg/day lovastatin	−45.4%	+8.1%
CAIUS	Placebo	40 mg/day pravastatin	−23%	+3%

The Atherosclerosis Risk in Communities Study has published IMT distributions for the general U.S. population (39). For reference, the average IMT for the common carotid artery for 55-year-old white men was reported to be 0.67 mm; the 90th percentile was reported to be 0.90 mm. These statistics indicate that ACAPS, PLAC-II, and CAIUS studied populations with advanced atherosclerosis.

F. Differences in IMT Progression Attributable to Lipid Lowering

Table 6 presents the treatment group differences in IMT progression for different carotid arterial beds among the six trials. It is noted that the reduction in progression attributable to lipid lowering is evident in each of the 11 arterial combinations presented in the papers for these trials. Only in the combination of the common+bifurcation+internal arteries in PLAC-II was the reduction in progression not statistically significant ($P = .44$). It is further noted that four of the six trials actually reported disease regression (CLAS, ACAPS, MARS, CAIUS), strongly suggesting that the atherosclerotic process can be reversed by lipid-lowering treatment.

Table 4 Baseline Descriptions

Trial	Mean age (years)	Percent male	Mean blood pressure (mm/Hg)	Mean LDL (mg/dL)	Mean HDL (mg/dL)	Percent smokers
CLAS[a]	54	100	124/81	170	44	0
ACAPS	62	52	131/77	156	46 (men) 58 (women)	12
PLAC II	63	85	130/77	166	40	12
KAPS	57	100	137/86	190	46	25
MARS	58	91	123/80	157	43	16
CAIUS	55	53	134/81	181	53	24

[a]On 46 men with a matching angiogram at 2 years of follow-up.

Table 5 Baseline Intimal-Medial Thicknesses by Trial (mm) (data not available for KAPS)

Trial	Artery	Treatment group	
		Active	Placebo
CLAS	Common carotid	0.65	0.61
ACAPS	Common carotid	1.15	1.14
	Common + bifurcation	1.36	1.38
	Common + bifurcation + internal	1.33	1.32
PLAC-II	Common carotid	1.01	1.01
	Common + bifurcation	1.32	1.38
	Common + bifurcation + internal	1.32	1.32
MARS	Common carotid	0.78	0.69*
CAIUS	Common carotid	0.89	0.85
	Common + bifurcation	1.11	1.08
	Common + bifurcation + internal	1.06	1.04

*$P < .001$ for difference between means.

1. Specific Results: CLAS (30)

In CLAS, the 39 participants who were randomized to placebo experienced increases in the mean IMT (right common carotid, far wall only) at 2 and 4 years postrandomization, going from 0.61 mm at baseline to 0.65 and 0.66 mm at 2 and 4 years, respectively. Conversely, the IMT among the 39 drug-treated participants decreased from 0.65 mm to 0.60 at both 2 and 4 years. The treatment group differences in the changes in these values from baseline to follow-up were statistically significant ($P < .001$).

Among the second subset of participants with the matching cervical angiogram, there was a statistically significant correlation between far wall IMT and average coronary stenosis at baseline ($r = .34$, $P < 0.05$).

The investigators also crudely examined the correlation between change in IMT and the change in lipids and lipoproteins in this second subset. In multivariate analyses combining both treatment groups, they reported that higher levels of on-trial HDL were associated with reduced wall thicknesses at 2 and 4 years. In univariate analyses, they reported that lower levels of LDL were statistically associated with smaller IMTs at 2 years.

The authors concluded that aggressive colestipol-niacin treatment reduces common carotid intimal-medial thickening.

2. Specific Results: ACAPS (32)

As noted in Table 6, lovastatin was associated with a statistically significant ($P = .001$) regression of IMT, which was defined as the average of the maximum

Table 6 Progression of Intimal-Medial Thicknesses by Treatment Group and by Trial

Trial	Artery	Treatment group		Percent reduction In rates	Absolute difference (mm/yr)	P value
		Active (mm/yr)	Placebo (mm/yr)			
CLAS	Common carotid (over 4 years)[a]	-0.013	0.013	200%	-0.026	<.001
ACAPS	Common + bifurcation + internal[a]	-0.009	0.006	250%	-0.0150	.001
PLAC-II	Common carotid	0.0295	0.0456	35%	-0.0161	.03
	Common + bifurcation	0.0616	0.0800	23%	-0.0184	.02
	Common + bifurcation + internal[a]	0.0593	0.0675	12%	-0.0082	.44
KAPS	Common carotid	0.0096	0.0285	66%	-0.0189	.0019
	Common + bifurcation[a]	0.0168	0.0309	46%	-0.0141	.0046
MARS	Common carotid (over 2 years)[a]	-0.038	0.019	300%	-0.0570	<.001[b]
CAIUS	Common carotid	-0.0032	0.0077	142%	-0.0109	.0047
	Common + bifurcation	-0.0082	0.0106	177%	-0.0188	.0001
	Common + bifurcation + internal[a]	-0.0043	0.0089	148%	-0.0132	.0007

[a]Primary ultrasonographic outcome measure for trial.
[b]Adjusted for baseline difference (see Table 5).

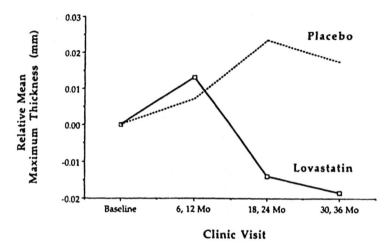

Figure 2 ACAPS results. Graph showing cross-sectional intimal-medial thicknesses for the lovastatin and placebo groups, excluding participants on warfarin. (From Ref. 32.)

thicknesses across 12 arterial walls. This result may also be noted in Figure 2, which plots the relative average cross-sectional IMT measurements across time. It is noted that the lovastatin IMT values initially increased, but decreased after a year, giving us the overall regression noted in Table 6.

ACAPS was not designed to be an events trial, and the power to detect a treatment group difference in events was not expected to be great because of the relatively small sample size and the short length of follow-up. However, both the incidence of fatal/nonfatal cardiovascular disease and the all-cause mortality rate were observed to be statistically reduced among the 460 participants assigned to the lovastatin groups. There were 14 cardiovascular events among these individuals compared to only five among the individuals assigned to the placebo groups ($P = .04$). Similarly, there were eight deaths among the lovastatin participants compared to only one among the placebo participants ($P = .02$).

The investigators concluded that lovastatin reverses the progression of carotid IMT and appears to reduce the risk of major cardiovascular events.

3. Specific Results: PLAC-II (33,47)

As noted in Table 6, the primary B-mode measure in PLAC-II (the change in IMT progression in the combination of the common, bifurcation, and internal arteries) did not reach statistical significance, although the trend was in the expected direction. As shown in Figure 3, this diminution of an effect was primarily due to there being no treatment effect at all in the internal carotid artery

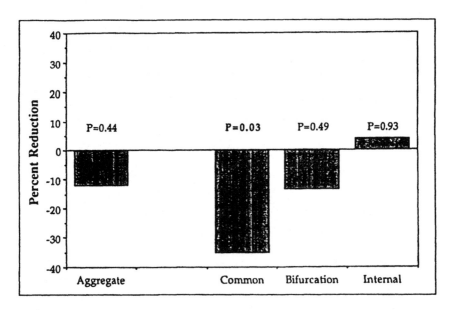

Figure 3 PLAC-II results. Adjusted percent reductions in aggregate intimal-medial thickness and in segment-specific intimal-medial thickness (pravastatin progression rate relative to placebo). (From Ref. 33.)

($P = .93$) (33). In fact, if that artery is excluded (not a prespecified hypothesis), a 23% reduction ($P = .02$) may be estimated.

PLAC-II was being conducted simultaneously with ACAPS, and as in ACAPS, the PLAC-II investigators discovered that the pravastatin participants experienced a 60% reduction in fatal/nonfatal coronary events compared to placebo (four events in the pravastatin group compared with 10 in the placebo). Although this difference did not reach statistical significance, when all-cause mortality was added, the resultant 61% reduction in events was statistically significant (adjusted $P = .04$) (47).

The investigators concluded that pravastatin use was associated with a reduction in the progression of carotid IMT (specifically in the common carotid artery and in the bifurcation) and with the incidence of any fatal event plus nonfatal MI.

4. Specific Results: KAPS (35)

As noted in Table 6, pravastatin was associated with a reduction in progression in both carotid segments. There was no statistical difference in the rates in the femoral artery.

Treatment effect was noted to be influenced by baseline IMT and by baseline smoking status (Fig. 4). In the placebo group, the investigators reported that the higher the baseline IMT, the higher the progression rate—a relationship not found in participants on pravastatin. Furthermore, placebo group smokers progressed more quickly than nonsmokers, although smokers in the pravastatin group had very similar progression rates compared to nonsmoking pravastatin participants. Taken together, the treatment benefit was found to be greater for smokers than nonsmokers.

5. Specific Results: MARS (31)

Overall, the lovastatin group experienced a regression of IMT disease compared to the progression in the placebo group (Table 6), at both 2 and 4 years ($P < .001$). (Follow-up statistics were adjusted for baseline values because of the difference noted at baseline [Table 5]).

The investigators concluded that lipid lowering with lovastatin reverses the progression of early, preintrusive carotid atherosclerosis.

6. Specific Results: CAIUS (34)

As noted in Table 6, pravastatin was associated with a regression in IMT whereas the placebo group experienced a progression of disease. This effect was noted for all of the specified outcome measures. In addition, the CAIUS investigators reported that the cross-sectional IMT values followed a pattern similar to that

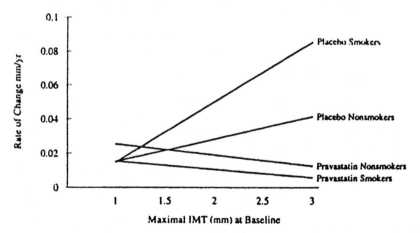

Figure 4 KAPS results. Annual estimated carotid atherosclerotic progression (change in intimal-medial thickness) for pravastatin and placebo groups, expressed as a function of the baseline mean maximum thickness. (From Ref. 35.)

noted in Figure 2 for ACAPS: whereas the IMT values in the placebo group had the tendency to increase over the follow-up period, the values in the pravastatin group first increased and subsequently decreased.

The authors concluded that pravastatin stops the progression of carotid IMT in asymptomatic men and women with moderately elevated lipid levels.

G.   Event Reduction Attributable to Lipid-Lowering

The six ultrasound trials noted in this review, five of which were statin trials, were part of the first generation of imaging trials examining statins and other treatment modalities with large lipid-lowering effects. None of the trials were conceived as event trials, although each monitored the occurrence of events for safety and scientific curiosity.

In the last half decade or so, however, as trials such as these have been published, it has been repeatedly reported that trials with large lipid-lowering effects have demonstrated large reductions in coronary events associated with the active treatment. This was noted above in the descriptions of the results from ACAPS and PLAC-II, and has been described in a number of meta-analyses. For example, the PLAC-II results have been pooled with the PLAC-I results and subsequently with the results from KAPS and REGRESS (an angiographic trial of pravastatin). In both of these analyses, pravastatin was shown to be associated with a reduction in clinical coronary events (48,49).

Table 7 presents the coronary event rates for the six B-mode/lipid-lowering trials discussed in this chapter. In both the primary and secondary prevention settings, the lipid-lowering therapies used in these trials are associated with a decrease in events. It is specifically noted that among the 609 patients in the three secondary prevention trials, there is a 41% reduction in coronary events attributable to lipid lowering over the relatively short 2.25 years of follow-up (95% confidence interval = 8% to 61% reduction, $P = .05$).

The question naturally arises, how can we be seeing these reductions in events from trials that are so much smaller and so much shorter than the four classic megatrials described in the beginning of this chapter? Also, if the absolute differences in IMT progression in Table 6 are examined, one is struck by the observation that the differences in progression attributable to treatment, although statistically significant, appear tiny (0.01 to 0.02 mm/year). Can these minuscule changes translate into the large reductions in the risk of an event?

Brown et al., in a review of angiographic trials of therapies with large lipid-lowering effects, noted the same phenomena: the angiographic changes, although statistically significant, were tiny, and yet clear reductions in clinical event rates were unexpectedly evident (50). In his discussion, Brown et al. present the argument that it is the large reductions in LDL that may account for these observations. A large reduction in blood lipids, they argue, would deplete the choles-

Table 7 Effect of Lipid-Lowering on Fatal/Nonfatal Coronary Events in B-Mode Trials

Clinical trial	Average length of follow-up (yrs)	Lipid-lowering group			Placebo group			% Relative reduction in rates	95% CI of % relative reduction
		Sample size	Events Number	Rate/1000/yr	Sample size	Events Number	Rate/1000/yr		
Primary prevention trials									
KAPS	3.00	224	5	7.4	223	8	12.0		
CAIUS	3.00	151	2	4.4	154	2	4.3		
ACAPS	3.00	410	5	4.1	459	9	6.5		
Total	3.00	785	12	5.1	836	19	7.6	34	−35, 67
Secondary Prevention Trials									
CLAS[a]	2.00	94	1	5.3	94	5	26.6		
PLAC II	3.00	75	4	17.8	76	10	43.9		
MARS[a]	2.00	134	22	82.1	136	31	114.0		
Total	2.25	303	27	39.6	306	46	66.9	41	8, 61
Grand total	2.80	1088	39	12.8	1142	65	20.3	39	11, 58

[a]The sample sizes and lengths of follow-up reflect the larger, parent trials of the B-mode substudies.

terol from the underlying core lipid pool of a plaque, decrease the number of lipid-filled macrophages, and thus stabilize the plaque. These effects, in turn, would reduce the likelihood of plaque fissuring, thrombosis, and subsequent clinical events.

Pathophysiology aside, it is evident in retrospect that we should not have been surprised that these smaller trials produced these large reductions in clinical events. Given the large LDL-lowering effects that were observed, Law et al. (51) and Holme (52) have shown that the observed event reductions in these new trials may have been predicted as an extrapolation of the results of the older, larger trials with their smaller lipid effects.

VI. CONCLUSION

The trials presented in this chapter consistently demonstrate that B-mode ultrasonography is a useful technique to measure and monitor the progression of carotid atherosclerosis, and is a useful technique to evaluate the effects lipid-lowering regimens.

Furthermore, each trial presented here demonstrated that the rate of carotid atherosclerosis could be slowed or even reversed by lipid-lowering therapy. Supporting this is the concomitant reduction in coronary events attributable to lipid-lowering therapy.

REFERENCES

1. Committee of Principal Investigators. A cooperative trial in the prevention of ischaemic heart disease using clofibrate. Br Heart J 1978; 40:1069–1118.
2. Committee of Principal Investigators. WHO cooperative trial on primary prevention of ischaemic heart disease using clofibrate to lower serum cholesterol: mortality follow-up. Lancet 1980; ii:379–385.
3. Committee of Principal Investigators. WHO cooperative trial on primary prevention of ischaemic heart disease using clofibrate to lower serum cholesterol: final mortality follow-up. Lancet 1984; ii:600–604.
4. Lipid Research Clinics Program: The Coronary Primary Prevention Trial: Design and implementation. J Chron Dis 1979; 32:609–631.
5. Lipid Research Clinics Program: The Lipid Research Clinics Coronary Primary Prevention Trial results. I. Reduction in incidence of coronary heart disease. JAMA 1984; 251:351–364.
6. Lipid Research Clinics Program: The Lipid Research Clinics Coronary Primary Prevention Trial results. II. The relationship of reduction in incidence of coronary heart disease to cholesterol lowering. JAMA 1984; 251:365–374.
7. Frick MH, Elo O, Haapa K, Heinonen OP, Heinsalmi P, Helo P, Huttunen JK,

Kaitaniemi P, Koskinen P, Manninen V, Mäenpää H, Mälkönen M, Mänttäri M, Norola S, Pasternack A, Pikkarainen J, Romo M, Sjöblom T, Nikkilä EA. Helsinki Heart Study: Primary prevention trial with gemfibrozil in middle-aged men with dyslipidemia. Safety of treatment, changes in risk factors, and incidence of coronary heart disease. N Engl J Med 1987; 317:1237–1245.

8. Coronary Drug Project Research Group. The Coronary Drug Project: Design, methods, and baseline results. American Heart Association monograph No. 38. Circulation 1973; 47(suppl I):I-1–I-79.

9. Coronary Drug Project Research Group. Clofibrate and niacin in coronary heart disease. JAMA 1975; 231:360–381.

10. Canner PL, Berge KG, Wenger NK, Stamler J, Friedman L, Prineas RJ, Friedewald W, for the Coronary Drug Project Research Group. Fifteen year mortality in Coronary Drug Project patients: Long-term benefit with niacin J Am Coll Cardiol 1986; 8:1245–1255.

11. Gould AL. Rossouw JF, Santanello NC, Heyse JF. Furberg CD. Cholesterol reduction yields clinical benefit. A new look at old data. Circulation 1995; 91:2274–2282.

12. Gordon DJ. Cholesterol lowering and total mortality. In: Rifkind BM, ed. Lowering Cholesterol in High-Risk Individuals and Populations. New York: Marcel Dekker, 1995.

13. Scandinavian Simvastatin Survival Study Group. Randomised trial of cholesterol lowering in 4444 patients with coronary heart disease: the Scandinavian Simvastatin Survival Study (4S). Lancet 1994; 344:1383–1389.

14. Shepherd J, Cobbe SM, Ford I, Isles CG, Lorimer AR, Macfarlane PW, McKillop JH, Packard CJ, for the West of Scotland Coronary Prevention Study Group. Prevention of coronary heart disease with pravastatin in men with hypercholesterolemia. N Engl J Med 1995; 333:1301–1307.

15. Glagov S, Weisenberg E, Zarins CK, Stankunavicius R, Kolettis GJ. Compensatory enlargement of human atherosclerotic coronary arteries. N Engl J Med 1987; 316:1371–1375.

16. Pignoli P, Tremoli E, Poli A, Oreste P, Paoletti R. Intimal plus medial thickness of the arterial wall: a direct measurement with ultrasound imaging. Circulation 1986; 74:1399–1406.

17. Bond MG, Wilmoth SK, Enevold GL, Strickland HL. Detection and monitoring of asymptomatic atherosclerosis in clinical trials. Am J Med 1989; 86(suppl 4A):33–36.

18. Touboul P-J, Crouse JR, eds. Intimal-Media Thickness and Atherosclerosis—Predicting the Risk. New York: Parthenon, 1997.

19. Craven TE, Ryu JE, Espeland MA, Kahl FR, McKinney WM, Toole JF, McMahan MR, Thompson CJ, Heiss G, Crouse JR. Evaluation of the associations between carotid artery atherosclerosis and coronary artery stenosis. A case-comparison study. Circulation 1990; 82:1230–1242.

20. Wofford JL, Kahl FR, Howard GR, McKinney WM, Toole JF, Crouse JR. Relation of extent of extracranial carotid artery atherosclerosis as measured by B-mode ultrasound to the extent of coronary atherosclerosis. Arterioscler Thromb 1991; 11:1786–1794.

21. Salonen R, Salonen JT. Progression of carotid atherosclerosis and its determinants: a population-based ultrasonography study. Atherosclerosis 1990; 81:33–40.

22. Salonen R. Risk Factors for Ultrasonographically Assessed Common Carotid Atherosclerosis—a Cross-Sectional and Longitudinal Population Based Study in Eastern Finnish Men. University of Kuopio (Finland) Community Health Original Reports, 1991.

23. Sutton KC, Wolfson SK, Kuller LH. Carotid and lower extremity arterial disease in elderly adults with isolated systolic hypertension. Stroke 1987; 18:817–822.

24. Heiss G, Sharrett AR, Barnes R, Chambless LE, Szklo M, Alzola C, for the ARIC Investigators. Carotid atherosclerosis measured by B-mode ultrasound in populations: associations with cardiovascular risk factors in the ARIC study. Am J Epidemiol 1991; 134:250–256.

25. Salonen R, Salonen JT. Determinants of carotid intima-media thickness—a population-based ultrasonography study in eastern Finnish men. J Intern Med 1991; 229:225–231.

26. Furberg CD, Byington RP, Riley W. B-mode ultrasound: a noninvasive method for assessing atherosclerosis. In: Willerson JT, Cohn JN, eds. Cardiovascular Medicine. New York: Churchill Livingstone; 1995.

27. Pignoli P. Ultrasound B-mode imaging for arterial wall thickness measurement. Atheroscler Rev 1984; 12:177–184.

28. Wikstrand J, Wiklund O. Frontiers in cardiovascular science—quantitative measurements in humans. Arterioscler Thromb 1992; 12:114–119.

29. Furberg CD, Byington RP, Craven TE. Lessons learned from clinical trials with ultrasound end-points. J Intern Med 1994; 236:575–580.

30. Blankenhorn DH, Selzer RH, Crawford DW, Barth JD, Liu C-r, Liu C-h, Mack WJ, Alaupovic P. Beneficial effects of colestipol-niacin therapy on the common carotid artery: two- and four-year reduction of intima-media thickness measured by ultrasound. Circulation 1993; 88:20–28.

31. Hodis HN, Mack WJ, LaBree L, Selzer RH, Liu R-r, Liu C-h, Alaupovic P, Kwong-Fu H, Azen SP. Reduction in carotid arterial wall thickness using lovastatin and dietary therapy. A randomized, controlled clinical trial. Ann Intern Med 1996; 124:548–556.

32. Furberg CD, Adams HP, Applegate WB, Byington RP, Espeland MA, Hartwell T, Hunninghake DB, Lefkowitz DS, Probstfield J, Riley WA, Young B, for the Asymptomatic Carotid Artery Progression Study (ACAPS) Research Group. Effect of lovastatin on early carotid atherosclerosis and cardiovascular events. Circulation 1994; 90:1679–1687.

33. Crouse JR, Byington RP, Bond MG, Espeland MA, Craven TE, Sprinkle JW, McGovern ME, Furberg CD. Pravastatin, Lipids and Atherosclerosis in the Carotid Arteries (PLAC-II). Am J Cardiol 1995; 75:455–459.

34. Mercuri M, Bond MG, Sirtori CR, Veglia F, Crepaldi G, Feruglio FS, Descovich G, Ricci G, Rubba P, Mancini M, Gallus G, Bianchi G, D'Alò G, Ventura A. Pravastatin reduces carotid intima-media thickness progression in an asymptomatic hypercholesterolemic Mediterranean population: the Carotid Atherosclerosis Italian Ultrasound Study. Am J Med 1996; 101:627–634.

35. Salonen R, Nyyssönen K, Porkkala E, Rummukainen J, Belder R, Park J-S, Salonen J, Kuopio Athersclerosis Prevention Study (KAPS)—A population-based primary preventive trial of the effect of LDL lowering on atherosclerotic progression in carotid and femoral arteries. Circulation 1995; 92:1758–1764.

36. Byington R. Carotid and femoral-iliac disease as models for the use of newer imaging techniques. In: Fuster V, ed. Syndromes of Atherosclerosis—Correlations of Clinical Imaging and Pathology. Armonk, NY: Futura, 1996.

37. Espeland MA, Byington RP, Hire D, Davis VG, Hartwell T, Probstfield J, for the ACAPS Study Group. Analysis strategies for serial multivariate ultrasonographic data that are incomplete. Stat in Med 1992; 11:1041–1056.

38. Blankenhorn DH, Johnson RL, Nessim SA, Azen SP, Sanmarco ME, Selzer RH, and the CLAS Investigators and Staff. The Cholesterol Lowering Atherosclerosis Study (CLAS): design, methods, and baseline results. Controlled Clin Trials 1987; 8:354–387.

39. Howard G, Sharrett AR, Heiss G, Evans GW, Chambless LE, Riley WA, Burke GL, for the ARIC Investigators. Carotid artery intimal-medial thickness distribution in general populations as evaluated by B-mode ultrasound. Stroke 1993; 24:1297–1304.

40. Blankenhorn DH, Azen SP, Kramsch DM, Mack WJ, Cashin-Hemphill L, Hodis HN, DeBoer LWV, Mahrer PR, Masteller MJ, Vailas LI, Alaupovic P, Hirsch LJ, and the MARS Research Group. Coronary angiographic changes with lovastatin therapy. The Monitored Atherosclerosis Regression Study (MARS). Ann Intern Med 1993; 119:969–976.

41. *Lipid Research Clinics Population Studies Data Book. Vol. I. The Prevalence Study.* Washington: Lipid Metabolism Branch, Division of Heart and Vascular Diseases, National Heart, Lung, and Blood Institute. U.S. Department of Health and Human Services, Public Health Service, National Institutes of Health. NIH Publication No. 80-1527, July 1980.

42. Blankenhorn DH, Nessim SA, Johnson RL, Sanmarco ME, Azen SP, Cashin-Hemphill L. Beneficial effects of combined colestipol-niacin therapy on coronary atherosclerosis and coronary venous bypass grafts. JAMA 1987; 257:3233–3240.

43. ACAPS Group. Rationale and design for the Asymptomatic Carotid Artery Plaque Study (ACAPS). Controlled Clin Trials 1992; 12:293–314.

44. Riley WA, Barnes RW, Applegate WB, Dempsey R, Hartwell T, Davis VG, Bond MG, Furberg CD. Reproducibility of noninvasive ultrasonic measurement of carotid atherosclerosis. The Asymptomatic Carotid Artery Plaque Study. Stroke 1992; 23:1062–1068.

45. Crouse JR, Byington RP, Bond MG, Espeland MA, Sprinkle JW, McGovern M, Furberg CD. Pravastatin, lipids, and atherosclerosis in the carotid arteries: design features of a clinical trial with carotid atherosclerosis outcome. Controlled Clin Trials 1992; 13:494–506.

46. Pitt B, Mancini GBJ, Ellis S, Rosman HS, Park J-S, McGovern ME, for the PLAC I investigators. Pravastatin Limitation of Atherosclerosis in the Coronary Arteries (PLAC I): reduction in atherosclerosis progression and clinical events. J Am Coll Cardiol 1995; 26:1133–1139.

47. Furberg CD, Byington RP, Crouse JR, Espeland MA. Pravastatin, lipids, and major coronary events. Am J Cardiol 1994; 73:1133–1134.
48. Pitt B, Furberg C, McGovern M, PLAC-I and PLAC-II Investigators. Reduction in cardiovascular events during treatment with pravastatin: pooled analysis from coronary and carotid atherosclerosis intervention trials. Eur Heart J 1994; 15(suppl):P2565. Abstract.
49. Byington RP, Jukema JW, Salonen JT, Pitt B, Bruscke AV, Hoen H, Furberg CD, Mancini GBJ. Reduction in cardiovascular events during pravastatin therapy. Pooled analysis of clinical events of the Pravastatin Atherosclerosis Intervention Program. Circulation 1995; 92:2419–2425.
50. Brown BG, Zhao X-Q, Sacco DE, Albers JJ. Lipid lowering and plaque regression. New insights into prevention of plaque disruption and clinical events in coronary disease. Circulation 1993; 87:1781–1791.
51. Law MR, Wald NJ, Thompson SG. By how much and how quickly does reduction in serum cholesterol concentration lower risk of ischaemic heart disease? BMJ 1994; 308:367–373.
52. Holme I. Relation of coronary heart disease incidence and total mortality to plasma cholesterol reduction in randomized trials: use of meta-analysis. Br Heart J 1993 69(suppl):S42–S47.

12

Cholesterol Lowering Reduces Mortality

The Statins

David J. Gordon
National Heart, Lung, and Blood Institute, National Institutes of Health, Bethesda, Maryland

I. INTRODUCTION

Although the etiologic role of cholesterol in coronary atherogenesis is well established, and clinical trials of cholesterol-lowering drugs have confirmed that this process can be slowed or even partially reversed by treatment, the net effect of cholesterol lowering on mortality was in doubt as little as 4 years ago (1). Clinical trials using bile acid sequestrant resins, fibrates, niacin, diet, and hormones to lower cholesterol suffered from a combination of modest cholesterol-lowering efficacy, side effects, poor compliance, and inadequate sample size, and often gave equivocal results, with favorable trends in the incidence of myocardial infarction (MI) and other clinical sequelae of atherosclerotic coronary heart disease (CHD) offset by unfavorable trends in non-CHD mortality.

In the early 1990s, several meta-analyses were performed to address quantitatively the issue of whether cholesterol lowering reduces mortality (2–10). Although the selection of trials included by the authors of these analyses and the interpretations they offered varied widely, certain common conclusions could be drawn:

1. Cholesterol lowering reduced CHD morbidity and mortality.
2. Non-CHD mortality was increased among patients receiving active treatment, and all-cause mortality was unchanged.
3. There were trends toward reduced mortality in secondary prevention trials (where most deaths were due to CHD) and toward increased mortality in primary prevention trials (where fewer than half the deaths were due to CHD).

4.   Trials in which the greatest degree of cholesterol lowering was attained had the most favorable results for CHD and all-cause mortality. The apparent adverse effect of treatment on non-CHD mortality did *not* increase in proportion to the degree of cholesterol lowering.

5.   Drug-specific toxicity with certain interventions used to lower cholesterol in these trials, i.e., clofibrate and hormones, may have obscured their ability to assess the effect of cholesterol lowering per se.

In the concluding paragraph of a meta-analysis addressing cholesterol-lowering trials published before October 1993 (1), it was suggested that "any real hope of resolving the unanswered questions about the impact of cholesterol-lowering on mortality should be invested in ongoing trials using potent cholesterol-lowering interventions without known major adverse non-CHD effects", i.e., the HMG CoA reductase inhibitors or "statins."

In the ensuing 4 years, 12 trials using these agents and fitting the same criteria used for the 22 prestatin trials (1) were completed (11–22) (Table 1). The Scandinavian Simvastatin Survival Study (4S) (12), West of Scotland Coronary Prevention Study (WOSCOPS) (14), Cholesterol and Recurrent Events (CARE) trial (18), Air Force-Texas Coronary Atherosclerosis Prevention Study (AFCAPS/TexCAPS) (21), Long-term Intervention with Pravastatin in Ischaemic Heart Disease (LIPID) study (22) were large multicenter trials, each containing 4000 to 9000 participants and designed to address the impact of treatment on clinical endpoints. The Asymptomatic Carotid Artery Progression Study (ACAPS) (11), Multicentre Anti-Atheroma Study (MAAS) (13), Kuopio Atherosclerosis Prevention Study (KAPS) (15), Pravastatin Limitation of Atherosclerosis in the Coronaries (PLAC I) study (16), Pravastatin, Lipids, and Atherosclerosis in the Carotid Arteries (PLAC-II) study (17), Carotid Atherosclerosis Italian Ultrasound Study (CAIUS) (19), and Post Coronary Artery Bypass Graft (Post-CABG) trial (20) were smaller studies (150 to 1400 patients) designed with angiographic or ultrasonographic endpoints in mind. The present paper reevaluates the impact of cholesterol lowering on mortality in the light of results from these statin trials.

II.   METHODS

As in the earlier meta-analysis (1), only randomized, unconfounded trials of at least 3 years' planned duration in which active treatment lowered cholesterol by at least 4% more than the control regimen are included. Multifactor trials in which interventions on nonlipid risk factors were administered along with cholesterol reduction to the active treatment group but not to the control group are excluded, but trials with a factorial design, in which the effects of cholesterol lowering are distinguishable from the concurrent effects of other intervention factors, are allowed. Trials in which a low dose of a statin or other cholesterol-lowering drug

Table 1 Randomized Trials of Statins and CHD Prevention[a]

Trial (Ref.)	Prior CHD	# Persons Trt.	# Persons Ctrl.	Follow-up (years)	Percent cholesterol reduction	Total Trt.	Total Ctrl.	Total OR	Mortality CHD Trt.	CHD Ctrl.	CHD OR	Non-CHD Trt.	Non-CHD Ctrl.	Non-CHD OR	CHD Incidence[b] Trt.	CHD Ctrl.	OR
ACAPS (11)	No	460	459	3.0	19.9	1	8	0.21[d]	0	4	0.13[d]	1	4	0.30	**5**	**9**	**0.56**
AFCAPS/TexCAPS (21)[c]	No	3,304	3,301	4.8	19.3	80	77	1.04	11	15	0.73	69	62	1.11	**57**	**95**	**0.60**[d]
CAIUS (19)	No	151	154	3.0	12.4	1	0	7.54	1	0	7.54	0	0	—	2	2	1.02
CARE (18)	Yes	2,081	2,078	5.0	20.0	180	196	0.91	96	119	0.80	84	77	1.09	**212**	**274**	**0.75**[d]
KAPS (15)	No	224	223	3.0	21.0	3	4	0.75	2	2	1.00	1	2	0.51	5	8	0.62
LIPID (22)	Yes	4,512	4,502	6.1	17.9	498	633	0.76[d]	287	373	0.75[d]	211	260	0.80[d]	557	715	**0.75**[d]
MAAS (13)	Yes	193	188	4.0	23.0	4	11	0.37	4	4	0.97	**0**	**7**	**0.13**[d]	14	9	1.54
PLAC1 (16)	Yes	206	202	3.0	19.0	4	6	0.65	3	3	0.98	1	3	0.36	10	19	0.50
PLAC2 (17)	Yes	75	76	3.0	21.6	3	5	0.60	1	2	0.52	2	3	0.67	4	10	0.40
Post CABG (20)	Yes	676	675	4.5	31.6	32	35	0.91	6	4	1.49	26	31	0.83	35	40	0.87
4S (12)	Yes	2,221	2,223	5.4	25.0	**182**	**256**	**0.69**[d]	**111**	**189**	**0.57**[d]	71	67	1.06	**431**	**622**	**0.62**[d]
WOSCOPS (14)	No	3,302	3,293	4.9	20.0	106	135	0.78	**41**	**61**	**0.67**[d]	65	74	0.87	**174**	**248**	**0.69**[d]
All statin trials (89,123 person-years)		17,405	17,374	5.1	20.0	**1,094**	**1,366**	**0.78**[d]	**563**	**776**	**0.71**[d]	531	590	0.89	**1,506**	**2,051**	**0.70**[d]

[a]Meeting the following criteria: analyzed by intent to treat; no nonlipid interventions; at least 3 years' duration; at lesat 4% cholesterol reduction.

[b]Combined incidence of CHD death and nonfatal myocardial infarction.

[c]The primary endpoint of this study, which included unstable angina as well as nonfatal myocardial infarction and CHD death, is not shown in this table.

Abbreviations: Trt. = Active treatment group, Ctrl. = Control group, OR = Odds Ratio (Trt. vs. Ctrl.).

[d]Statistically significant endpoint results (P < .05) are indicated in boldface type.

is administered to the control group are allowed, provided that the 4% minimum cholesterol difference between active treatment and control groups is exceeded. All results are analyzed by "intent to treat."

Meta-analyses of mortality due to CHD, non-CHD causes, and all causes combined and of the combined incidence of CHD death and nonfatal MI (CHD incidence) were performed using the Mantel-Haenszel procedure as modified by Yusuf et al. (23). This procedure, which can accommodate even small trials with few events, entails a comparison of observed versus "expected" (by chance) numbers of deaths in each trial. The difference between observed and expected events (O-E) was divided by its estimated variance (V) to obtain an estimate of the log odds ratio for each trial. The values of O-E and V for each trial were summed over all trials to estimate the log of the overall odds ratio (Σ(O-E)/ΣV) and its variance (ΣV). For each trial that compared more than one qualifying active treatment arm with a common control group, the active treatment arms were combined before this summation to avoid double-counting their control groups. However, the active treatment arms of multiarmed trials were considered separately when subgroups defined by intervention were examined. The nominal two-tailed P values based on these Mantel-Haenszel odds ratios do not take into account the follow-up time at which the events in each trial occurred; thus, significance levels for individual trials may not agree precisely with those that appeared in the original trial reports.

III. RESULTS

A. Prestatin Trials

As previously reported (1), analysis of the 22 treatment arms from 18 trials published before October 1993 that meet the criteria specified above, which included 15,847 participants and 86,660 person-years of follow-up on treatment with a 10% mean net cholesterol reduction, showed a significant 17% reduction in CHD incidence ($P < .001$) but essentially no change in all-cause mortality (Table 2). A 9% decrease in CHD mortality ($P = .03$) was offset by a 24% increase in non-CHD mortality ($P = .001$). Significant reductions in CHD incidence were seen for all classes of intervention except hormones (two dextrothyroxine trials and a small estrogen trial). Significant adverse trends in non-CHD mortality, leading to nonsignificant increases in total mortality, were seen in the trials using hormones and fibrates. Favorable trends in total mortality were seen for the other classes of cholesterol-lowering interventions, but approached significance only in the single trial that used partial ileal bypass surgery, which lowered cholesterol at least twice as much as the other classes of interventions using diets or drugs. Subgroup analyses of the 22 trials by degree of cholesterol lowering showed more favorable effects of cholesterol lowering on non-CHD as well as

Table 2 Cholesterol-Lowering Trials by Intervention Class

Intervention	# Trials	# Treated	Person-Years	Mean cholesterol reduction	Percent change in risk[a]			CHD Incidence[b]
					Total	Mortality		
						CHD	Non-CHD	
Prestatin								
Surgery	1	421	4,084	22%	-24%	-30%	-7%	**-43%**[b]
Resins	3	1,992	14,491	9%	-11%	-32%	33%	**-21%**[b]
Niacin	2	1,264	7,365	8%	-4%	-7%	8%	**-17%**[b]
Diet	6	1,200	6,356	11%	-6%	-21%	0%	**-24%**[b]
Fibrates	7	9,669	50,333	9%	3%	-8%	**32%**[b]	**-18%**[b]
Hormones	3	1,301	4,031	11%	18%	5%	**77%**[b]	7%
Subtotal	22	15,847	86,660	10%	1%	-9%	**24%**[b]	**-17%**[b]
Statins	12	17,405	89,123	20%	**-22%**[b]	**-29%**[b]	-11%	**-30%**[b]
Total	34	33,252	175,782	15%	**-10%**[b]	**-17%**[b]	5%	**-24%**[b]

[a]Statistically significant endpoint results (P < .05) are indicated in boldface type.
[b]Combined incidence of CHD death and nonfatal myocardial infarction.

CHD mortality in the trials in which cholesterol was lowered by >12% (the median) (1).

The four primary prevention trials and 18 secondary prevention trials showed similar favorable trends toward reduced CHD mortality and morbidity and toward increased non-CHD mortality (Table 3). Due to the higher proportion of CHD deaths in the secondary prevention trials, the net impact of cholesterol lowering on total mortality was more favorable than in the primary prevention trials; however, neither effect was statistically significant.

B. Statin Trials

The 12 statin trials reported since 1993 contained 17,405 persons randomized to active treatment. Combined, they represent 89,123 person-years of experience on statin drugs, slightly more person-years than represented by the 22 prestatin trials, and with double the cholesterol lowering (20% vs. 10%). Trials using pravastatin (14–19, 22), lovastatin (11,20,21), and simvastatin (12,13) obtained similarly strong positive results (Table 1). Statistically significant reductions in the primary endpoint (total mortality for 4S, CHD death for LIPID, CHD death plus nonfatal MI for WOSCOPS and CARE, and CHD plus nonfatal MI plus unstable angina for AFCAPS/TexCAPS) were found in each of the five large trials. Significant reductions in atherosclerosis progression (by coronary angiography or carotid ultrasonography), accompanied by mostly favorable trends in clinical CHD endpoints, were observed in the seven smaller trials. With the exception of an increased number of breast cancers among pravastatin-treated women in the CARE trial (18) (which was not reproduced in the LIPID trial [22], there was no evidence of adverse noncardiovascular effects in any statin trial.

Meta-analysis of the twelve statin trials (Table 2) showed approximately 30% reductions ($P < .0001$) in both CHD mortality and incidence. These reductions in CHD mortality and morbidity are 3.2 and 1.8 times those seen in the prestatin trials, respectively, as might be expected when cholesterol levels are lowered by 20% rather than 10%. Non-CHD mortality, which was significantly increased in the 22 prestatin trials, was reduced by 11% ($P = .07$) in the statin trials. Overall mortality was reduced in the statin trials by 22% ($P < .0001$).

When primary and secondary prevention trials are analyzed separately, the reductions in CHD mortality (33%, 28%) and CHD incidence (34%, 29%) are actually slightly greater in the primary prevention trials (Table 3). The reduction in all-cause mortality was still greater in the secondary prevention trials (23%, $P < .0001$) than in the primary prevention trials (15%, $P = .10$), due mainly to the smaller proportion of CHD deaths in the latter trials. This is in marked contrast to the prestatin trials, in which the adverse effects of treatment on non-CHD mortality resulted in a net 13% *increase* in all-cause mortality in primary prevention trials.

Table 3 Cholesterol-Lowering Trials—Primary Versus Secondary Prevention

Population	# Trials	# Treated	Person-Years	Mean cholesterol reduction	Percent change in risk			
					Total	Mortality CHD	Non-CHD	CHD Incidence[a]
Primary prevention	9	17,154	90,133	14%	3%	-15%	**15%**[b]	**-28%**[b]
Prestatin	4	9,713	55,589	9%	13%	-7%	**26%**[b]	**-23%**[b]
Statins	5	7,441	34,544	20%	-15%	**-33%**[b]	-4%	**-34%**[b]
Secondary prevention	25	16,098	85,649	17%	**-13%**[b]	**-18%**[b]	-1%	**-22%**[b]
Prestatin	18	6,134	31,071	11%	-3%	**-10%**[b]	22%	**-14%**[b]
Statins	7	9,964	54,579	20%	**-23%**[b]	**-28%**[b]	-12%	**-29%**[b]
Total	34	33,252	175,782	15%	**-10%**[b]	**-17%**[b]	5%	**-24%**[b]

[a]Combined incidence of CHD death and nonfatal myocardial infarction.
[b]P < .05.
Note: Statistically significant endpoint results (P < .05) are indicated in boldface type.

There are no obvious differences in the results for lovastatin, pravastatin, and simvastatin in terms of risk reduction and safety. While it is reasonable to assume that significant reductions in CHD risk may be expected with the newer statin drugs (fluvastatin, atorvastatin, cerivastatin) in proportion to their degree of cholesterol lowering, it should be recognized that these agents differ structurally from the older statins and that their long-term safety must still be established.

C. All Trials

Clearly in terms of their far greater potency, acceptability to patients, and freedom from adverse noncardiovascular effects, the statins stand apart from the interventions used in the 22 pretstatin trials. Nevertheless it is instructive to examine the overall experience of 34 cholesterol-lowering trials in which 33,252 patients have been treated with a variety of drug and nondrug regimens for an average duration of 4.8 years (Table 2). Highly significant reductions ($P \leq .0001$) have now been observed in CHD incidence (24%), CHD mortality (17%), and all-cause mortality (10%). Non-CHD mortality is still 5% greater in treated than control patients ($P = .3$) due to the adverse experience in the early hormone and fibrate trials, but this increase is no longer statistically significant. The overall mortality experience is still better in secondary than in primary prevention trials despite similar reductions in CHD incidence and mortality, due to the 15% increase in non-CHD mortality ($P = .05$) in the primary prevention trials (Table 3).

IV. WHERE DO WE STAND NOW?

The overwhelmingly positive results of the statin trials have demonstrated unambiguously that, given a class of drugs that can lower cholesterol safely and effectively, CHD morbidity and mortality can be reduced without an offsetting increase in non-CHD mortality. The failure of prestatin trials to demonstrate a reduction in all-cause mortality reflects the suboptimal characteristics of cholesterol-lowering interventions used to test that hypothesis in these studies. However, some important questions about cholesterol lowering remain unanswered:

   1.   *At what level of initial risk should cholesterol-lowering drugs be used?* While relative risk is most germane to questions of etiology and mechanism, absolute reductions in risk, which are the product of relative risk reduction and initial risk, are paramount in quantifying the potential benefit of an intervention on its target population. This benefit must exceed the adverse impact of any drug toxicity and, in a world where health care resources are finite, must be substantial relative to the cost of intervention. Other things being equal, the absolute benefit of an intervention, and thus its ratio to risk and to cost, can be expected

to be most favorable when initial risk is highest. The trend toward increased all-cause mortality in prestatin primary prevention trials is an obvious example of this principle; even low-level drug toxicity may negate a beneficial effect on CHD mortality when the initial risk (i.e., in the absence of treatment) of CHD death is relatively modest.

The advent of the statins, with their greater CHD reduction and apparent freedom from adverse effects (at least in the 3- to 6-year time frame of clinical trials), broadens the indications for cholesterol-lowering drugs to lower-risk groups. The statins are clearly beneficial in hypercholesterolemic men without prior clinical evidence of CHD (WOSCOPS [14]) and in what we would consider "average" cholesterol ranges in patients with prior CHD (CARE [18], LIPID [22]). But can we say this about persons with "average" cholesterol levels and no prior CHD? In AFCAPS/TexCAPS (21), relatively low-risk persons like these derived significant benefit from 5 years of lovastatin treatment in terms of nonfatal CHD events, but not in terms of mortality. However, the mortality rate in this trial was only 0.5% per year, and only 17% of those deaths were due to CHD. While this trial confirms the hypothesis that cholesterol lowering reduces CHD risk in this target population, one would need a much longer and/or larger trial to extend the indication for statin drugs to such a large segment of the population.

2. *How low should our cholesterol target be?* Initial levels of LDL cholesterol in some of the statin trials extended well below the population mean (130 to 150 mg/dL above age 45 [24]), and LDL cholesterol levels <100 mg/dL were often attained in the active treatment groups. However, the effect of cholesterol lowering on CHD events was not uniform in the CARE trial, in which LDL cholesterol levels ranged from 115 to 174 mg/dL at entry and ranged as low as 65 mg/dL on treatment. Posthoc subgroup analyses of this study showed (1) no reduction in CHD events in patients with initial LDL cholesterol levels below 125 mg/dL, and (2) no incremental reduction in CHD event rates for lowering LDL cholesterol <125 mg/dL with pravastatin (25). Moreover, among hypercholesterolemic men receiving pravastatin in the WOSCOPS, no additional benefit appeared to accrue from cholesterol reductions >25% (26). This disassociation between cholesterol lowering and reduction of CHD event rates was not observed in the 4S (27) and is somewhat at odds with epidemiologic observations that cholesterol is a continuous risk factor at least down to *total* cholesterol levels <150 mg/dL (28,29). Ongoing trials, particularly those comparing high versus standard doses of simvastatin and those using atorvastatin, a potent new statin drug, may help clarify whether there is a point of diminishing return for cholesterol-lowering treatment.

3. *Are benefits of cholesterol lowering generalizable to all segments of the population?* The prestatin trials were done almost exclusively in middle-aged male Caucasians. While the majority of statin trials do include women and ex-

tend the upper age limit beyond age 60, randomized trial data in women, men above age 70, and racial and ethnic minorities are still sparse. Subgroup analyses of the four large trials (12,18,21,22) that included women all suggest that they benefit at least as much as men from cholesterol lowering, but women comprise only 10% to 20% of the patients in these studies. Subgroup analyses of these trials and WOSCOPS (14) by age suggest that the older participants derive significant benefit from treatment, although the relative risk reduction tends to be smaller than in younger patients. However, the upper age limits in these trials ranged from 64 to 75 years at entry, and they lacked power to address whether the benefits of cholesterol lowering extend beyond age 70.

More data are also needed on the efficacy of cholesterol lowering in conditions conferring high risk of atherosclerotic CHD (diabetes, hypertension, renal failure), which have often been grounds for exclusion from randomized trials of cholesterol lowering. If cholesterol lowering confers as great a relative risk reduction in the presence of these conditions as it does in their absence, then it may be a particularly valuable and cost-effective adjunct to the primary treatment of these conditions. Several ongoing statin trials are now looking at these subgroups.

4. *How early in life should cholesterol-lowering treatment begin?* While a heart-healthy diet in which <30% of calories are derived from fat and <10% from saturated fat is considered appropriate for everyone above age 2, and further dietary restrictions are commonly recommended for hypercholesterolemic children, physicians are understandably reluctant to use cholesterol-lowering drugs in children, except in extreme cases like familial hypercholesterolemia. The efficacy and safety of statins is difficult to establish in children, because both the CHD events one is trying to prevent and the adverse effects of therapy may be decades away. However, short-term studies in high-risk children examining the effects of statin therapy on intermediate endpoints that may be assessed noninvasively may be a useful start.

V. CONCLUSION

In an earlier meta-analysis of prestatin trials (1) it was suggested that "the new generation of cholesterol-lowering trials, using far more potent cholesterol-lowering agents with (as yet) no known major adverse effects on non-CHD mortality, may provide a more pure and powerful test of the effect of cholesterol lowering on mortality." The results of the 12 qualifying statin trials reported since 1993 now provide a clear and positive resolution of this issue. Lowering cholesterol using statin drugs not only reduces CHD risk, but it does so without offsetting increases in noncardiovascular mortality. Furthermore, meta-analysis of all cholesterol-lowering trials meeting the specified criteria for duration and specificity now demonstrates a significant reduction in all-cause mortality.

REFERENCES

1. Gordon DJ. Cholesterol Lowering and Total Mortality. In: Rifkind BM, ed. *Contemporary Issues in Cholesterol Lowering: Clinical and Population Aspects*. New York: Marcel Dekker, 1995:33–48.
2. Holme I. An analysis of randomized trials evaluating the effect of cholesterol reduction on total mortality and coronary heart disease incidence. Circulation 1990; 82:1916–1924.
3. Muldoon MF, Manuck SB, Mathews KA. Lowering cholesterol concentrations and mortality: a quantitative review of primary prevention trials. Br Med J 1990; 301:309–314.
4. Rossouw JE, Lewis B, Rifkind BM. The value of lowering cholesterol after myocardial infarction. N Engl J Med 1990; 323:1112–1119.
5. Ravnskov U. Cholesterol lowering trials in coronary heart disease: frequency of citation and outcome. BMJ 1992; 305:15–19.
6. Davey Smith G, Song F, Sheldon TA. Cholesterol lowering and mortality and mortality: the importance of considering initial level of risk. BMJ 1993; 306:1367–1373.
7. Holme I. Relation of coronary heart disease incidence and total mortality to plasma cholesterol reduction in randomised trials: use of meta-analysis. Br Heart J 1993; 69(suppl):S42–S47.
8. Law MR, Wald NJ, Thompson SG. By how much and how quickly does reduction in serum cholesterol concentration lower risk of ischaemic heart disease? BMJ 1994; 308:367–373.
9. Law MR, Thompson SG, Wald NJ. Assessing possible hazards of reducing serum cholesterol. BMJ 1994; 308:373–379.
10. Gould AL, Rossouw J, Santanello NC, Heyse, JF, Furberg C. Cholesterol lowering confers clinical benefit: a new look at old data. Circulation 1995; 91:2274–2282.
11. Furberg CD, Adams HP Jr, Applegate WB, Byington RP, Espeland MA, Hartwell T, Hunninghake DB, Lefkowitz DS, Probstfield J, Riley WA, Young B, for the Asymptomatic Carotid Artery Progression Study (ACAPS) Research Group, Effect of lovastatin on early carotid atherosclerosis and cardiovascular events. Circulation 1994; 90:1679–1687.
12. Scandinavian Simvastatin Survival Study Group. Randomised trial of cholesterol lowering in 4444 patients with coronary heart disease: the Scandinavian Simvastatin Survival Study (4S). Lancet 1994; 344:1383–1389.
13. MAAS Investigators. Effect of simvastatin on coronary atheroma: the Multicentre Anti-Atheroma Study (MAAS). Lancet 1994; 344:633–638
14. Shepherd J, Cobbe S, Ford I, Isles C, Lorimer AR, Macfarlane PW, McKillop JH, Packard CJ, for the West of Scotland Coronary Prevention Study Group. Preven-

tion of coronary heart disease with pravastatin in men with hypercholesterolemia. N Engl J Med 1995; 333:1301–1307.

15. Salonen R, Nyyssönen K, Porkkala E, Rummukainen J, Belder R, Park JS, Salonen J. Kuopio Atherosclerosis Prevention Study (KAPS). A population-based primary preventive trial of the effect of LDL lowering on atherosclerotic progression in carotid and femoral arteries. Circulation 1995; 92:1758–1764.

16. Pitt B, Mancini GBJ, Ellis SG, Rosman HS, Park JS, McGovern ME, for the PLAC I investigators. Pravastatin Limitation of Atherosclerosis in the Coronaries (PLAC I): reduction in atherosclerosis progression and clinical events. J Am Coll Cardiol 1995; 26:1133–1139.

17. Crouse JR, Byington RP, Bond MG, Espeland MA, Craven TE, Sprinkle JW, McGovern ME, Furberg CD, Pravastatin, Lipids, and Atherosclerosis in the Carotid Arteries (PLAC-II). Am J Cardiol 1995; 75:455–459.

18. Sacks FM, Pfeffer MA, Moye LA, Rouleau JL, Rutherford JD, Cole TG, Brown L, Warnica JW, Arnold JMO, Wun CC, Davis BR, Braunwald E, for the Cholesterol and Recurrent Events Investigators. The effect of pravastatin on coronary events in patients with average cholesterol levels. N Engl J Med 1996; 335:1001–1009.

19. Mercuri M, Bond G, Sirtori CR, Veglia F, Crepaldi G, Feruglio FS, Descovich G, Ricci G, Rubba P, Mancini M, Gallus G, Bianchi G, D'Alo G, Ventura A. Pravastatin reduces intima-media thickness in an asymptomatic hypercholesterolemic Mediterranean population: the Carotid Atherosclerosis Italian Ultrasound Study. Am J Med 1996; 101:627–634.

20. Post Coronary Artery Bypass Graft Trial Investigators. The effect of aggressive lowering of low-density lipoprotein cholesterol levels and low-dose anti-coagulation on obstructive changes in saphenous-vein coronary-artery bypass grafts. N Engl J Med 1997; 336:153–162.

21. Downs JR, Clearfield M, Weis S, Whitney E, Shapiro DR, Beere PA, Langendorfer A, Stein EA, Kruyer W, Gotto AM, for the AFCAPS/TexCAPS Research Group. Primary prevention of acute coronary events with lovastatin in men and women with average cholesterol levels. Results of the AFCAPS/TexCAPS Trial. JAMA 1998; 279:1615–1622.

22. Long-term Intervention with Pravastatin in Ischaemic Disease (LIPID) Study Group. Prevention of cardiovascular events and death with pravastatin in patients with coronary heart disease and a broad range of initial cholesterol levels. N Engl J Med 1998; 339:1349–1357.

23. Yusuf S, Peto R, Collins R, Sleight P. Beta blockade during and after myocardial infarction: an overview of randomized trials. Prog Cardiovasc Dis 1985; 27:335–371.

24. Johnson CL, Rifkind BM, Sempos CT, Carroll MD, Bachorik PS, Briefel RR, Gordon DJ, Burt VL, Brown CD, Lippel K, Cleeman JI. Declining serum total cholesterol levels among U.S. adults: the National Health and Nutrition Examination Surveys. JAMA 1993; 269:3002–3008.

25. Sacks FM, Moyé LA, Davis BR, Cole TG, Rouleau JL, Nash DT, Pfeffer MA, Braunwald E. Relationship between plasma LDL concentrations during treatment

with pravastatin and recurrent coronary events in the Cholesterol and Recurrent Events trial. Circulation 1998; 97: 1446–1452.

26. West of Scotland Coronary Prevention Study Group. Influence of pravastatin and plasma lipids on clinical events in the West of Scotland Coronary Prevention Study (WOSCOPS). Circulation 1998; 97:1440–1445.

27. Pedersen TR, Olsson AG, Færgeman O, Kjekshus J, Wedel H, Berg K, Wilhelmsen L, Haghfelt T, Thorgeirsson G, Pyörälä K, Miettinen T, Christophersen B, Tobert JA, Musliner TA, and Cook TJ. Lipoprotein changes and reduction in the incidence of major coronary heart disease events in the Scandinavian Simvastatin Survival Study (4S). Circulation 1998; 97:1453–1460.

28. Stamler J, Wentworth D, Neaton JD. Is the relationship between serum cholesterol and risk of premature death from coronary heart disease continuous and graded? Findings in 356,222 primary screenees of the Multiple Risk Factor Intervention Trial (MRFIT). JAMA 1986;256:2823–2828.

29. Chen Z, Peto R, Collins R, MacMahon S, Lu J, Li W. Serum cholesterol and coronary heart disease in a low-cholesterol population: positive relationship with no evidence of a threshold. BMJ 1991; 303:276–282.

Index

ABOUT THE EDITOR

SCOTT M. GRUNDY is a Professor of Internal Medicine, and the Director and Chairman for the Center for Human Nutrition and Department of Clinical Nutrition, at the University of Texas Southwestern Medical Center at Dallas. The author or coauthor of 450 journal articles, he is a member of the American Society of Clinical Investigation, the American Association of Physicians, and the Institute of Medicine of the National Academy of Sciences. He was a recipient of a MERIT Award from the National Institutes of Health (1987–1997). Dr. Grundy received the B.S. degree (1955) in chemistry from Texas Technological College, Lubbock, Texas, the M.S. degree (1960) in biochemistry, the M.D. degree (1960) from Baylor College of Medicine, Houston, Texas, and the Ph.D. degree (1968) in biochemistry from Rockefeller University, New York, New York.